Heat Transfer:
Heat Exchangers, Steam Generators Cooling Towers

by D. James Benton

Copyright © 2018-2021 by D. James Benton, all rights reserved.

Trademarks

Excel® is a registered trademark of Microsoft®, as is the company name. GateCycle™ is a trademark of the General Electric Company® and the name of a very useful thermal cycle modeling tool.

Preface

This is a combination of three books: *Heat Exchangers, Heat Recovery Steam Generators*, and *Evaporative Cooling*. In this text we cover performance prediction and evaluation, as well as classical analytical and numerical methods for computation. Evaporative cooling is vital for life and industry, yet is rarely covered in a course on heat transfer. Cooling towers are widely used, making analysis of the evaporative processes essential for completeness of this subject.

Programming

All of the examples presented in this book are implemented either in the C programming language or Excel® macros. The code, spreadsheets, data files, and other material are arranged in folders within a single ZIP archive accompanying each of the three texts. All three can be freely downloaded at the address below.

All of the examples contained in this book,
(as well as a lot of free programs) are available at:
https://www.dudleybenton.altervista.org/software/index.html

All of the color figures can be found here (click on cover):
https://djamesbenton.altervista.org/

Typical Tube Bundle and Tube Sheet

Table of Contents

	page
Preface	i
Chapter 1. Log-Mean Temperature Difference	1
Chapter 2. The NTU-Effectiveness Method	3
Chapter 3. The F-LMTD Method	5
Chapter 4. Water-Cooled Steam Surface Condenser	10
Chapter 5. Air-Cooled Steam Surface Condenser	17
Chapter 6. Simple Shell-and-Tube Heat Exchanger	24
Chapter 7. Feedwater Heater	29
Chapter 8. Heat Recovery Steam Generator	33
Chapter 9. Moisture Separator/Reheater	40
Chapter 10. The TEMA Designs	44
Chapter 11. Simple Numerical Methods	48
Chapter 12: Variable Properties	54
Chapter 13: Variable Conductance	58
Chapter 14. Two-Phase Flow Inside Tubes	61
Chapter 15. Condensation in Crossflow	69
Chapter 16. Operational Data	74
Chapter 17. Monte Carlo Methods	80
Chapter 18. Heat Recovery Steam Generators	85
Chapter 19. Analytical vs. Numerical	86
Chapter 20. Expected Variations	104
Chapter 21. Actual Designs	107
Chapter 22. Heat Release Diagram	113
Chapter 23. Expansion Line	124
Chapter 24. Element Arrangement	127
Chapter 25. Attemperation	142
Chapter 26. Performance Testing	143
Chapter 27. GT and HRSG Heat Balances	151
Chapter 28. Correcting Test Results	161
Chapter 29. Psychrometrics	166
Chapter 30. Thermodynamic Properties of Moist Air	172
Chapter 31. Merkel's Equation	176
Chapter 32. Counterflow Demand Curves	179
Chapter 33. Crossflow Demand Curves	182
Chapter 34. A Closer Look at Crossflow	184
Chapter 35. Natural-Draft Cooling Towers	187
Chapter 36. Heat Transfer from Falling Droplets	195
Chapter 37. Spray Cooling Systems	202
Chapter 38. Cooling Ponds	208
Chapter 39. The Lewis Number	213
Chapter 40. Air into Water	219
Appendix A. Crossflow Program	225
Appendix B. Moisture Separator/Reheater Program	229
Appendix C: Monte Carlo Codes	239
Appendix D: Steam Properties	245
Appendix E: Exhaust Properties	247
Appendix F: Combustion Calculations	250
Appendix G. Moist Air Properties	252

Appendix H. Heat Transfer Coefficients 255
Appendix I. Flow Measurement 260
Appendix J. Cooling Tower Terms 263
Appendix K: Moist Air Property Functions 264
Appendix L. Runge-Kutta 2D for Crossflow Calculations 268
Appendix M. Falling Droplet Trajectory and Mass Transfer 270
Appendix N. Nuclear Plant Thermal Performance 275
Appendix O. Cooling Pond Performance 288
Appendix P. Power Plant Thermal Simulation 294
Appendix Q. Nitrogen Supersaturation Models 297

Typical Spray and Splash Bars

Chapter 1. Log-Mean Temperature Difference

Unlike companion fields, such as thermodynamics and fluid flow, there are relatively few governing equations in heat transfer and even fewer in heat exchanger analysis. The fundamental one being:

$$dQ = U\Delta T dA \tag{1.1}$$

The differential heat transfer rate (per unit area), dQ, is equal to the overall conductance (i.e., series combination of all heat transfer coefficients), U, times the local temperature difference, ΔT, times the differential area, dA. All of these variables may vary over the surface of the heat exchanger, thus integrating this differential equation requires different approaches. We will begin with the most simple.

The first consideration is geometry. The simplest geometry would be a double pipe, where one fluid flows through the inner pipe and a second fluid flows through the annulus formed between the two pipes. There are two flow configurations: co-current and counter-current. A second consideration is thermal properties, primarily specific heat. At this point, we will consider both fluids to have constant specific heats and use the notation, C_P.

The mass flow rates will be given the symbol, m, and the subscripts H and C will be used to designate the hot and cold streams, respectively. The conservation of energy leads to:

$$dQ = \dot{m}_C C_{PC} dT_C = -\dot{m}_H C_{PH} dT_H \tag{1.2}$$

The negative sign before the third term in Equation 1.2 arises from the fact that as the temperature of the cold stream increases the temperature of the hot stream decreases. The temperature difference is given by:

$$\Delta T = T_H - T_C \tag{1.3}$$

The differential of the temperature difference is given by:

$$d\Delta T = dT_H - dT_C \tag{1.4}$$

Equation 1.2 can be rearranged to form:

$$dT_H - dT_C = -\frac{dQ}{\dot{m}_H C_{PH}} - \frac{dQ}{\dot{m}_C C_{PC}} \tag{1.5}$$

Equations 1.4 and 1.5 can be substituted into Equation 1.1 and both sides divided by ΔT to yield:

$$\left(\frac{\dot{m}_H C_{PH} \dot{m}_C C_{PC}}{\dot{m}_H C_{PH} + \dot{m}_C C_{PC}} \right) \frac{d\Delta T}{\Delta T} = U dA \tag{1.6}$$

Both sides of Equation 1.6 are easily integrated to arrive at:

$$\left(\frac{\dot{m}_H C_{PH} \dot{m}_C C_{PC}}{\dot{m}_H C_{PH} + \dot{m}_C C_{PC}}\right) \ln\left(\frac{\Delta T_{out}}{\Delta T_{in}}\right) = UA \quad (1.7)$$

The heat transfer is also equal to:

$$Q = \dot{m}_H C_{PH} \left(T_{H,in} - T_{H,out}\right) = \dot{m}_C C_{PC} \left(T_{C,out} - T_{C,in}\right) \quad (1.8)$$

These last two equations can be combined (substitute $mC_P = Q/\Delta T$ hot and cold from Eqn. 1.8 into Eqn. 1.7) to form:

$$Q = UA \frac{\Delta T_{out} - \Delta T_{in}}{\ln\left(\frac{\Delta T_{out}}{\Delta T_{in}}\right)} \quad (1.9)$$

The last group in Equation 1.9 is called the log-mean temperature difference or LMTD. Several cardinal rules of calculus were broken in the process of arriving at this result, including:

- U may depend on temperature, in which case, it should have been on the left side of Equation 1.6.
- The specific heats may not have been constant, in which case, they shouldn't have been taken outside the implied integral.
- ΔT might have been zero, in which case we shouldn't have divided by it when taking it to the left side of the differential equation.
- ΔT might not be zero, but $d\Delta T$ might be. If $m_H C_{PH} = m_C C_{PC}$ and the configuration is counter-current, the difference in temperatures of the two streams is constant. Although dQ is not zero, by using Equation 1.5 we inadvertently multiplied both sides of Equation 1.1 by zero. The LMTD becomes 0/ln(1), which is undefined, but $\Delta T_{in} = \Delta T_{out}$.

The LMTD approach can fail in more complex arrangements with variable specific heats. Several examples will be given after other methods are introduced so that these may also be evaluated side-by-side. It is worth noting at this point that no method is adequate for every possible application and circumstance.

Know which approach is best
for your particular application
and what its limitations are.

Chapter 2. The NTU-Effectiveness Method

This method is similar to the LMTD method, but takes a slightly different approach to solving the same differential equation (viz., 1.1). The effectiveness, ε, is defined as the ratio of the actual to the maximum heat transfer:

$$\varepsilon = \frac{Q}{Q_{MAX}} \tag{2.1}$$

The heat transfer, Q, for both the hot and cold streams is given by Equation 1.8. The maximum possible heat transfer would be equal to the difference in the two inlet temperatures times the minimum product of the mass flow and specific heat or:

$$Q_{MAX} = (\dot{m}C_P)_{MIN}(T_{H,in} - T_{C,in}) \tag{2.2}$$

where the minimum could be the hot or cold side. The ratio of the minimum to maximum product of mass and specific heat is given the symbol, R:

$$R = \frac{(\dot{m}C_P)_{MIN}}{(\dot{m}C_P)_{MAX}} \tag{2.3}$$

In the special case of phase change (most often this occurs in a condenser) CP is infinite, making R=0. The number of transfer units, NTU, is defined as:

$$NTU = \frac{UA}{(\dot{m}C_P)_{MIN}} \tag{2.4}$$

Through the same steps used to arrive at Equations 1.5, 1.6, and 1.7, we can arrive at the following effectiveness for co-current flow in a double pipe:

$$\varepsilon = \frac{1 - e^{[-NTU(1+R)]}}{1+R} \tag{2.5}$$

and for counter-current:

$$\varepsilon = \frac{1 - e^{[-NTU(1-R)]}}{1 - Re^{[-NTU(1-R)]}} \tag{2.6}$$

which reduces to the following when $R=1$:

$$\varepsilon = \frac{NTU}{1+NTU} \tag{2.7}$$

Of course, the same cardinal rules of calculus were violated in arriving at these equations. Analytical solutions for many other heat exchanger geometries have been worked out and can readily be found on-line. These will only be covered here as they compare to more general numerical solutions. Computer solutions are so convenient and flexible that the analytical solutions have lost some of the utility they once had when computers were not as readily available.

The NTU-ε method leads to an even more useful approach by considering a more general definition of effectiveness:

$$P_I = \frac{q}{C_I \left(T_{hot,in} - T_{cold,in}\right)} \quad (2.8)$$

where the subscript, *I*, can be either the hot or cold stream. The capacity of the stream is given by:

$$C_I = (\dot{m} C_P)_I \quad (2.9)$$

and the number of transfer units associated with stream, *I*, is given by:

$$NTU_I = \frac{UA}{(\dot{m} C_P)_I} \quad (2.10)$$

The capacitance ratio for the stream is given by:

$$R_I = \frac{C_I}{C_{II}} \quad (2.11)$$

where R_{II} is the non-reference (i.e., the other) stream. It also follows that:

$$P_I = \left(\frac{C_h}{C_c}\right)\left(\frac{T_{h,i} - T_{h,o}}{T_{h,i} - T_{c,i}}\right) = \left(\frac{C_c}{C_h}\right)\left(\frac{T_{c,o} - T_{c,i}}{T_{h,i} - T_{c,i}}\right) \quad (2.12)$$

Once cast in this form, the performance of many heat exchangers performance may be expressed analytically by *P* as a function of *NTU$_I$*, *R*, and the arrangement. Alternately, *NTU$_I$* may be expressed as a function of *P*, *R*, and the arrangement. The premier reference text on this subject is *Heat Transfer* by L. C. Thomas. There are several editions, but the Professional Edition is by far the most useful. One source to find this would be the following link:

https://www.amazon.com/Lindon-C.-Thomas/e/B001HOVPI4

This present text is merely a source of examples and illustrations. The reader is directed to Lindon's text for comprehensive details on the P-NTU$_I$ method and many other things, including: convection, radiation, boiling, condensation, and heat transfer coefficients.

Chapter 3. The F-LMTD Method

It should be evident that a double pipe cannot adequately approximate any but the most simplistic heat exchanger geometries. One approach has been to multiply by a fudge factor, **F**, to account for the geometry and still use the log-mean temperature difference. The F-LMTD method is somewhat useful, at least for simple geometries and constant properties plus it has some historical interest. Only the crossflow geometry will be considered here.

$$UA = \frac{Q}{F \times LMTD} \quad (3.1)$$

Performance Prediction

The method is most often presented in the form of parametric curves based on analytical solutions to the steady-state conduction equation. Separate analytical solutions must be obtained if either of the sides (i.e., hot/cold fluids) is mixed across the direction of flow. Before the advent of the microcomputer, these solutions were evaluated on mainframes and drawn on pen plotters—that held an actual pen and moved it or the paper around! The solution for both sides unmixed is illustrated in the following figure for values of **R**=0.2, 0.4, ... 4.0:

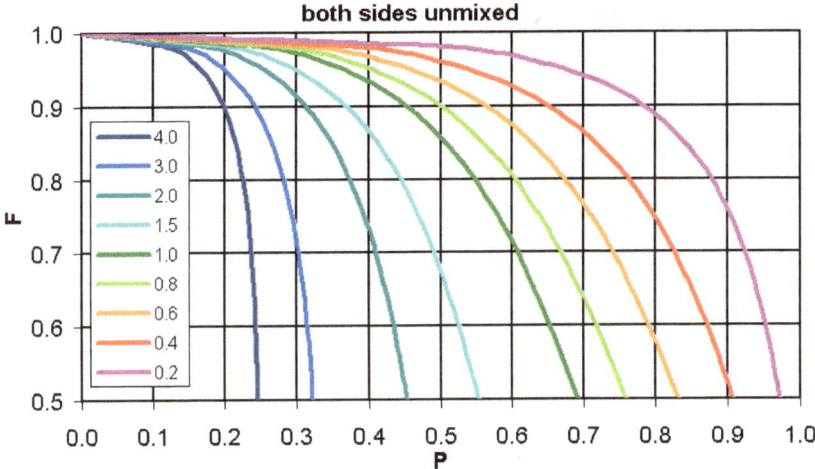

These crossflow plus similar analytical solutions may be cast in the form of $P_I(NTU_I, R_I,$ arrangement) and are strictly applicable in the case of constant properties. A variety of such solutions are provided in Lindon Thomas' book previously mentioned. Only one will be included here–that corresponding to the preceding figure (unmixed):

$$F = \frac{1}{NTU_I(1-R)} \ln\left[\frac{R\exp(-\Gamma R)}{R-1+\exp(-\Gamma R)}\right] \quad R < 1 \quad (3.2)$$

$$F = \frac{1}{NTU_I}\left[\frac{1-\exp(-\Gamma)}{\exp(-\Gamma)}\right] \quad R=1 \qquad (3.3)$$

$$P = \sum_{m=0}^{\infty}\left\{\frac{\Phi_m \Psi_m}{R\,NTU_I}\right\} \qquad (3.4)$$

$$\Phi_m = 1-\exp(-NTU_I)\sum_{j=0}^{m}\frac{NTU_I^{\,j}}{j!} \qquad (3.5)$$

$$\Psi_m = 1-R\exp(-R\,NTU_I)\sum_{j=0}^{m}\frac{(R\,NTU_I)^j}{j!} \qquad (3.3)$$

$$P = \frac{1-\exp(-\Gamma R)}{R} \qquad (3.7)$$

$$\Gamma = 1-\exp(-NTU_I) \qquad (3.8)$$

The crossflow solution for one side mixed is shown in the following figure:

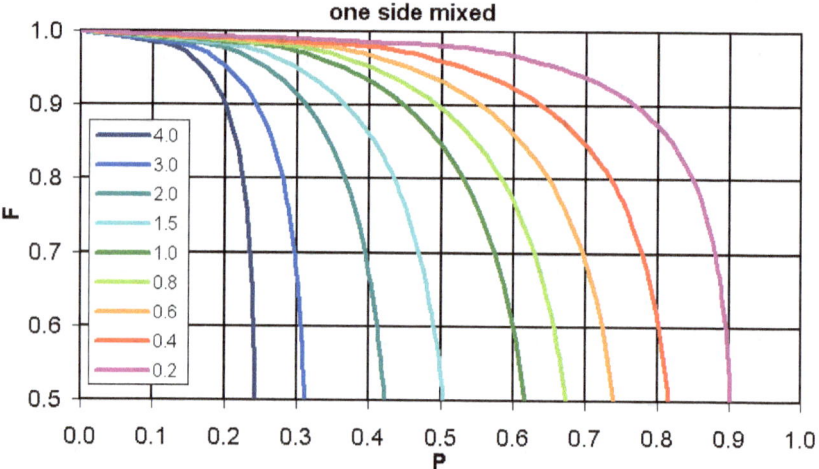

Notice that all of the curves in the second figure fall below and to the left of the corresponding curve in the preceding figure. For example, in the first figure (unmixed) the magenta **R**=0.2 curve intersects the horizontal axis (**F**=0.5) at **P**=0.97, while this same curve in the second figure (one side mixed) intersects at **P**=0.90. The both sides mixed solution yields even lower values of **F**, as shown in this next figure, where this same curve intersects at **P**=0.89. An examination of Equation 3.1 reveals that a smaller value of **F** will require a larger **UA** and more heat exchange surface, most likely costing more.

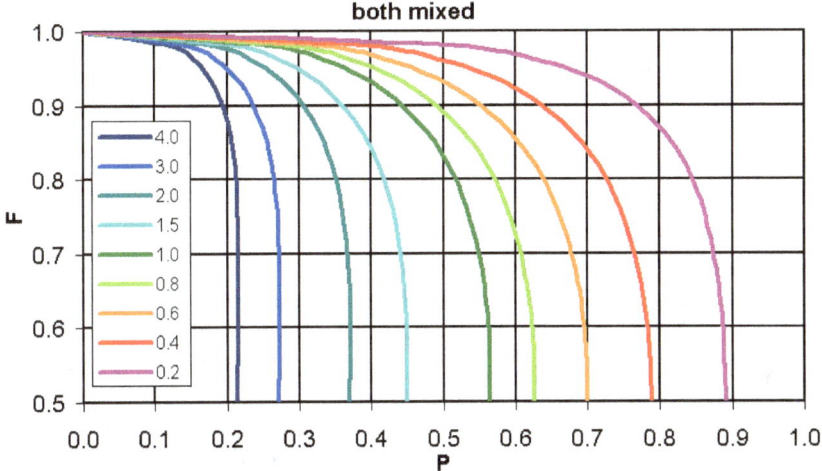

The F-LMTD method for all three crossflow configurations is illustrated in the Excel® spreadsheet named crossflow_analytical.xls, which is contained in the on-line archive. Macros (i.e., VBA® functions) are provided for all the preceding curves plus several more to facilitate their use in problem solving.

	A	B	C	D	E
1		crossflow example			
2	symbol	units	hot	cold	
3	U	W/m²/°C	150		
4	Cp	kJ/kg/°C	4.0	1.0	
5	m	kg/s	100	500	
6	Tin	°C	50	15	
7	Tout	°C	20	39.0	
8	q	kW		12,000	
9	LMTD	°C		7.61	
10	P	-		0.686	
11	R	-		0.800	
12	symbol	units	unmix	1 mix	mixed
13	F	-	0.667	0.352	0.183
14	UA	kW/°C	2362	4481	8622
15	A	m²	15.7	29.9	57.5
16	user inputs in blue				
17	calculations in orange				

If any of the blue values are changed, the orange ones will automatically update, yielding the F factor and required surface area. In the case shown here, mixing either of the sides increases the required surface area from 15.7 to 29.9 m². Mixing both sides increases the required area to 57.5 m²! This example illustrates why mixing either of the two streams is not recommended from a thermal perspective, although from a maintenance or mechanical perspective it might be.

In general, it's not a good idea to mix either of the streams—think: Second Law of Thermodynamics. Mixing always results in the production of entropy, which is an irreversible process. Only mix the hot and/or cold sides in such a heat exchanger if there's some other reason, such as: to avoid fouling or plugging of small channels, to simplify fabrication, or to provide greater structural strength and eliminate the need for additional material.

Performance Evaluation

This unmixed design is presented as a performance test in the same spreadsheet. Test measurements include: the hot and cold inlet and outlet temperatures along with the hot and cold side flows. The uncertainty for each instrument is listed along with the average value over the test period. The instrument uncertainty is divided into two components: temporal and persistent.

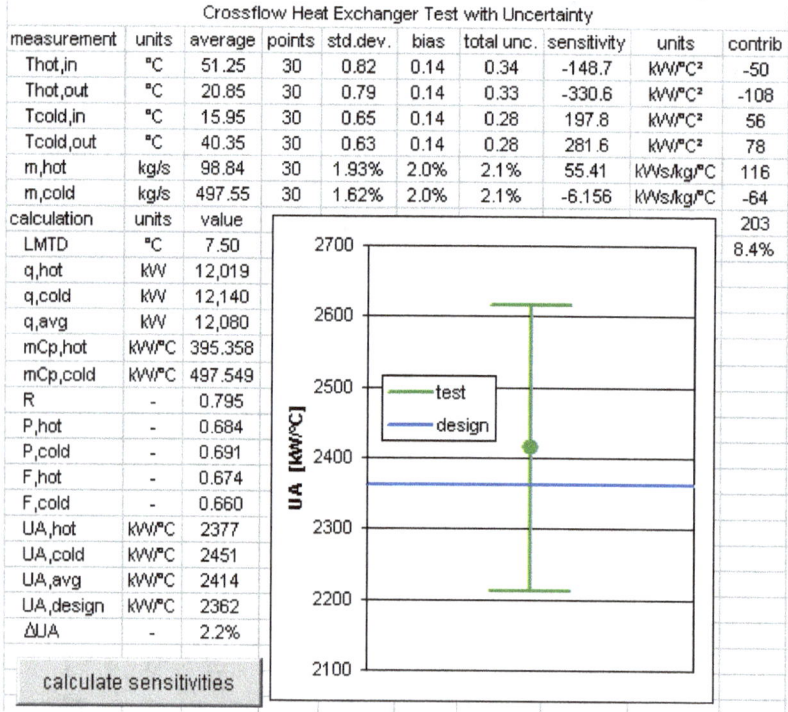

The temporal component of the uncertainty is assumed to be random and proportional to the standard deviation of the average. This is multiplied by Student's-T value at 95% to account for the number of readings. The time-independent component is often called *bias* or *systematic* uncertainty. This represents the accuracy of the instrument and its calibration.

The systematic uncertainty of the temperature measurements is listed as 0.26/1.8°C, which is typical for precision RTDs[1] with NIST-traceable pedigree[2]. The uncertainty of the flow measurement is listed as 2%. If someone tells you they have a flow meter that's more accurate than this, don't believe them. It is evident from the following equation that the measurements cannot possibly be any more accurate than the systematic uncertainty of the device.

$$u_i = \sqrt{b_i^2 + \left(t_{95} \frac{s_X}{\sqrt{N}}\right)^2} \tag{3.9}$$

Here u_i is the uncertainty, b is the bias (or systematic uncertainty), s_X is the standard deviation of the measured values, t_{95} is Student's-T value at the 95% confidence level[3], and N is the number of points. The sensitivities, θ_I, are partial derivatives of the result, UA_{AVG}, with respect to each measurement. A macro is provided that perturbs each value and calculates the sensitivity, which is multiplied by u_{TOTAL} to obtain each uncertainty contribution. The total uncertainty is the root-sum-square of the components, in this case 203 kW/°C or 8.4% of the average 2414 kW/°C.[4]

$$u_{TOTAL} = \sqrt{\sum_{i=1}^{m} (\theta_i u_i)^2} \tag{3.10}$$

The average test UA is 2.2% above the design value of 2362 kW/°C, but the test uncertainty is much larger than this at ±8.4%, as shown by the green bars in the figure, so that it is not possible to say that the heat exchanger is performing better than design. This result would, however, constitute adequate proof that the heat exchanger has *passed* the performance test.

---

[1] Resistance Temperature Detector of the 4-wire Platinum element variety.
[2] The calibration process meets the requirements of and can be traced back to the National Institute of Standards and Technology.
[3] Use TINV(0.05,count) to calculate this value in Excel®.
[4] More details on uncertainty calculations may be found in the standard reference for Test Uncertainty, ASME PTC 19.1-2013. I must warn you that this document is very difficult to understand, even when you're already quite familiar with the material.

Chapter 4. Water-Cooled Steam Surface Condenser

A water-cooled steam surface condenser (WCC) is basically a box with a bunch of tubes running through it. These range in size from a couch to a motel. Practical considerations put the condensing steam on the outside of the tubes and the cooling water on the inside. Condensing heat transfer coefficients (on the vapor side) are quite high and don't require large velocities; whereas, convective heat transfer coefficients (on the liquid side) very much depend on velocity to achieve higher values. A typical condenser waterbox and tubesheet is depicted below:

Performance Prediction

The two key references for WCCs are: ASME PTC-12.2 (2010) and Heat Exchange Institute Standards for Steam Surface Condensers 11th Ed. (2012). These two documents are quite different in approach and content, the former is more academic and the latter is more practical. PTC-12.2 presents dimensionless correlations involving Reynolds and Nusselt numbers and the HEI document contains lots of interesting tables and curves.

The primary concern of these documents and various formulas therein is to obtain a value for the overall heat transfer coefficient, U. In practice, there are only three possibilities (arranged in order of decreasing likelihood): 1) you will buy a condenser from somebody who makes them, 2) you will work for somebody who makes them, 3) you will start your own company to make condensers.

If you are buying a condenser, the manufacturer will calculate U using their own secret proprietary formula that the founder (who died long ago) came up with. If you work for that company, you'll walk past the founder's picture on the way to your desk, where you'll use that secret formula to design condensers. If you're starting your own company, you'll average all of the formulas, hoping that at least some of them work, add a 2% margin just to be safe, worry, add another 2% margin, worry some more, and add another 1% margin. Over the years you'll develop your own formula to leave behind along with a picture to inspire future employees.

Since, in the vast majority of cases, you will be given a set of condenser performance curves, we will begin there. Shown below is a typical set of such curves for a large power plant condenser:

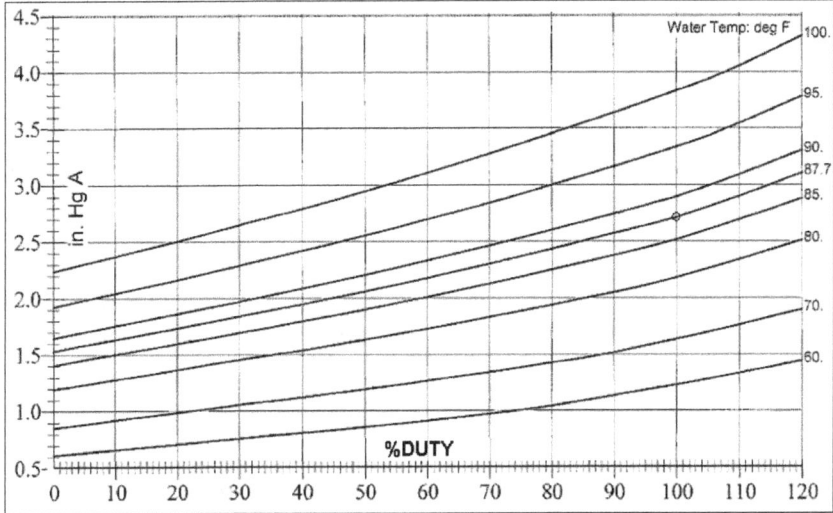

These most often have heat load or percent duty along the X-axis and absolute pressure in inches of mercury (or kPa) on the Y-axis. The curves are at different values of inlet cooling water temperature. There will be a curve that runs through the design point, in this case, 87.7°F, indicated by the little circle.

These curves are often provided in mixed units. The modern engineer should be comfortable working with many systems of units. The impression is often given in academia that the formal SI system is used consistently throughout the world, but this is not the case. My advice is to get comfortable with it and don't fuss. There are way too many people fussing already.

These curves show what the customer is paying for (viz., pressure), but are not the most efficient representation of the information. The upward curvature of the lines in the preceding graph are entirely due to the relationship between the saturation pressure and temperature of steam and are not an artifact of the heat transfer. If, instead of plotting pressure on the Y-axis, we plot saturation temperature, the curvature is completely eliminated, as illustrated in this next figure:

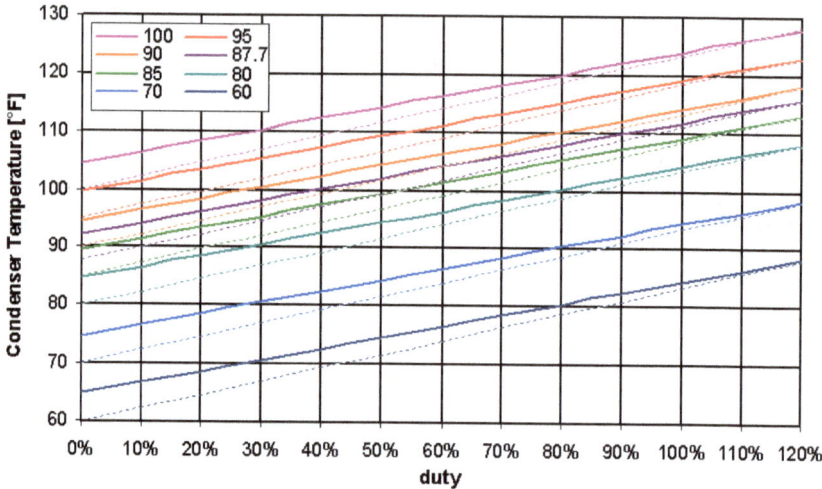

Ignore the dotted lines for the moment. The curves from the previous graph have been reduced to a series of parallel straight lines. While the previous curves could be represented by a multi-variable regression with six to ten terms, this same information as drawn above can be represented by a simple equation with only three terms:

$$T_{SAT} = \alpha + \beta\, duty + \gamma T_W \tag{4.1}$$

This is to be expected, considering these curves were derived from the relationship, with the product, **UA**, remaining essentially constant:

$$Q = UA \cdot LMTD \tag{4.2}$$

The dotted lines in the graph above are an interesting detail. Some manufacturers provide curves with a gradual uplift on the left side, while others have an abrupt change in slope at low duty. In either case, these represent the same thing. Clearly, if there were no heat transfer (i.e., 0% duty) and there were no non-condensable gases, each line should intersect the Y-axis at the point corresponding to the inlet water temperature, hence, the dotted lines.

The dotted lines are the heat transfer that would be achieved in a *perfect world*. Since we don't live on that planet, we use the solid lines. No one will guarantee the performance of a condenser at this low duty. Manufacturers only provide curves that go down this far because customers fuss if they don't. Each manufacturer has their own way of "adjusting" curves at low duty (or very low inlet water temperatures). No one cares about condenser pressures under these conditions, at least from a performance perspective. This is one of the reasons why plants are designed with more than one circulating water pump. If the pressure gets too low in the winter, resulting in sonic velocity at the exit of the steam turbine, the operator simply turns off one of the pumps. The performance of the steam turbine improves plus there is less auxiliary load: a win-win.

> There's a spreadsheet (water_cooled_condenser.xls) in the on-line archive that contains these figures plus macros that illustrate WCC calculations.

Performance Evaluation

Accurately testing a water-cooled steam surface condenser is a lot more complicated than it might seem. Below are some of the difficult questions you must answer in order to do so:

Q1: What is the actual steam flow during the test (not the amount shown on the twenty-year-old heat balance that wasn't accurate on the day it was printed)?

Q2: How are you going to account for all of the other miscellaneous (un-instrumented) flows into and out of the condenser during the test as well as the change in the level of the oddly shaped hotwell?

Q3: What is the quality of the steam as it hits the top of the tube bundle (generally assumed to be at the UEEP or used energy end point)? You're going to need everything from generator loss curves to annulus area, last stage blade length, and an exhaust loss curve, which you'll have to assume is accurate.[5]

Q4: How are you going to figure out how much water is flowing through the condenser?

Q4a: Are you planning to use an acoustic flow meter? Then don't bother testing the condenser. Just scribble down some numbers from the control room on a napkin and come up with an estimate over a few beers, because that's all the accuracy you'll end up with.

Q4b: Are you planning to use the pump curves? Better to rent an acoustic flow meter. Those pump curves were made from a plastic 1:16 scale model and never verified on a prototype.

Q4c: You're going to need access to a straight run of pipe where you can perform a pitot traverse. There aren't many laboratories that can calibrate your pitot. There's Alden, TVA Norris, Iowa Institute of Hydraulic Research, and the U. S. Naval facility. It's unlikely the latter will be responsive to inquiries, so don't bother asking.

Q4d: Your only other practical option for accurate condenser flow measurement is dye dilution. There are a *few* people who can perform an

---

[5] In order to calculate the UEEP you start with the generator output and power factor, obtained from a precision power meter, and add the losses, which are usually of adequate precision. Then you subtract the exhaust loss to get the ELEP, including any adjustment, such as the moisture correction. You use this to build an expansion line along with the previous extractions, which must be dry steam. The whole process is iterative. There's a spreadsheet in the *Thermodynamics* package that does this for you.

accurate dye dilution test. This technique is replete with subtleties and surprisingly easy to mess up. Look for experience and you'll be pleased with the results. If you don't have good pitot access, dye dilution is your only option.

Q5: What pressure(s) are you going to use? The pressure varies all over above the tube bank up into the hood all the way to the annulus of the turbine. Are you going to use the static or the stagnation pressure?

Q5a: You might try to measure the stagnation pressure if the velocity doesn't rip the end off your probe or bend the shaft so you can't get it out when the test is over.

Q5b: Are you going to measure the static pressure with a basket? There are several types. The one shown below is omni-directional. Some aren't. I've seen many of the directional ones installed backwards. Granted, it may not be obvious which is the front and back face on these and I've even heard arguments that they work better backwards.

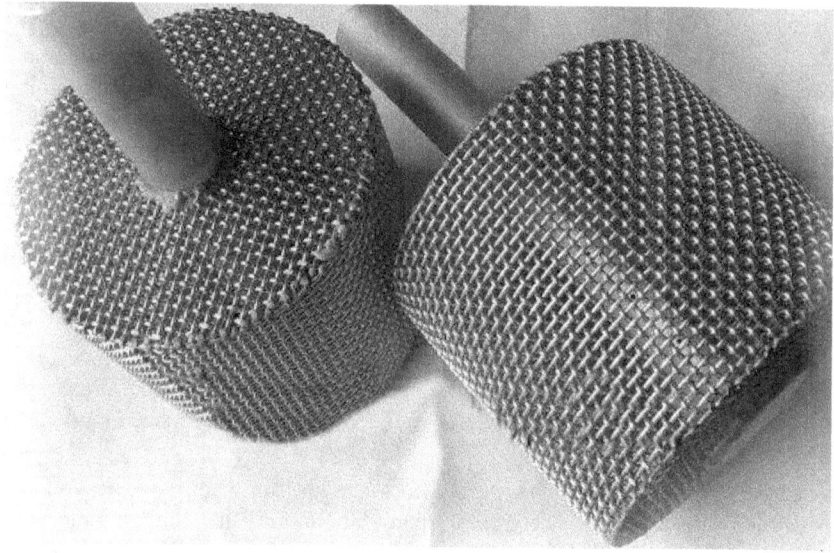

Q5c: Where are you going to measure the pressure? Underneath a flange on the side of the hood where it will be nicely protected but also inside a separation zone, such as the one depicted here behind an airfoil?

I trust you know that the steam exiting the turbine can reach sonic velocity, even faster than that represented by the flow over the airfoil in the following picture. The separation zone shown in this figure and the distorted pressure field is no exaggeration for what can occur in the region where you're trying to measure pressure. You will get as many opinions as to where best locate your pressure sensors as there are people with an interest in the outcome of the test.

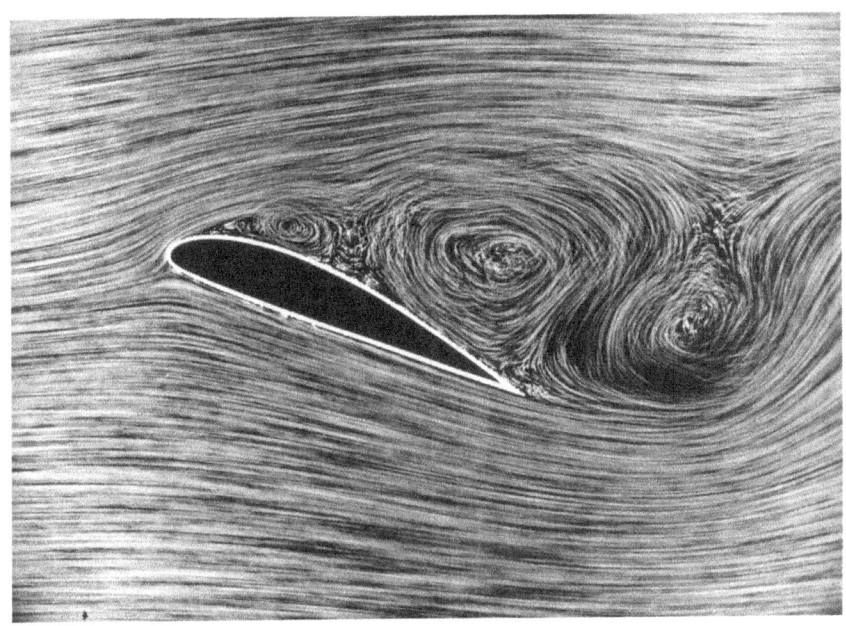

The ideal for such a test would be to measure the temperature of the condensate dripping off the top row of tubes. Sadly, this isn't possible. The next best thing would be to measure the temperature of the condensate running off the lowest row of tubes that aren't flooded. I'm sorry to say that I've never had this measurement either.

> The truth is that I'm suspicious of any and all condenser pressure measurements and am skeptical of any that deviate very far from that indicated by the hotwell temperature. When there is a lot of inventory and a large hotwell sub-cooling, this number is useless and I'm skeptical of the whole lot. I only feel good about a test when the pressure lines up fairly well with the hotwell temperature. It's like the testimony of two witnesses, neither one of which is quite believable by themselves, but together are just barely adequate.

Once you have decided where to collect the data and with what instruments and have agreed upon a way to correct for the inflows, outflows, and hotwell inventory, you'll need at least three hours of steady operation. It is also essential to maintain adequate removal of non-condensables during the test. If there is excessive in-leakage, there is no point testing the condenser. You must fix that first.

The following table summarizes the results of an actual test:

	A	B	C	D	E	F	G	H	I
1			CONDENSER THERMAL PERFORMANCE TEST						
2	GEOMETRY	Units	Symbol	Design	Test 1	Test 2	Test 3	Avg.	Notes
3	tubes	-		20,500	20,500	20,500	20,500	20,500	
4	surface area	ft²	A	975,000	975,000	975,000	975,000	975,000	
5	outer diameter	in	Do	1.000	1.000	1.000	1.000	1.000	1"
6	wall thickness	in	wt	0.028	0.028	0.028	0.028	0.028	22g
7	inside diamer	in	Di	0.944	0.944	0.944	0.944	0.944	
8	TEST INPUTS	Units		Design	Test 1	Test 2	Test 3	Avg.	
9	steam pressure	in HgA	Ps	2.50	2.63	2.62	2.67	2.64	meas.
10	water flow rate	gpm	W	335,000	338,147	338,389	338,087	338,208	pitot4xtr
11	water inlet temp.	°F	T1	90.0	90.8	90.8	91.3	91.0	RTDx2
12	water outlet temp.	°F	T2	108.0	109.4	109.3	109.9	109.5	RTDx4
13	PROPERTIES								
14	density	lbm/ft³	ρ	62.00	61.98	61.98	61.98	61.98	
15	specific heat	BTU/lbm/°F	Cp	0.9981	0.9981	0.9981	0.9981	0.9981	
16	CALCULATIONS	Units		Design	Test 1	Test 2	Test 3	Avg.	
17	duty	MBTU/hr	Q	2993	3121	3106	3120	3121	mCpΔT
18	steam temp.	°F	Ts	108.6	110.4	110.3	110.9	110.5	Tsat
19	LMTD	°F	ΔT	5.3	6.3	6.3	6.3	6.3	
20	U	BTU/hr/ft²/°F	U	576.0	510.2	508.4	509.6	509.4	Q/A/ΔT
21	RESULTS	Units		Design	Test 1	Test 2	Test 3	Avg.	
22	Cleanliness	-	Cf	90.0%	88.6%	88.3%	88.5%	88.4%	U/Udes

The spreadsheet (WCC_test.xls) is contained in the archive.

Chapter 5. Air-Cooled Steam Surface Condenser

An air-cooled steam surface condenser (ACC) is basically a car radiator the size of a Walmart on stilts, such as the one pictured below. They're quite expensive to buy and operate. Their chief advantage is that they don't consume water like an evaporative cooling tower. An ACC is like a WCC combined with a dry cooling tower only with fewer moving parts to break—a good choice in arid regions.

Performance Prediction

The performance curves for ACCs are very similar to the ones for WCCs with the addition of fan speed as a variable. The condensation and separating material (e.g., tube wall) conductances are much higher than the convective heat transfer coefficient on the air side, making this the dominating factor. This is even more pronounced than for the water on the inside of the tubes in a WCC. Because air has a much lower thermal conductivity and specific heat than water, it takes a lot more of it and a lot more surface area to carry away the heat of condensation.

Water temperature correction factors (e.g., HEI's) are of some value with a WCC, but empirical air properties are more important with an ACC. The following figure is typical of such curves.

ACCs fan power consumption can be considerable so that variable-speed motors and/or drives are often used to manage auxiliary load. It is, therefore, necessary to quantify performance in terms of both duty and fan speed. This relationship varies with many factors and must be supplied by the ACC manufacturer. The only alternative would be to establish it through a formal testing process using ASME PTC-30.1 (2007), "Air-Cooled Steam Condensers."

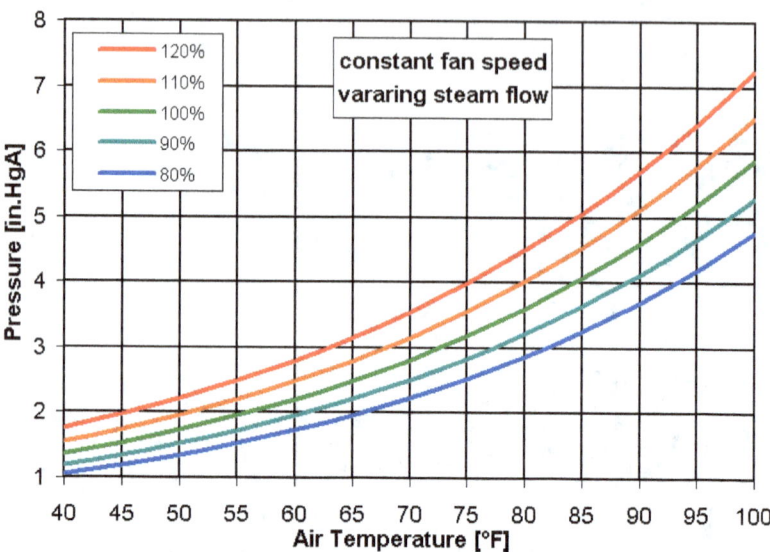

This next figure shows the impact of fan speed at constant steam flow:

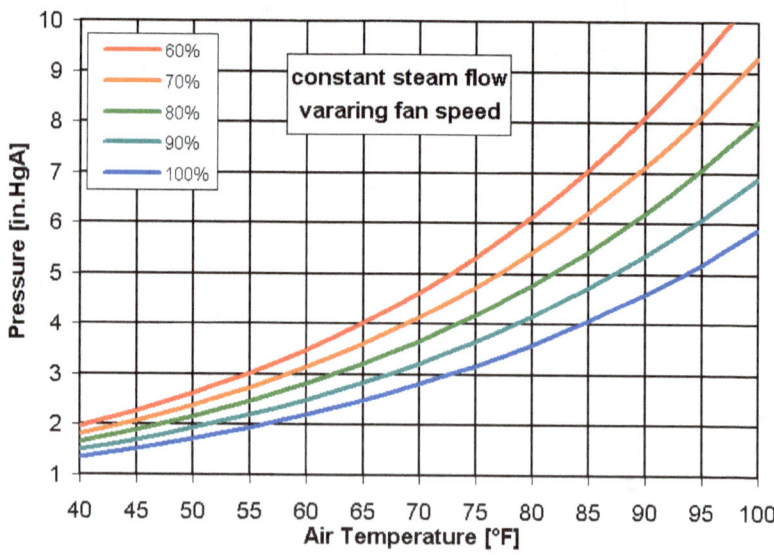

A subtle difference between WCCs and ACCs is that typically the performance of the former is given in terms of percent duty or heat load; whereas, the performance of the latter is most often given in terms of steam flow. The difference being a result of either the entering quality and/or the slight change in enthalpy with the vertical axis, as the corresponding saturation temperature changes.

The preceding two curve sets were reproduced here using multi-variable regression. The manufacturer generated these using empirical correlations for heat transfer coefficients, which means that these can also be inferred from the curves and then a regression performed on the result. Once the product of the effective surface area, A, and overall conductance, U, have been "extracted" from the curves, the same may be used to regenerate the curves from Q=UA*LMTD. The spreadsheet contains macros that perform the calculation both ways so that you can compare the two.

Multi-variable regression on ACC curves is a little more challenging than WCC curves, but far from overwhelming. First, get my digitizing tool if you don't already have it. If the curves are in color, reduce the colors to Window's 16 before copying the image onto the clipboard so you can take advantage of the automatic feature (Ctl-A per selected color or Ctl-Alt-A for all 16 colors).

If the image is already on the clipboard before you launch Digitize, it will load it for you and you won't have to open a file. I usually get the images onto the clipboard using the Acrobat snapshot tool, since the graphs usually come in a PDF.

I always define 4 control points. If the R^2 isn't close to 1.00, then one or more of your points is bad. Go back and check them all.

When you exit Digitize, the results are automatically copied onto the clipboard. Just paste them into Excel. You can save it in a file if you want to, but you don't have to. The first two columns are the pixels and the third and fourth are x and y. After you arrange all of the data, use my curve-fitting tool to perform the regression. It will copy the resulting VB macro onto the clipboard for you to paste into Excel.

There's a spreadsheet (air_cooled_condenser.xls) in the on-line archive that contains these figures plus macros that illustrate ACC calculations.

<u>Performance Evaluation</u>

Testing an ACC is in some ways less complicated than testing a WCC—but that's not to say that it's cheaper. First of all, there are lots of places to measure the steam pressure and there are frequently already fittings where you can readily attach instruments. You will have the same problem determining the quality of the entering steam (or the UEEP), but it's often easier to quantify the flow rate, because the condensate is all coming out of a pipe at the bottom.

Don't even try measuring the air flow through an ACC. There are holes and gaps all over it and you'll never account for all of them. Just calculate it from the heat load. You'll have enough trouble getting a velocity-weighted average exiting air temperature. On a big ACC it costs way too much to traverse all of the fans. Pick a geometric pattern (center, edge, lateral, corner, etc.), traverse several fans, and apply the results across the rest based on location. It's just like traversing the fans atop a cooling tower and you use the same equipment.

The biggest problem with ACCs is sub-cooling. You neither pay for nor care about sub-cooling. The standpipe will provide more than adequate net positive suction head (NPSH) for the condensate booster pump(s). Sub-cooling means that a bunch of the expensive heat exchange surface that you paid for is flooded, which is about as useful as having a finished basement full of water.

> *During an ACC performance test, the manufacturer's representative will make a really big deal about how wonderful all the water in your basement is and go to great lengths to explain how they calculated the amount down to the last drop. They must give this spiel. Just smile and listen.*

Here's the problem with sub-cooling... If you test when it's hot and you still have sub-cooling, then your ACC is more than adequate. If you test when it's cold and you have sub-cooling, you don't know that you'll have adequate capacity when the weather gets hot. The manufacturer will try to tell you that X degrees of sub-cooling in the winter will translate into Y inches of mercury less pressure in the summer. The trouble is, they never gave you that graph and they can't provide it to you now, because it's just too complicated to draw.

Of course, the test conditions aren't on any of the graphs that you got. Oh, yeah, there's also the fact that half of the fans were off during the test and they never gave you that graph either. They could give you a graph with one-third of the fans turned off, but their software won't let them turn half of the fans off.

Welcome to the world of performance testing!

An accurate velocity-weighted exit air temperature is important to an ACC performance test for two reasons: 1) it is not likely that you will be testing at the guarantee conditions, so you will need to correct the performance back to the base reference conditions, and 2) you will be interested in the fan power consumption, as this is the second most important aspect of the performance.

The following table summarizes an actual fan test and includes all of the calculations:

	A	B	C	D	E	F	G	H	I	J	K
1	ACC Fan Test (two traverses at right angles)										
2	fan dia.		ft	34.4							
3	hub dia.		ft	3.44							
4	net area		ft²	921.2							
5	baro		psia	14.77							
6	pitot coef.		-	835.5	includes calibration and units						
7		6	point along each radius								
8	pnt	X	Y	temp	p	ΔP	angle	Vtang	Vnorm	Area	ρVA
9	-	ft.	ft.	°F	lb/ft³	in.H2O	deg*	fpm	fpm	ft²	lb/min
10	1	-11.6	11.6	86.5	0.073	0.48	32	2142	1817	38.4	5091
11	2	-10.5	10.5	88.5	0.073	0.46	31	2101	1801	38.4	5028
12	3	-9.3	9.3	88.5	0.073	0.52	11	2234	2193	38.4	6122
13	4	-7.9	7.9	89.5	0.073	0.36	18	1860	1769	38.4	4931
14	5	-6.1	6.1	89.5	0.073	0.07	26	820	737	38.4	2055
15	6	-3.2	3.2	85.0	0.073	0.00	0	49	49	38.4	137
16	7	3.2	-3.2	85.0	0.073	0.00	0	39	39	38.4	110
17	8	6.1	-6.1	85.0	0.073	0.00	0	202	202	38.4	569
18	9	7.9	-7.9	88.0	0.073	0.25	23	1548	1425	38.4	3982
19	10	9.3	-9.3	88.0	0.073	0.45	13	2077	2024	38.4	5656
20	11	10.5	-10.5	87.5	0.073	0.57	10	2337	2301	38.4	6436
21	12	-11.6	11.6	87.0	0.073	0.40	11	1957	1921	38.4	5377
22	pnt	X	Y	temp	p	ΔP	angle	Vtang	Vnorm	Area	ρVA
23	-	ft.	ft.	°F	lb/ft³	in.H2O	deg	fpm	fpm	ft²	lb/min
24	13	11.6	11.6	87.0	0.073	0.35	25	1830	1659	38.4	4644
25	14	10.5	10.5	88.5	0.073	0.37	14	1884	1828	38.4	5105
26	15	9.3	9.3	89.0	0.073	0.38	12	1911	1869	38.4	5213
27	16	7.9	7.9	90.0	0.073	0.32	19	1755	1659	38.4	4620
28	17	6.1	6.1	90.0	0.073	0.15	24	1201	1098	38.4	3056
29	18	3.2	3.2	85.0	0.073	0.00	0	44	44	38.4	123
30	19	-3.2	-3.2	85.0	0.073	0.00	0	31	31	38.4	87
31	20	-6.1	-6.1	85.0	0.073	0.00	0	195	195	38.4	549
32	21	-7.9	-7.9	92.0	0.072	0.26	23	1585	1459	38.4	4047
33	22	-9.3	-9.3	91.5	0.072	0.47	12	2130	2083	38.4	5784
34	23	-10.5	-10.5	90.0	0.073	0.57	16	2342	2251	38.4	6268
35	24	11.6	11.6	88.0	0.073	0.44	40	2054	1573	38.4	4397
36	*Note: The air doesn't flow straight up. You must								88.7		89,383
37	measure the angle and multiply by cos(θ).										

The points of a fan traverse must correspond to equal areas. This is described in several ASME and CTI test codes. The relationships are as follows. The net flow area, excluding the hub, is given by:

$$A = \frac{\pi}{4}\left(D_{FAN}^2 - D_{HUB}^2\right) \tag{5.1}$$

Each of the *n* equal annular areas corresponding to the measurement points is given by:

$$\frac{A}{n} = \pi\left(R_{i+1}^2 - R_i^2\right) \qquad (5.2)$$

The first (inner most) and last (outer most) radii correspond to the hub and fan, respectively:

$$R_1 = \frac{d}{2} \quad R_N = \frac{D}{2} \qquad (5.3)$$

The measurement is taken at the center of each annulus. The outer radius of each annulus is given by:

$$R_O = \sqrt{\frac{i}{n}D_{FAN}^2 + \frac{n-i}{n}D_{HUB}^2} \qquad (5.4)$$

This test consisted of two perpendicular traverses with 12 points along each diameter (6 per radius), spaced to be representative of equal areas. Note that the air does not flow vertically upward. You must measure the angle at each point and then multiply by the $\cos(\theta)$. The total mass flow rate and mass weighted average temperature are easily calculated with functions built into Excel®. The measurement points are illustrated in this next figure:

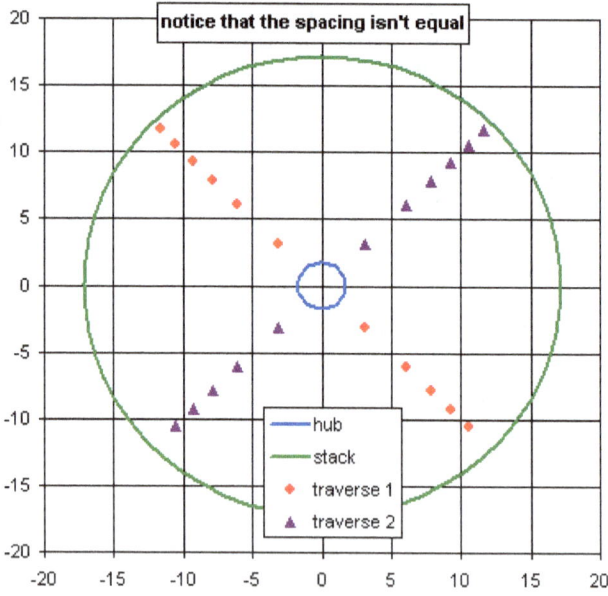

The spreadsheet (ACC_fan_test.xls) is contained in the archive. This is only part of the overall test. There is also the thermal performance on another tab in the same spreadsheet as the performance curves. In this case, two tests were conducted: one in the winter and a second in the summer:

	A	B	C	D	E	F
1	Winter Test					
2	Inputs	Units	Test 1	Test 2	Test 3	Avg.
3	fan spd.	%	75.0%	75.0%	75.0%	75.0%
4	stm. flow	%	102.1%	102.4%	102.6%	102.4%
5	amb. air	°F	48.2	49.5	51.1	49.6
6	stm. pres.	in.HgA	2.19	2.27	2.39	2.28
7	stm. temp.	°F	104.1	105.3	107.1	105.5
8	Calculations	Units	Test 1	Test 2	Test 3	Avg.
9	exp. pres.	in.HgA	2.22	2.31	2.41	2.31
10	exp. pres.	Δ%	-1.3%	-1.5%	-0.9%	-1.2%
11	exp. temp.	°F	104.6	105.9	107.4	105.9
12	exp. temp.	Δ°F	-0.4	-0.5	-0.3	-0.4
13	Corrections	Units	Test 1	Test 2	Test 3	Avg.
14	at design	in.HgA	1.70	1.70	1.71	1.71
15	guarantee	in.HgA	1.73	1.73	1.73	1.73
16	as-tested	Δ%	-1.3%	-1.5%	-0.9%	-1.2%

This ACC passed the winter test by approximately 1.2%, in spite of the fans running at 70% speed and condensing 2.4% additional flow. The summer test was not so impressive:

	A	B	C	D	E	F
18	Summer Test					
19	Inputs	Units	Test 1	Test 2	Test 3	Avg.
20	fan spd.	%	100.0%	100.0%	100.0%	100.0%
21	stm. flow	%	98.0%	97.9%	97.5%	97.8%
22	amb. air	°F	97.3	98.0	98.7	99.3
23	stm. pres.	in.HgA	5.49	5.57	5.66	5.57
24	stm. temp.	°F	137.3	137.9	138.5	137.9
25	Calculations	Units	Test 1	Test 2	Test 3	Avg.
26	exp. pres.	in.HgA	5.40	5.48	5.56	5.48
27	exp. pres.	Δ%	1.7%	1.6%	1.8%	1.7%
28	exp. temp.	°F	136.6	137.2	137.8	137.2
29	exp. temp.	Δ°F	0.6	0.6	0.7	0.7
30	Corrections	Units	Test 1	Test 2	Test 3	Avg.
31	at design	in.HgA	1.76	1.76	1.76	1.76
32	guarantee	in.HgA	1.73	1.73	1.73	1.73
33	as-tested	Δ%	1.7%	1.6%	1.8%	1.7%

The same ACC failed a second performance test ten months later by 1.7% with the fans at 100% and condensing only 97.8% of the design steam flow. There are more variables impacting the performance of ACCs than WCCs and this should not be overlooked when considering a performance testing plan.

Chapter 6. Simple Shell-and-Tube Heat Exchanger

Condensers were introduced before simple shell-and-tube heat exchangers, as their analysis is often less complicated. For one thing, the LMTD and NTU-ε methods are the same for condensers because C_P on the condensing side is infinite. Because the temperature on one side of a condenser is constant, geometry isn't an issue either.

The simplest type of shell-and-tube heat exchanger to analyze is one that has sufficient baffles to approximate a counter-current double-pipe geometry, as illustrated schematically in this next figure:

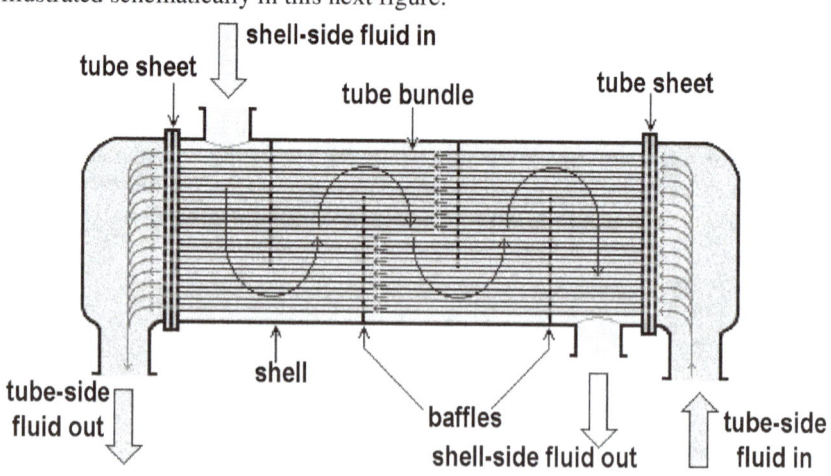

Performance Prediction

The heat transfer coefficient inside the tubes is fairly simple to calculate and there are several equations to choose from. No one should ever design a heat exchanger where the flow inside the tubes isn't turbulent, so there's no point developing a correlation for anything but turbulent flow. The classic equation was developed by Dittus&Boelter[6]:

$$Nu = 0.023 \, Re^{\frac{4}{5}} \, Pr^n \quad (6.1)$$

where n=0.4 for heating and n=0.3 for cooling. A more recent correlation has been developed by Sieder&Tate[7]:

$$Nu = 0.027 \, Re^{\frac{4}{5}} \, Pr^{\frac{1}{3}} \left(\frac{\mu_{BULK}}{\mu_{SURFACE}} \right)^{0.14} \quad (6.2)$$

Yet another correlation has been developed by Rabas&Cane[8]:

$$Nu = 0.0158 \, Re^{0.835} \, Pr^{0.462} \quad (6.3)$$

Resistance of the tube wall can be found in any heat transfer text:

$$R_W = \frac{D_O}{2k_W} \ln\left(\frac{D_O}{D_I}\right) \quad (6.4)$$

There are numerous correlations for the shell side heat transfer coefficients. You'll have to find one that is appropriate for your particular application. These vary with the fluids, baffles, and tube spacing. Fouling is often used as a *fudge* factor. The overall heat transfer coefficient is found by summing the resistances:

$$\frac{1}{U} = \left(\frac{D_O}{D_I}\right)(R_{FI} + R_T) + R_M + R_S + R_{FO} \quad (6.5)$$

Here R_{FI} and R_{FO} are the inside and outside fouling resistance, respectively. R_T, R_W, and R_S are the tube side, tube material, and shell side resistances, respectively. The D_O/D_I term accounts for the fact that the inner and outer surface areas of a tube aren't the same per unit length.

---

[6] Dittus, P. W. and L. M. Boelter, L.M., University of California Publications in Engineering, Vol. 1, No. 13, pp. 443-461 1930 (reprinted in *International Communications in Heat and Mass Transfer*, Vol. 12, pp. 3-22, 1985).

[7] Sieder, E. N. and G. E. Tate, "Heat Transfer and Pressure Drop of Liquids in Tubes," Industrial Engineering Chemistry, Vol. 28, p. 1429, 1936.

[8] Rabas, T. J., and D. Cane, "An Update of Intube Forced Convection Heat Transfer Coefficients of Water," *Desalinization*, Vol. 44, pp. 109–119, 1983.

There are also several methods for estimating the shell side heat transfer coefficient. One of the earliest methods was developed by Kern[9], based on industrial heat exchangers and is quite similar to the Sieder-Tate.

$$Nu = 0.36 \, Re^{0.55} \, Pr^{\frac{1}{3}} \left(\frac{\mu_{BULK}}{\mu_{SURFACE}} \right)^{0.14} \tag{6.6}$$

This correlation could also be cast in the same form as Dittus-Boelter for convenience:

$$Nu = 0.36 \, Re^{0.55} \, Pr^{n} \tag{6.7}$$

The length in both the Nusselt and Reynolds number is an equivalent diameter (D_E), that takes into account the tube spacing, pitch (P), and bundle alignment. This first equation (6.7) is for a square tube arrangement:

$$D_E = \frac{4P^2}{\pi D_O} - D_O \tag{6.8}$$

and this second (6.8) is for a triangular arrangement:

$$D_E = \frac{2\sqrt{3}P^2}{\pi D_O} - D_O \tag{6.9}$$

The shell-side velocity, V_s, is given by:

$$V_S = \frac{\dot{m}_S}{\rho N_R (P - D_O) \left(\dfrac{L_T}{N_B} \right)} \tag{6.10}$$

where N_R is the number of tubes per row (across the flow), L_T is the tube length, and N_B is the number of baffles. You may want to adjust the effective shell-side area to better represent the actual tube bundle and baffle arrangement in your heat exchanger.

Performance Evaluation

We will first consider a side-stream oil cooler. As usual, we are only given minimal information by the manufacturer and have been asked to conduct a performance test to determine if it is fouled and how badly. We are given the design operating conditions, from which we will infer the overall heat transfer coefficient and as-new fouling resistance. For this simple arrangement and almost constant properties the LMTD and NTU-ε methods yield essentially the same results.

---

[9] Kern, D. Q., *Process Heat Transfer*, McGraw-Hill, 1950.

The following table shows the geometry, properties, and performance calculations at the design conditions:

Side-Steam Oil Cooler - Design Calculations					
geometry	units	value	dimensions	units	result
tubes	-	16	inside dia.	mm	9.40
tubes/row	-	4	tube clear.	mm	4.45
outside dia.	mm	12.7	baffle spacing	m	0.46
tube wall	mm	1.7	shell eff. dia.	mm	12.8
tube pitch	mm	17.1	calculations	units	result
tube len.	m	3.66	heat transfer	kW	50.8
baffles	-	8	water flow	kg/hr	7,521
operating point	units	value	LMDT	°C	84.3
oil flow	kg/hr	3,402	UA	W/C°	602
oil inlet temp.	°C	115.6	A	m²	2.33
oil exit temp.	°C	90.6	U	W/m²/°C	258
water in tmp.	°C	15.6	tube vel.	m/s	0.99
water out tmp.	°C	21.1	tube Re	-	2,546
properties	units	value	tube Nu	-	40.1
tube cond.	W/m/°C	25.9	tube side h	W/m²/°C	546
oil density	kg/m³	859	tube side R	°C-m²/W	0.0018
oil spec. ht.	kJ/kg/°C	2.15	tube wall h	W/m²/°C	4732
oil viscosity	cp	3.14	tube wall R	°C-m²/W	0.0002
oil cond.	W/m/°C	0.128	shell vel.	m/s	0.26
oil Prandtl	-	52.8	shell Re	-	3,145
water density	kg/m³	998	shell Nu	-	68
water sp. ht.	kJ/kg/°C	4.38	shell side h	W/m²/°C	3179
water visco.	cp	1.05	shell side R	°C-m²/W	0.0003
water cond.	W/m/°C	0.597	fouling h	W/m²/°C	1140
water Prandtl	-	7.68	fouling R	°C-m²/W	0.0009

The properties are calculated using interpolation and tables that are also provided in the same spreadsheet. The geometry (number and size of tubes), inlet and outlet temperatures, and oil flow are provided. The heat transfer is calculated from the oil side and the water flow rate from that. The overall heat transfer coefficient is calculated from $U=Q/A/LMTD$, the tube side, wall, and shell side heat transfer coefficients are calculated, leaving only the fouling resistance, R_W.

The four one-hour test periods are summarized in the following table:

Side-Stream Oil Cooler - Test Calculations						
Measurements	units	Test 1	Test 2	Test 3	Test 4	Average
oil flow	kg/hr	3,241	3,241	3,235	3,245	3,240
oil inlet temp.	°C	112.9	113.2	113.0	113.9	113.3
oil exit temp.	°C	89.8	90.5	91.4	91.6	90.8
water in tmp.	°C	14.4	15.5	17.5	15.0	15.6
water out tmp.	°C	18.9	20.1	22.0	19.5	20.1
properties	units	value				
tube cond.	W/m/°C	15.0	16.0	17.0	18.0	16.5
oil density	kg/m³	860.38	860	860	860	860
oil spec. ht.	kJ/kg/°C	2.147	2.15	2.15	2.15	2.15
oil viscosity	cp	3.24	3.21	3.19	3.16	3.20
oil cond.	W/m/°C	0.128	0.128	0.128	0.128	0.128
oil Prandtl	-	54.4	53.9	53.6	53.1	53.8
water density	kg/m³	998.77	999	998	999	999
water sp. ht.	kJ/kg/°C	4.38	4.38	4.38	4.38	4.38
water visco.	cp	1.09	1.06	1.01	1.08	1.06
water cond.	W/m/°C	0.594	0.596	0.599	0.595	0.596
water Prandtl	-	8.05	7.80	7.38	7.92	7.79
calculations	units	result				
heat transfer	kW	44.6	44.0	41.8	43.3	43.4
water flow	kg/hr	8,054	7,873	7,510	8,062	7,875
LMDT	°C	84.4	83.8	82.2	85.2	83.9
UA	W/C°	529	526	508	508	518
U	W/m²/°C	227	225	218	218	222
tube vel.	m/s	0.94	0.94	0.94	0.94	0.94
tube Re	-	2,349	2,372	2,383	2,415	2,380
tube Nu	-	38.0	38.1	38.2	38.5	38.2
tube side h	W/m²/°C	517	519	520	524	520
tube side R	°C-m²/W	0.0019	0.0019	0.0019	0.0019	0.0019
tube wall h	W/m²/°C	4732	4732	4732	4732	4732
tube wall R	°C-m²/W	0.0002	0.0002	0.0002	0.0002	0.0002
shell vel.	m/s	0.28	0.27	0.26	0.28	0.27
shell Re	-	3,231	3,247	3,256	3,283	3,254
shell Nu	-	71	70	68	71	70
shell side h	W/m²/°C	3271	3250	3201	3283	3251
shell side R	°C-m²/W	0.0003	0.0003	0.0003	0.0003	0.0003
fouling H	W/m²/°C	781	758	680	666	721
fouling R	°C-m²/W	0.0013	0.0013	0.0015	0.0015	0.0014
					-14%	59%

As shown at the bottom of the table, the fouling resistance has increased from 0.0009 °C-m²/W to 0.0014 (59%) for a reduction in the overall conductance from 258 W/m²/°C to 222 (14%).

Chapter 7. Feedwater Heater

There are three factors that can greatly complicate heat exchanger analysis. We have already discussed two: varying properties and geometry. The third is zones of phase change within the heat exchanger. Each of these factors is ignored (glossed over, set aside) during the process of separation and integration of the governing differential equation in order to obtain a closed-form solution.

Feedwater heaters are perhaps the most common type of heat exchanger that has all three zones (de-superheating, condensing, and sub-cooling) within the same shell. Heat Recovery Steam Generators (HRSGs) used in combined cycle power plants also have these three zones, as to all conventional boilers, but the three are separated into discrete sections in those designs.

Performance Prediction

In order to explore this type of heat exchanger, it is helpful to consider a graph with heat transfer along the horizontal axis and temperature along the vertical, as in the following figure:

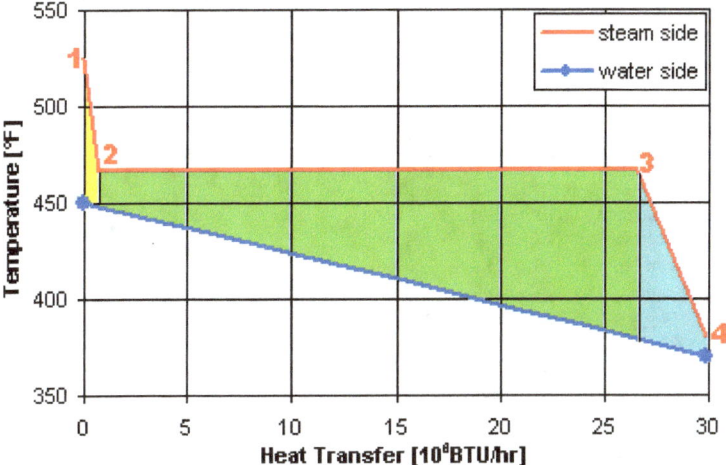

The yellow area represents the de-superheating zone, in which the steam is cooled from a superheated state to saturated vapor. The red process line has a sharp downward slope between points 1 and 2. The green area represents the condensing zone, in which the steam condenses on the outside of the tube bundle. The red process line is flat over this zone because the saturation temperature is constant. The cyan area represents the sub-cooling zone, in which the condensate is further cooled, as it approaches the inlet feedwater temperature. The red process line from point 3 to 4 has a downward slope that is not as steep as the line from point 1 to 2 because the specific heat of the condensate is larger than the specific heat of the vapor and the rate of change of temperature along the process line is given by:

$$\frac{dT}{dq} = \frac{-1}{\dot{m}C_P} \tag{7.1}$$

Neither the F-LMTD nor the P-NTU method can be applied to this heat exchanger as a whole because there are three thermally discontinuous zones set apart by the phase change. Inside the shell there are often partitions plus there is always an interface below which the tube bundle is flooded. However, these methods can be used effectively on each of the three sections separately, as illustrated in the spreadsheet, feed_water_heater_analytical.xls, which is included in the archive.

	A	B	C	D	E	F	G	H	I
1				Feed Water Heater Example					
2	INPUTS	units		CALCULATIONS	units				
3	feed water inlet			feed water inlet					
4	flow	lbm/hr	350,000	enthalpy	BTU/lbm	345.9			
5	pressure	psia	2250	feed water outlet					
6	temperature	°F	370	enthalpy	BTU/lbm	431.4			
7	feed water outlet			extraction					
8	pressure	psia	2200	saturation	°F	467.0			
9	temperature	°F	450	TTD	°F	17.0			
10	extraction			superheat	°F	58.0			
11	pressure	psia	500	enthalpy	BTU/lbm	1249.7			
12	temperature	°F	525	flow	lbm/hr	33,382			
13	drain			drain					
14	temperature	°F	380	DCA	°F	10.0			
15				enthalpy	BTU/lbm	354.0	U	A	area
16	heat trans. zone	Qx10⁶	UA	zone color	units	ΔT	BTU/hr/ft²/°F	ft²	frac
17	de-superhtng	1.5	0.035	yellow	°F	42.4	100	354	23%
18	condensing	25.2	0.537	green	°F	46.9	600	895	58%
19	subcooling	3.2	0.089	cyan	°F	36.0	300	295	19%
20	process line		0.661	calculations		T		1544	
21	point 1	0	525	sat. vapor	BTU/lbm	1204.7			
22	point 2	1.5	467	sat. liquid	BTU/lbm	449.5			
23	point 3	26.7	467	desup. exit temp	°F	446.0			
24	point 4	29.9	380	cond. exit temp.	°F	378.5			

The LMTD in the de-superheating, condensing, and sub-cooling (yellow, green, and cyan) zones is 42.4, 46.9, and 36.0°F, respectively. The heat transfer in these three zones is 1.5, 25.2, and 3.2x10⁶ BTU/hr, respectively. The conductance, **UA=Q/LMTD**, in these three zones is 0.035, 0.537, and 0.089x10⁶ BTU/hr/°F, respectively for a total of 0.661x10⁶ BTU/hr/°F.

Typical overall heat transfer coefficients, *U*, for the de-superheating, condensing, and sub-cooling zones would be: 100, 600, and 300 BTU/hr/ft²/°F. This would indicate required surface areas of 354, 895, and 295 ft², respectively, for a total of 1544 ft². As there is typically a single tube U-shaped bundle in these heat exchangers, this means that 23% of the tube area at the top near the steam inlet is surrounded by dry steam on the shell side, 58% of the tube area is dripping with condensate, and 19% at the bottom is flooded.

Feedwater heater performance is most often specified in terms of Terminal Temperature Difference (TTD) and Drain Cooler Approach (DCA). By industry convention, TTD, rather than being the actual terminal difference, is the saturation temperature of the steam in the condensing zone minus the feedwater outlet temperature. This value can be zero and even negative if there is sufficient superheat of the extraction steam. TTD is not used in the sense of $Q \neq UA*\Delta T$, so a value less than or equal to zero does not indicate a violation of the Second Law of Thermodynamics. The following figure shows the relative impact of TTD and DCA on the required size of the heat exchanger in this example.

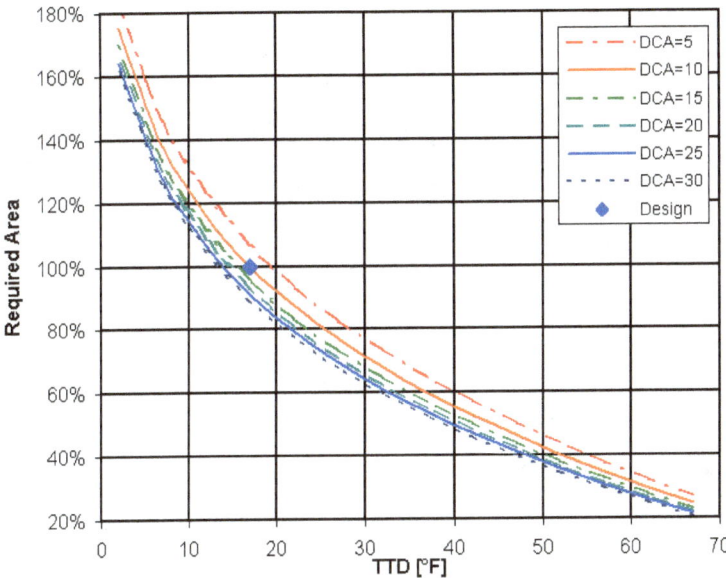

Performance Evaluation

Performance test calculations for a feedwater heater are similar to the crossflow and TEMA-E examples except that the process must be separated into the three zones (de-superheating, condensing, and sub-cooling).

Feed Water Heater Test with Uncertainty

measurement	units	average	points	std.dev.	bias	total unc.	sensitivity	units	contrib
Tfw,in	°F	371.5	30	0.63	0.14	0.28	-213.021	BTU/hr/°F/°F	-58.74
Tfw,out	°F	451.3	30	0.47	0.14	0.23	19948.34	BTU/hr/°F/°F	4530
Textract	°F	524.8	30	0.58	0.14	0.26	-964.167	BTU/hr/°F/°F	-250.7
Tdrain	°F	380.3	30	0.71	0.14	0.30	-3949.15	BTU/hr/°F/°F	-1191
Pfw,in	psia	2248	30	0.93%	0.75%	0.01	-11.601	BTU/hr/°F/psia	-0.096
Pfw,out	psia	2201	30	0.86%	0.75%	0.01	5.699826	BTU/hr/°F/psia	0.046
Pextract	psia	499	30	0.74%	0.75%	0.8%	-2985.15	BTU/hr/°F/psia	-23.86
flow,fw	lbm/hr	337,213	30	0.98%	2.0%	2.0%	1.966478	BTU/hr/°F/lbm/hr	0.04
calculation	units	value							4691
hfw,in	BTU/lbm	347.5							5.2%
hfw,out	BTU/lbm	432.8							
hextract	BTU/lbm	1249.6							
hg	BTU/lbm	1204.7							
hf	BTU/lbm	449.3							
hdrain	BTU/lbm	354.3							
flow,extract	lbm/hr	32,123							
Qdesuper	BTU/hr	1.4E+06							
Qconds	BTU/hr	2.4E+07							
Qsubcool	BTU/hr	3.1E+06							
Tsat	°F	466.8							
TTD	°F	15.5							
DCA	°F	8.8							
Tfw,desup	°F	447.3							
Tfw,subco	°F	380.0							
LMTD,desup	°F	40.7							
LMTD,cond	°F	45.1							
LMTD,subc	°F	34.1							
UA,desup	BTU/hr/°F	3.5E+04							
UA,conds	BTU/hr/°F	5.4E+05							
UA,subco	BTU/hr/°F	9.0E+04							
UA,total	BTU/hr/°F	6.6E+05							

Chapter 8. Heat Recovery Steam Generator

Heat Recovery Steam Generators (HRSGs) are common in twenty-first century power plants. Gas Turbines (GTs) are fairly efficient; yet continuously belch out large quantities of hot exhaust. Run this hot air into a heat exchanger, make steam, pass that through a turbine attached to a generator, and achieve even higher efficiency. The combination (GT+HRSG) is called a combined cycle power plant (CCPP) and is even more efficient than a coal-fired power plant (CFPP). The following is a typical HRSG: The GT exhaust enters on the bottom left and the stack gas leaves at the top right.

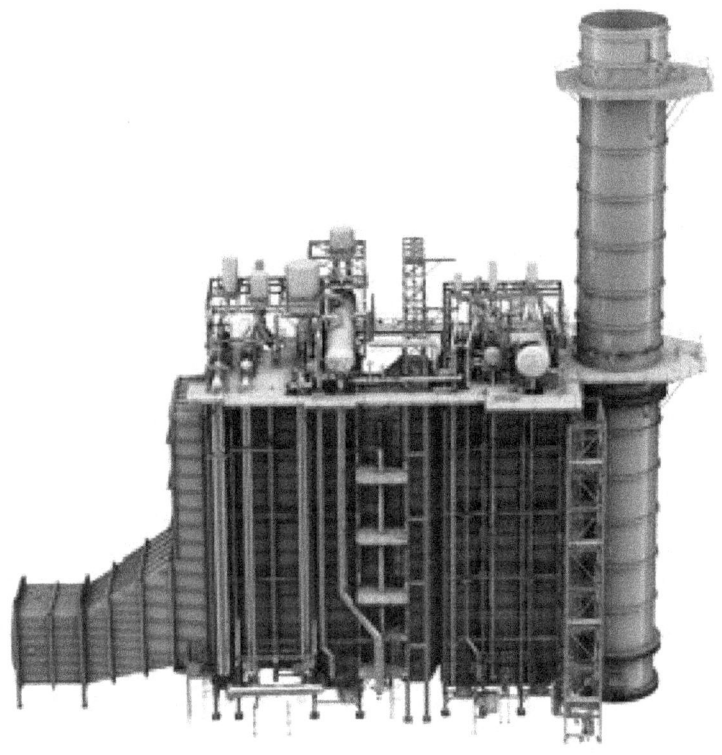

In the current fuel market with cheap natural gas and regulators putting the squeeze on emitters of what are perceived to be greenhouse gases, combined cycle power plants are far more economical than ones burning coal. Of course, some day the gas wells may run out or wars may impact the supply or some celebrity's house may fall into a sink hole near a gas field where there's been cracking so that the price of gas goes through the roof and we may be sorry to have torn down all the coal plants.

While combined cycle plants are thermodynamically efficient, they are by no means cheap. Turbines, heat exchangers, condensers, and cooling towers all cost a hundred million dollars. There are two basic types of HRSGs: single and triple pressure. The former deliver steam to a turbine without reheat or for process (e.g., a paper mill or desalination plant) and the latter to a steam turbine with reheat.

The three pressures in the HRSG are necessitated by the steam turbine design, not the aspects of the heat exchanger. Details of the reheat Rankine cycle are beyond the scope of this book, but suffice it to say that this is to keep the back end of the turbine out of the wet steam zone as much as possible. In a HRSG there are three distinct zones of differing heat exchange: superheating, evaporating, preheating (equivalent of de-sub-cooling or bringing the sub-cooled liquid up to the saturated liquid state).

The three heat exchange processes in a HRSG are accomplished in different types of heat exchangers: superheater, evaporator, and economizer. Most often, these will be three separate devices. Water flows inside tubes and exists only as a single phase in a superheater (vapor) and economizer (liquid), but transitions from liquid to vapor in an evaporator. The GT exhaust is always in a gaseous phase and flows through the shell in all three types. A typical single-pressure design is shown below:

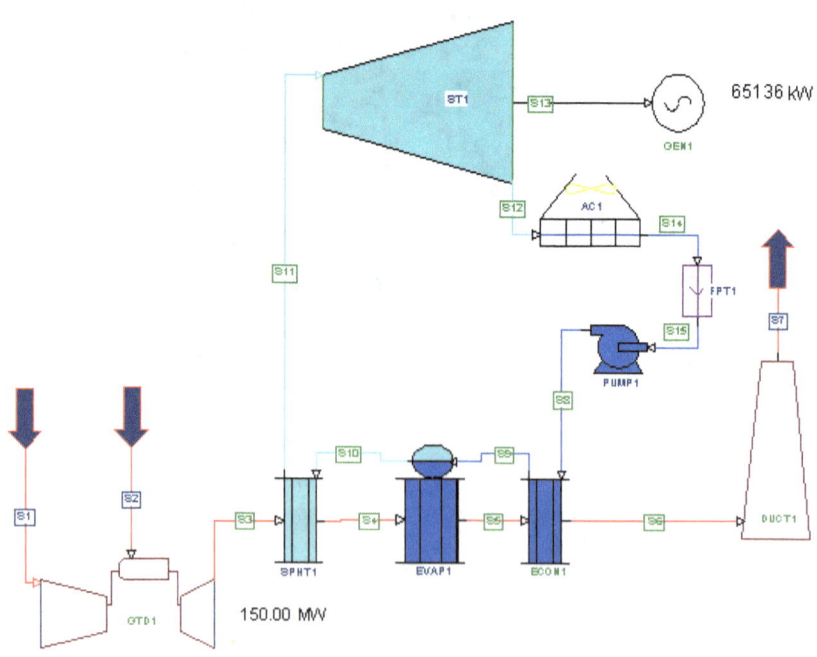

Performance Prediction

Thermal performance of a HRSG is illustrated by what is called a *heat release diagram*, which is a graph showing cumulative heat transferred from the gas to the steam on the horizontal axis and temperature on the vertical axis. The hot gas process line is drawn in red and the steam process line is drawn in blue.

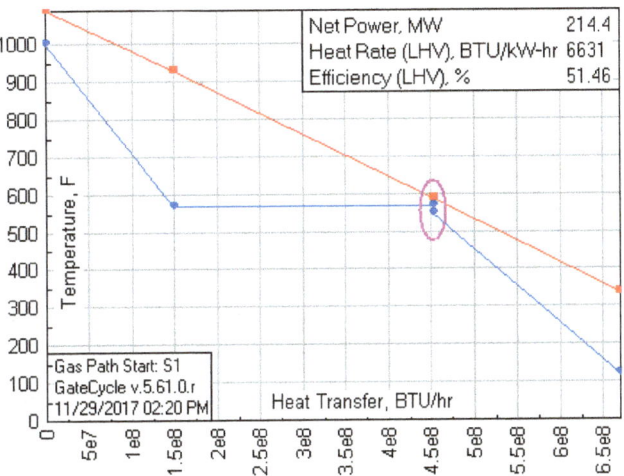

The slope of the hot process line is continuous and almost linear because the specific heat of the GT exhaust gas varies only slightly over this range of temperatures. The steam process line is discontinuous, owing to the fact that the first section is the superheater, the second is the evaporator, and the third is the economizer. This figure is very similar to the one in the preceding chapter on feedwater heaters and for the same reasons.

The area identified by the magenta ellipse is called a *pinch point* and occurs only this once in a single-pressure HRSG. As with the feedwater heater, each section can be analyzed using either the LMTD or P-NTU methods. In either case, the temperature difference at the pinch point is small, resulting in a large required UA. In fact, the performance of HRSGs is dominated by pinch points.

The pinch point shown above will limit the amount of steam that can be generated. A *tighter* (i.e., smaller) pinch will produce more steam produced and a *looser* (i.e., larger) pinch will produce less steam. The amount of steam produced is almost linearly proportional to the pinch, as illustrated in the following figure. The required surface area (or more accurately required UA) is quite a different matter. A tighter pinch would require a much larger UA, which would mean much more expensive heat exchangers.

Designing a cost-effective HRSG is an art and requires juggling more than just surface areas.

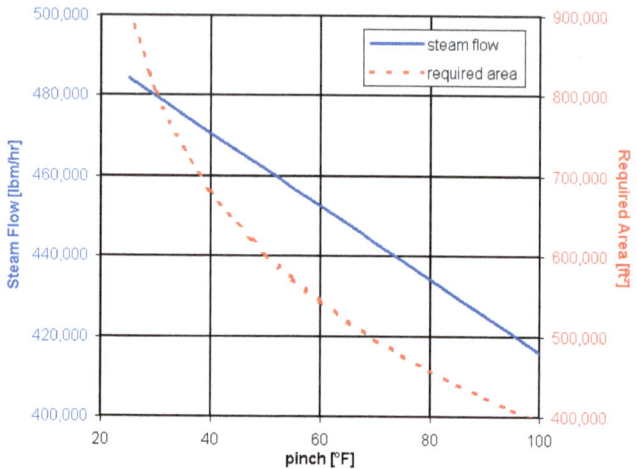

The following is a typical three-pressure HRSG heat release diagram:

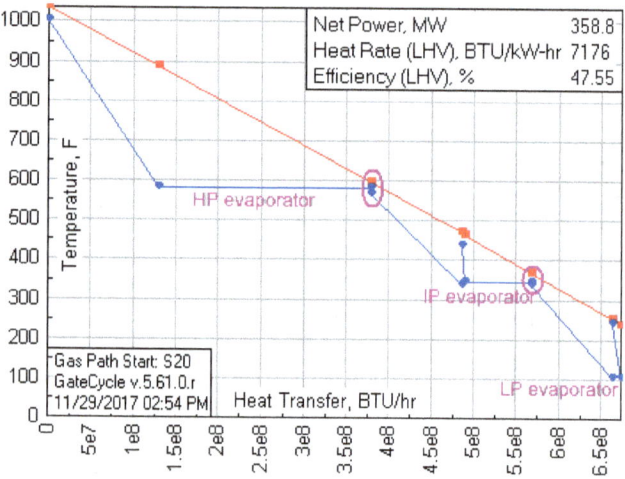

Thermal analysis of the three components that are combined to create a HRSG (superheater, evaporator, and economizer) is the same as for the individual components or a feedwater heater. The arrangement, construction specifics, and properties may be different, but these are all crossflow heat exchangers, as the water and steam flows inside vertical tubes and the exhaust gas flows across the outside of the tubes.

Performance Evaluation

The following is typical of data collected for a HRSG performance test as conducted in accordance with ASME PTC 4.4-2008 (see HRSG_test.xls):

	A	B	C	D	E	F	G	H	I	J	K	L
1		EGW	EGT	STACK	STEAM	FW	EVAP	SPHT	FW	ECON	EVAP	SPHT
2		lb/hr	°F	°F	lb/hr	psia	psia	psia	°F	°F	°F	°F
3	7/11/2017 11:00	3560009	1062.184	347.2219	474780.8	1410.52	1268.351	1204.69	109.8975	554.5939	574.2497	994.1239
4	7/11/2017 11:01	3563845	1062.229	347.3358	475826.9	1413.847	1271.057	1207.279	110.0538	554.7798	574.5233	994.1703
5	7/11/2017 11:02	3556583	1062.518	347.5338	475015.8	1412.153	1269.844	1206.209	110.5083	554.6401	574.4008	996.0175
6	7/11/2017 11:03	3560494	1062.37	347.8043	475418.8	1412.609	1270.061	1206.332	110.2606	554.6644	574.4227	994.2704
7	7/11/2017 11:04	3552742	1062.557	347.6749	474764.9	1410.732	1268.572	1204.929	110.5626	554.5766	574.2722	994.5612
8	7/11/2017 11:05	3564397	1062.12	348.0719	475776.4	1414.592	1271.832	1208.113	109.8982	554.8629	574.6016	995.9677
9	7/11/2017 11:06	3568099	1062.197	348.0178	476385.4	1415.531	1272.409	1208.563	110.021	554.8383	574.6599	993.8743
10	7/11/2017 11:07	3549199	1062.758	347.8878	474066.5	1408.799	1267.053	1203.505	110.8345	554.4473	574.1184	995.289
11	7/11/2017 11:08	3560778	1062.615	347.8424	474816.8	1411.941	1269.751	1206.163	110.5964	554.8085	574.3914	996.7969
12	7/11/2017 11:09	3559480	1062.706	348.1149	475709.6	1413.505	1270.785	1207.022	110.8229	554.6433	574.4958	994.2205
13	7/11/2017 11:10	3558904	1062.693	348.0472	475702.3	1414.393	1271.677	1207.969	110.836	554.7864	574.5859	996.053
14	7/11/2017 11:11	3560813	1062.862	348.0524	475932.4	1414.32	1271.467	1207.685	111.0774	554.7416	574.5647	994.3447
15	7/11/2017 11:12	3552305	1063.119	347.7818	475405	1413.676	1271.136	1207.475	111.5319	554.6536	574.5313	996.6328
16	7/11/2017 11:13	3552606	1062.882	348.3405	474236.8	1409.904	1268.057	1204.523	111.0212	554.6046	574.22	996.4935
17	7/11/2017 11:14	3558228	1062.961	348.1659	474855.1	1411.91	1269.697	1206.096	111.1537	554.7639	574.3859	996.5723
18	7/11/2017 11:15	3552335	1062.948	348.3035	474776	1411.633	1269.467	1205.873	111.1917	554.6356	574.3625	996.5461
19	7/11/2017 11:16	3556490	1063.007	348.0992	475358.4	1412.652	1270.14	1206.432	111.2825	554.7296	574.4307	994.7662
20	7/11/2017 11:17	3550507	1063.192	347.9697	474352.8	1410.104	1268.189	1204.632	111.5227	554.6177	574.2333	996.0248
21	7/11/2017 11:18	3546866	1063.384	348.0148	473955.2	1409.336	1267.656	1204.173	111.8187	554.495	574.1794	997.0692
22	7/11/2017 11:19	3562278	1062.778	348.6291	475352.5	1413.35	1270.735	1207.016	110.8908	554.8507	574.4908	994.9742
23	7/11/2017 11:20	3551808	1063.544	347.9311	475004.1	1412.144	1269.843	1206.211	112.1186	554.6127	574.4006	996.0909

EGW and EGT are the industry-standard acronyms for GT exhaust flow and temperature, respectively. The GT exhaust flow cannot be measured, rather it is calculated from an energy balance as part of a simultaneous performance test in accordance with ASME PTC 22-2005.

The steam flow is determined from the feedwater flow, as measured through a precision nozzle, and may be adjusted to account for makeup and/or blowdown. Flow measurement of the feedwater in the liquid state is far more accurate than steam in the vapor state, which is why this measurement point is preferred.

> *Flow* meters *are simply* **not** *accurate, regardless of what you might have been told. For flow measurements consult ASME PTC 19.5-2004, MFC-3M (2004), or MFC-11M (2003). The only accurate* meter *is a Coriolis and you will not find one on the feedwater or steam line in a HRSG. That's why there should always be a precision nozzle somewhere on the feedwater line.*

In accordance with the test code, data are divided into one-hour test periods. Averages are calculated using =AVERAGE(INDIRECT(ADDRESS(...))) and defining the appropriate rows and columns on the data tab. The analysis tab is divided into sections: data, properties, and calculations. The input section is:

	A	C	D	E	F	G	H
1			HRSG Performance Test				
4	Test Inerval		Test 1	Test 2	Test 3	Test 4	average
5	begin		11:00:00	12:00:31	12:58:57	14:00:31	
6	end		12:00:31	12:58:57	14:00:31	15:00:00	
7	INPUT DATA	units					
8	GT Exhaust						
9	flow	lbm/hr	3,541,849	3,502,088	3,502,019	3,543,139	3,522,274
10	temperature	°F	1064.0	1067.7	1067.8	1064.0	1065.9
11	Stack						
12	temperature	°F	348.4	349.9	349.9	348.5	349.2
13	Feedwater						
14	flow	lbm/hr	474,318	472,121	472,133	474,353	473,231
15	pressure	psia	1410.2	1404.7	1404.6	1410.3	1407.5
16	temperature	°F	112.8	118.3	118.3	112.8	115.6
17	Economizer						
18	temperature	°F	554.5	553.9	553.9	554.5	554.2
19	Evaporator						
20	pressure	psia	1268.3	1264.1	1264.0	1268.4	1266.2
21	temperature	°F	574.2	573.8	573.8	574.3	574.0
22	Superheater						
23	pressure	psia	1204.7	1200.9	1200.8	1204.9	1202.8
24	temperature	°F	996.5	999.6	999.4	996.6	998.0

Steam properties used are the same ones in feedwater_heater_analytical.xls plus three additional functions have been provided as macros for the GT exhaust gas properties appropriate for this particular test. These could be calculated based on the information in the appendix of PTC 4.4 or from the NASA Glenn tables (NASA TP-2002-211556). The property section is:

	A	C	D	E	F	G	H
25	PROPERTIES	units					
26	GT Exhaust						
27	enthalpy	BTU/lbm	263.9	264.9	264.9	263.8	264.4
28	Stack						
29	enthalpy	BTU/lbm	72.7	73.1	73.1	72.7	72.9
30	Feedwater						
31	enthalpy	BTU/lbm	84.5	89.9	89.9	84.4	87.2
32	Economizer						
33	enthalpy	BTU/lbm	554.7	554.0	554.0	554.8	554.4
34	Evaporator						
35	enthalpy	BTU/lbm	1181.8	1182.0	1182.0	1181.8	1181.9
36	Superheater						
37	enthalpy	BTU/lbm	1497.2	1499.1	1499.0	1497.2	1498.1

The calculation section is:

38	CALCULATIONS	units					
39	Heat to Steam						
40	Economizer	10⁶BTU/hr	223.1	219.1	219.1	223.1	221.1
41	Evaporator	10⁶BTU/hr	297.4	296.5	296.5	297.4	297.0
42	Superheater	10⁶BTU/hr	149.6	149.7	149.7	149.6	149.7
43	Total	10⁶BTU/hr	670.1	665.3	665.3	670.2	667.7
44	Heat from Gas						
45	Total	10⁶BTU/hr	677.0	671.7	671.8	677.2	674.4
46	Loss	10⁶BTU/hr	6.9	6.4	6.5	7.0	6.7
47	Loss	%	1.03%	0.95%	0.96%	1.03%	0.99%
48	Gas Enthalpy						
49	Superheater	BTU/lbm	221.2	221.7	221.7	221.2	221.5
50	Evaporator	BTU/lbm	136.4	136.3	136.3	136.4	136.3
51	Gas Temp.						
52	Superheater	°F	910.0	912.0	912.1	910.0	911.0
53	Evaporator	°F	594.1	593.8	593.8	594.2	593.9
54	LMTDs						
55	Superheater	°F	167.3	168.6	168.8	167.1	167.9
56	Evaporator	°F	111.7	112.4	112.5	111.8	112.1
57	Economizer	°F	87.2	86.3	86.3	87.3	86.8
58	Heat Capacity						
59	Superheater						
60	Gas Side	10⁶BTU/hr/°F	0.981	0.971	0.971	0.982	0.976
61	Steam Side	10⁶BTU/hr/°F	0.354	0.352	0.352	0.354	0.353
62	R	-	0.361	0.362	0.362	0.361	0.362
63	P	-	0.311	0.312	0.312	0.311	0.312
64	F	-	0.991	0.991	0.991	0.991	0.991
65	Economizer						
66	Gas Side	10⁶BTU/hr/°F	0.918	0.907	0.907	0.918	0.913
67	Steam Side	10⁶BTU/hr/°F	0.908	0.899	0.898	0.908	0.903
68	R	-	0.990	0.990	0.990	0.990	0.990
69	P	-	0.505	0.508	0.508	0.505	0.506
70	F	-	0.843	0.838	0.838	0.843	0.841
71	UAs						
72	Superheater	10⁶BTU/hr/°F	0.903	0.896	0.895	0.904	0.900
73	Evaporator	10⁶BTU/hr/°F	2.662	2.637	2.636	2.661	2.649
74	Economizer	10⁶BTU/hr/°F	3.034	3.029	3.029	3.031	3.030
75	Total	10⁶BTU/hr/°F	6.598	6.562	6.561	6.596	6.579

All three heat exchangers are crossflow; however, the evaporator has a heat capacity ratio, R, of zero, simplifying the calculations. The superheater, evaporator, and economizer are analyzed separately. The uncertainty for each test period, if desired, would be calculated as before.

Chapter 9. Moisture Separator/Reheater

The steam generators in nuclear plant of the pressurized water reactor variety produce saturated steam at best. The point is, they don't produce superheated steam. Wet steam is very hard on a steam turbine. While steam turbines in a nuclear plant have special moisture extraction blades, this alone isn't enough. The steam must be somewhat dried and reheated. This is what a moisture separator/reheater (MSR) does.

A MSR looks like an oblong tank but contains several sections: a moisture separator plus one or two tube bundles. The low-pressure wet steam may enter at the top or bottom, but always flows through a stack of chevrons to remove water droplets, then upward across the tube bundle(s) to heat and dry what's left. The tube bundles are fed by high-pressure steam. Chevrons are depicted below:

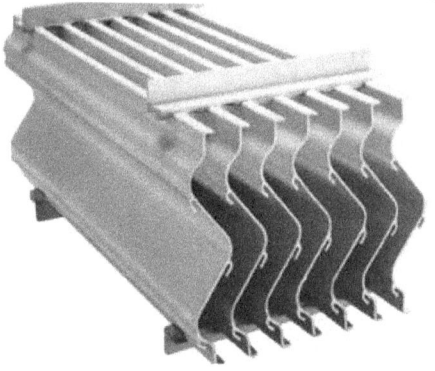

An MSR is not employed to improve thermal efficiency. On the contrary, it reduces thermal efficiency, sacrificing high-pressure steam to heat low-pressure steam, which generates entropy. The following is a typical schematic:

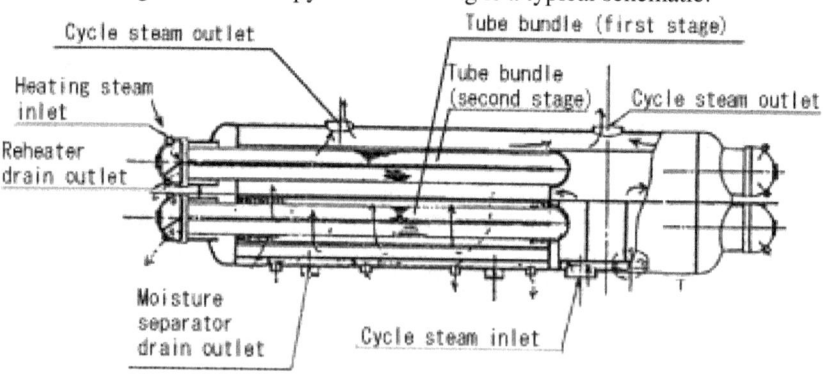

A simple analysis is presented here using the LMTD method. A more detailed, numerical example is presented in Chapter 14. The entering and exiting steam is most often not superheated; therefore, it is necessary to either specify enthalpy or quality. The pressure drop is also important and will be specified here as an input rather than calculating it, as in the more detailed example.

Performance Prediction

	A	B	C	D	E	F
1		Moisture Separator Reheater Example				
2	INPUTS	units	value	CALCS.	units	value
3	HP Tubes			LP Tubes		
4	Flow	lbm/hr	100,000	Tinlet	°F	443.6
5	Pinlet	psia	834.5	Toutlet	°F	438.7
6	Hinlet	BTU/lbm	1295.6	Houtlet	BTU/lbm	1054.9
7	Poutlet	psia	824.2	Shell Side		
8	Xoutlet	%	97%	Tinlet	°F	370.8
9	LP Tubes			Xinlet	%	88.7%
10	Flow	lbm/hr	75,000	Tchevr	°F	369.2
11	Pinlet	psia	396.1	Xchevr	%	98.3%
12	Hinlet	BTU/lbm	1150.1	Hchevr	BTU/lbm	1181.7
13	Poutlet	psia	376.4	Hmidl	BTU/lbm	1196.0
14	Xoutlet	%	81%	Tmidl	°F	368.7
15	Shell Side			Houtlet	BTU/lbm	1219.4
16	Flow	lbm/hr	500,000	Toutlet	°F	385.8
17	Pinlet	psia	175.0	superheat	°F	17.6
18	Hinlet	BTU/lbm	1100.0	LP Tubes		
19	chevrons	%	85%	Q	BTU/hr	7.14E+06
20	Pchevr	psia	171.7	LMTD	°F	72.2
21	Poutlet	psia	169.4	FUA	BTU/hr/°F	9.90E+04
22	CALCS.	units	value	HP Tubes		
23	HP Tubes			Q	BTU/hr	1.17E+07
24	Tinlet	°F	596.0	LMTD	°F	63.7
25	Toutlet	°F	521.7	FUA	BTU/hr/°F	1.84E+05
26	Houtlet	BTU/lbm	1178.2			

There is often condensation inside the tube bundles, which reduces the complexity and necessity of estimating F. In this simplified example, it will be left in the product, **FUA**. The most challenging problems facing the designer of MSRs are mechanical. The original design used in all of the Westinghouse® Pressurized Water Reactors (PWRs) had finned copper-nickel tubes. While these are great for heat transfer, they have a very large coefficient of thermal expansion and a low ultimate stress. This combination in a heat exchanger is a recipe for failure—in this case, a colossal one, costing millions of dollars.

In order to avoid these mechanical problems, condensation inside the tubes of this design must be minimized. Several approaches were taken, including flow-restricting orifice plates at the tube bundle outlet, but these were marginally successful at best. All of the MSRs of this original design have been replaced with improved designs having other materials (e.g., stainless or titanium) and sheets that allow the tubes to slide back-and-forth.

In addition to the removal of moisture droplets, superheating of the steam is of primary importance. The impacts of extraction steam flow rates on superheating and required FUA are shown in the next three figures.

Impact of Extraction Flow on Superheat

	50,000	60,000	70,000	80,000	90,000	100,000	110,000	120,000	130,000	140,000	150,000
40,000	3.80	5.57	7.33	9.09	10.85	12.61	14.37	16.14	17.90	19.66	21.42
48,000	4.95	6.71	8.47	10.23	11.99	13.76	15.52	17.28	19.04	20.80	22.56
56,000	6.09	7.85	9.61	11.38	13.14	14.90	16.66	18.42	20.18	21.95	23.71
64,000	7.23	8.99	10.76	12.52	14.28	16.04	17.80	19.57	21.33	23.09	24.85
72,000	8.38	10.14	11.90	13.66	15.42	17.19	18.95	20.71	22.47	24.23	25.99
80,000	9.52	11.28	13.04	14.80	16.57	18.33	20.09	21.85	23.61	25.38	27.14
88,000	10.66	12.42	14.19	15.95	17.71	19.47	21.23	22.99	24.76	26.52	28.28
96,000	11.81	13.57	15.33	17.09	18.85	20.61	22.38	24.14	25.90	27.66	29.42
104,000	12.95	14.71	16.47	18.23	20.00	21.76	23.52	25.28	27.04	28.80	30.57
112,000	14.09	15.85	17.62	19.38	21.14	22.90	24.66	26.42	28.19	29.95	31.71
120,000	15.23	17.00	18.76	20.52	22.28	24.04	25.81	27.57	29.33	31.09	32.85

Impact of Extraction Flow on Required FUA

	50,000	60,000	70,000	80,000	90,000	100,000	110,000	120,000	130,000	140,000	150,000
40,000	0.1357	0.1535	0.1718	0.1906	0.2099	0.2297	0.2502	0.2712	0.2929	0.3152	0.3384
48,000	0.1469	0.1649	0.1834	0.2023	0.2218	0.2419	0.2625	0.2838	0.3058	0.3284	0.3518
56,000	0.1582	0.1763	0.1950	0.2141	0.2338	0.2541	0.2750	0.2965	0.3187	0.3417	0.3654
64,000	0.1694	0.1877	0.2066	0.2259	0.2458	0.2663	0.2875	0.3092	0.3317	0.3550	0.3791
72,000	0.1807	0.1992	0.2182	0.2377	0.2579	0.2786	0.3000	0.3220	0.3448	0.3684	0.3928
80,000	0.1928	0.2116	0.2309	0.2507	0.2712	0.2923	0.3140	0.3365	0.3597	0.3837	0.4086
88,000	0.2058	0.2250	0.2446	0.2649	0.2857	0.3072	0.3294	0.3524	0.3761	0.4007	0.4262
96,000	0.2191	0.2386	0.2586	0.2793	0.3005	0.3225	0.3452	0.3687	0.3930	0.4182	0.4443
104,000	0.2326	0.2524	0.2729	0.2940	0.3157	0.3381	0.3613	0.3854	0.4102	0.4361	0.4629
112,000	0.2464	0.2666	0.2874	0.3090	0.3312	0.3541	0.3778	0.4024	0.4279	0.4544	0.4819
120,000	0.2604	0.2810	0.3023	0.3243	0.3470	0.3704	0.3947	0.4199	0.4460	0.4732	0.5015

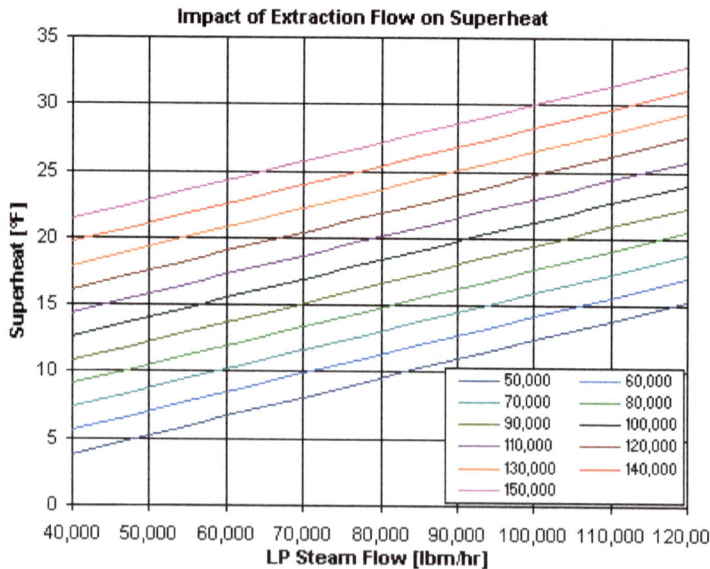

Impact of Extraction Flow on Superheat

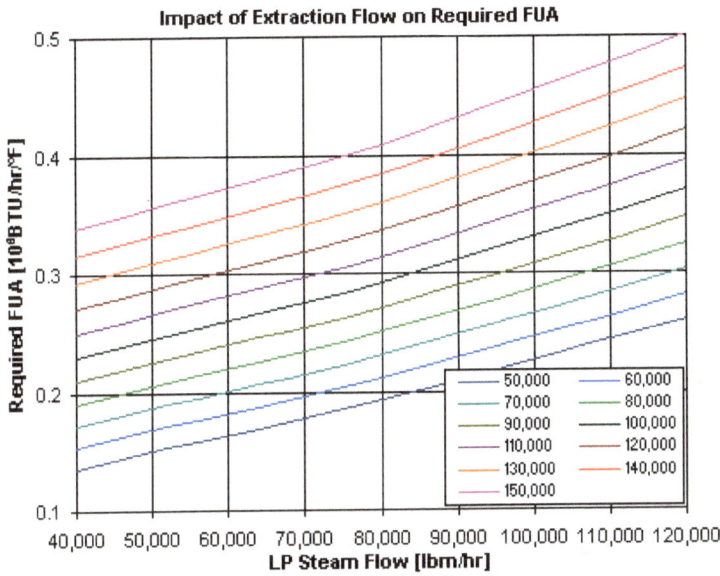

Performance Evaluation

It is virtually impossible to measure the thermal performance of a moisture separator/reheater. Not only are the flows two-phase (vapor plus liquid), but also the temperatures are not uniform across the flow stream. Besides being a useless exercise, traversing the flow with a probe is completely impractical. No one is going to let you stick a probe into anything inside an operating nuclear plant.

> *I can't say enough about the difficulties involved in measuring a two-phase flow. It is a known fact—thoroughly established by experiment—that the two phases travel at different speeds. Which one do you intend to measure? Presumably the vapor phase. Will you then calculate, estimate, or ignore the liquid phase? If so, why bother measuring anything? Resign yourself to calculating it from other things that you can measure; because that's as close as you will come.*

Not only are the flows impractical to measure, so are the steam qualities. Even if you were to pass the steam through a calorimeter, you cannot assure that your sample is representative of the whole. Such a test would be pointless.

Chapter 10. The TEMA Designs

TEMA stands for Tubular Exchangers Manufacturers Association, a group of leading shell-and-tube designers and fabricators. TEMA has identified seven arrangements for the shell and tubes. TEMA has also identified a number of head/end types, which will not be discussed here, as these are mechanical considerations and we are only considering thermal ones. The seven arrangements are given letter designations (E, F, G, H, J, K, X) and are shown below:

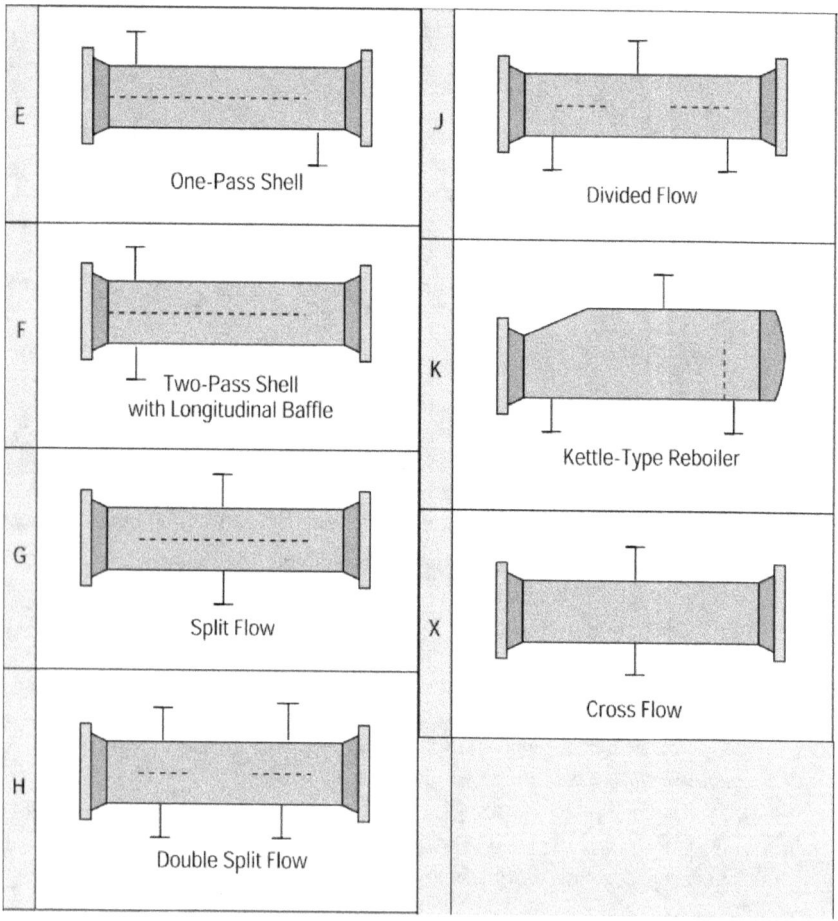

TEMA and its members have provided many resources to guide in the selection of the best design for most applications. These are readily available on the web. Tube-side and shell-side heat transfer coefficients for these designs are similar to those briefly discussed in Chapter 7. More details on the specifics (i.e., convective heat transfer inside tubes) may be found in any heat transfer textbook, such as the one referenced previously.

Performance Prediction

There are two ways of analyzing these designs: the F-LMTD/P-NTU method or numerically. The former will be presented here and the latter in subsequent chapters. Type-X is simply crossflow, which was covered in the previous chapter. Analytical solutions exist for many of these and separate solutions are required for variants, such as 1-2 and 1-4 arrangements, as illustrated in the following figure.

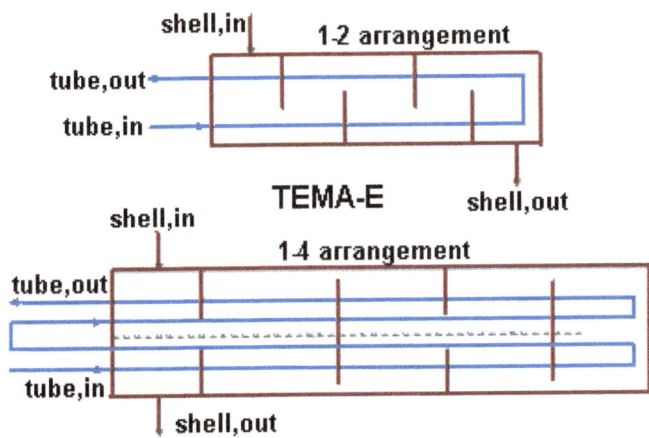

Only the 1-2 arrangement will be presented here. The NTU-ε method is readily applicable, and given in the following equations:

$$\varepsilon = \frac{2}{1+R+\sqrt{1+R^2}\left(\frac{1+e^{-\Gamma}}{1-e^{-\Gamma}}\right)} \tag{10.1}$$

$$\Gamma = NTU\sqrt{1+R^2} \tag{10.2}$$

$$NTU = \frac{1}{\sqrt{1+R^2}}\ln\left[\frac{2-\varepsilon\left(1+R-\sqrt{1+R^2}\right)}{2-\varepsilon\left(1+R+\sqrt{1+R^2}\right)}\right] \tag{10.3}$$

$$R = \frac{(\dot{m}C_P)_{MIN}}{(\dot{m}C_P)_{MAX}} \tag{10.4}$$

An illustration of this calculation may be found in the spreadsheet Eshell_analytical.xls, as shown below:

	A	B	C	D
1	**TEMA E-Shell Example**			
2	symbol	units	tube	shell
3	U	W/m²/°C	65	
4	Cp	kJ/kg/°C	2.5	1.0
5	m	kg/s	1000	8000
6	Tin	°C	250	50
7	Tout	°C	100	96.9
8	q	kW	375,000	
9	mCp	kW/°C	2,500	8,000
10	R	-	0.313	
11	ε	-	0.750	
12	NTU	-	1.965	
13	UA	-	4913	
14	A	m²	75.6	
15	user inputs in blue			
16	calculations in orange			

The NTU-ε method is used to calculate NTU and UA, the required conductance. It should be apparent by now that the product UA occurs quite frequently in thermal analyses of heat exchangers, much more so than either U or A alone. The overall conductance, U, is dominated by fluid and material properties as well as design and operation (i.e., Reynolds numbers, convective, evaporative, and condensive heat transfer coefficients). The heat exchange area, A, along with the overall design and selection of materials drives the cost.

Performance Evaluation

Evaluation of performance test data for this type of heat exchanger is the same as for the crossflow except for a few formulas in the calculation of *UA*. The uncertainty is calculated in exactly the same way.

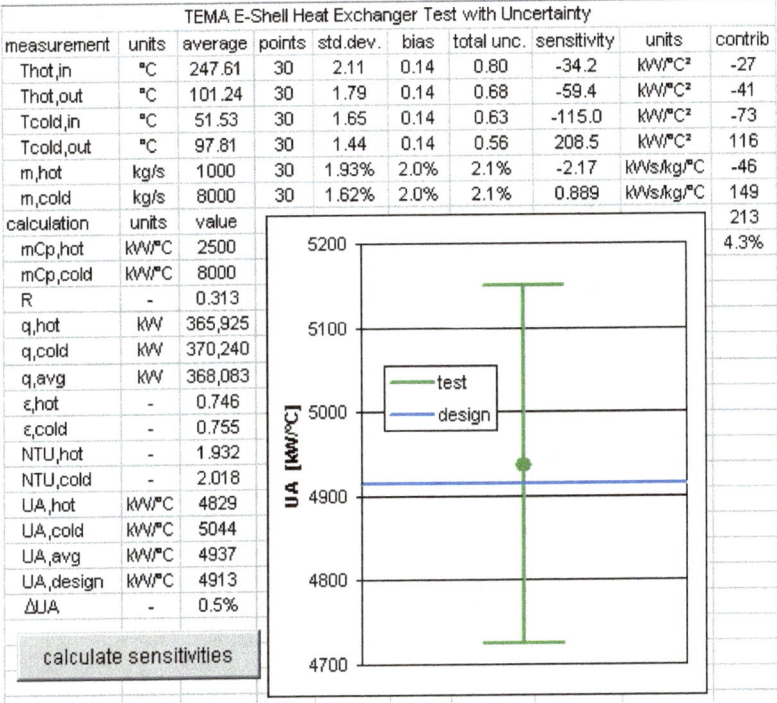

measurement	units	average	points	std.dev.	bias	total unc.	sensitivity	units	contrib
Thot,in	°C	247.61	30	2.11	0.14	0.80	-34.2	kW/°C²	-27
Thot,out	°C	101.24	30	1.79	0.14	0.68	-59.4	kW/°C²	-41
Tcold,in	°C	51.53	30	1.65	0.14	0.63	-115.0	kW/°C²	-73
Tcold,out	°C	97.81	30	1.44	0.14	0.56	208.5	kW/°C²	116
m,hot	kg/s	1000	30	1.93%	2.0%	2.1%	-2.17	kWs/kg/°C	-46
m,cold	kg/s	8000	30	1.62%	2.0%	2.1%	0.889	kWs/kg/°C	149
calculation	units	value							213
mCp,hot	kW/°C	2500							4.3%
mCp,cold	kW/°C	8000							
R	-	0.313							
q,hot	kW	365,925							
q,cold	kW	370,240							
q,avg	kW	368,083							
ε,hot	-	0.746							
ε,cold	-	0.755							
NTU,hot	-	1.932							
NTU,cold	-	2.018							
UA,hot	kW/°C	4829							
UA,cold	kW/°C	5044							
UA,avg	kW/°C	4937							
UA,design	kW/°C	4913							
ΔUA	-	0.5%							

The analysis proceeds as for the crossflow example with instrument uncertainties, a button to calculate sensitivities, and an overall test result.

Chapter 11. Simple Numerical Methods

With the availability of personal computers there is little point developing complicated closed-form analytical solutions to such mundane problems. Numerical solutions can be easily set up in Excel® and used over and over again. These can be adapted to account for cross mixing, varying properties, and property-dependent heat transfer coefficients. The need for numerical solutions in such cases arises from the fact that the assumptions made in order to obtain the analytical solutions (e.g., the LMTD and P-NTU methods) are not valid.

We will begin with the simplest useful application: a crossflow arrangement without mixing. The heat exchanger is divided into cells. The number of cells required depends on how much the temperatures and properties vary. The strategy here is to simplify the problem so that the aforementioned assumptions are reasonably valid over a single cell, hoping that the combination of cells will accurately capture the whole.

> *Ensemble Hypothesis - the whole may be represented by an ordered assemblage of distinguishable, yet simpler parts, no one of which exhibits all the characteristics of the whole.*

In order to illustrate this approach, the first example problem in Chapter 3 will be divided into a 10x10 grid of cells. This and the following two examples may be found in crossflow_numerical.xls.

crossflow heat exchanger - fully explicit - 10x10 grid

1724						hot side inlet					
		50	50	50	50	50	50	50	50	50	50
		34.9	40.1	43.5	45.8	47.2	48.2	48.8	49.2	49.5	49.7
		26.3	32.2	36.9	40.5	43.1	45.1	46.5	47.5	48.3	48.8
		21.4	26.5	31.1	35.1	38.5	41.2	43.3	45.0	46.3	47.3
		18.7	22.5	26.5	30.4	33.9	37.0	39.7	41.9	43.7	45.2
		17.1	19.8	23.0	26.4	29.8	33.0	35.9	38.5	40.7	42.6
		16.2	18.0	20.5	23.3	26.3	29.3	32.2	35.0	37.4	39.6
		15.7	16.9	18.7	20.9	23.4	26.1	28.9	31.6	34.2	36.5
		15.4	16.2	17.4	19.1	21.2	23.5	26.0	28.5	31.0	33.5
	Texit	15.2	15.7	16.6	17.9	19.5	21.4	23.5	25.8	28.2	30.6
	20.0	15.1	15.4	16.0	17.0	18.2	19.7	21.5	23.5	25.7	27.9
						hot side outlet					
	15	27.1	35.0	40.2	43.6	45.8	47.2	48.2	48.8	49.2	49.5
	15	21.9	28.2	33.5	37.7	41.0	43.5	45.3	46.7	47.6	48.3
cold side inlet	15	18.9	23.5	28.1	32.4	36.1	39.2	41.7	43.7	45.3	46.5
	15	17.2	20.4	24.1	27.9	31.6	34.9	37.8	40.3	42.4	44.1
	15	16.3	18.4	21.2	24.4	27.7	30.9	33.9	36.7	39.1	41.2
	15	15.7	17.1	19.2	21.7	24.5	27.4	30.3	33.1	35.7	38.1
	15	15.4	16.3	17.8	19.7	22.0	24.5	27.2	29.9	32.5	34.9
	15	15.2	15.8	16.8	18.2	20.0	22.1	24.5	26.9	29.4	31.9
	15	15.1	15.5	16.2	17.2	18.6	20.3	22.2	24.4	26.7	29.0
	15	15.1	15.3	15.8	16.5	17.5	18.8	20.5	22.3	24.3	26.5
										Texit	39.0

In this case, the hot side temperatures form an 11x10 block (11 rows and 10 columns). There is an extra row at the top, as this serves as the hot side inlet boundary condition. The hot side exit temperature is the average of the bottom row in this block. The result (20.0) is shown in bold.

The cold side temperatures form a 10x11 block. There is an extra column on the left, as this serves as the cold side inlet boundary. The cold side exit temperature is the average of the right column in this block. The result (39.0) is shown in bold.

This first example is a fully explicit temperature difference, that is, using only the hot and cold temperatures entering each cell, in this case, above and to the left, respectively.

$$\Delta T_{i,j} = (T_H)_{i,j} - (T_C)_{i,j} \qquad (11.1)$$

This is implemented in the spreadsheet by typing =MAX(0,G3-F15) into cell R4 and dragging it to fill the block. MAX() is used to prevent overshoot. The temperature differences and the heat transfer each form a 10x10 block:

ΔT in each cell										
use max(0,dT) to prevent overshoot										
35.0	22.9	15.0	9.8	6.4	4.2	2.8	1.8	1.2	0.8
19.9	18.2	15.4	12.3	9.5	7.2	5.3	3.9	2.8	2.0
11.3	13.3	13.4	12.3	10.7	9.0	7.3	5.8	4.6	3.5
6.4	9.3	10.7	11.0	10.6	9.6	8.5	7.2	6.0	4.9
3.7	6.2	8.1	9.2	9.6	9.4	8.8	8.0	7.0	6.1
2.1	4.1	5.9	7.3	8.1	8.5	8.5	8.1	7.5	6.8
1.2	2.6	4.2	5.5	6.6	7.4	7.8	7.8	7.6	7.1
0.7	1.7	2.9	4.1	5.2	6.1	6.8	7.1	7.2	7.1
0.4	1.1	2.0	3.0	4.0	4.9	5.7	6.3	6.6	6.8
0.2	0.7	1.3	2.1	3.0	3.9	4.7	5.4	5.9	6.2
ΔQ in each cell										
603	395	259	170	111	73	48	31	20	13
343	315	265	212	164	124	92	67	49	35
195	230	231	213	185	155	126	100	78	60
111	160	185	190	182	166	146	124	104	85
63	108	140	158	165	162	152	137	121	104
36	71	102	125	140	147	146	140	130	117
20	45	72	96	114	127	134	134	131	123
12	29	50	71	90	106	117	123	125	123
7	18	34	51	69	85	99	108	114	117
4	11	22	36	52	67	81	92	101	107
								q	12,000
								error	0.0%

If Nx is the number of cells perpendicular to the hot stream (in the direction of the cold stream) and Ny is the number of cells perpendicular to the cold stream (in the direction of the hot stream) then the heat transfer in each cell is simply:

$$Q_{i,j} = \frac{UA}{NxNy}\Delta T_{i,j} \qquad (11.2)$$

This is implemented by typing =$E$2*R4/10/10 into cell R15 and dragging it to fill the block. The temperature change of the cold stream in a cell is:

$$(T_C)_{i,j+1} = (T_C)_{i,j} + \frac{Q_{i,j}}{\left(\dfrac{\dot{m}_C}{Ny}\right)C_{PC}} \qquad (11.3)$$

This is implemented by typing =F15+R15/$D$4/($D$5/10) into cell G15 and dragging it to fill the block. The temperature change of the hot stream in a cell is:

$$(T_H)_{i+1,j} = (T_H)_{i,j} - \frac{Q_{i,j}}{\left(\dfrac{\dot{m}_H}{Nx}\right)C_{PH}} \qquad (11.4)$$

This is implemented by typing =G3-R15/$C$4/($C$5/10) into cell G4 and dragging it to fill the block. The total heat transfer is the sum of all $Q_{i,j}$ and is shown in bold (12,000). The required conductance, *UA*, is placed in cell E2 and shown in orange (1724) and is assumed constant throughout. This requires iterative solution and a button is provided to accomplish this.

How many grid cells is enough?

The answer is not as simple as it might seem. First we increase the number of grid cells to see if this changes the result. It's not convenient to do this in Excel®, so a simple program written in C is used that dynamically allocates memory and runs much faster than a spreadsheet. The code is described in Appendix A and is included in the on-line archive along with the spreadsheet.

The result converges to 2800 kW/°C at about 400x400=160,000 cells (not something you'd want to try in Excel®). The analytical answer is 2362, which occurs at 33x33=1089 cells. Sometimes having too many cells leads to round-off errors, but this is not what's happening here. The problem is more subtle than this.

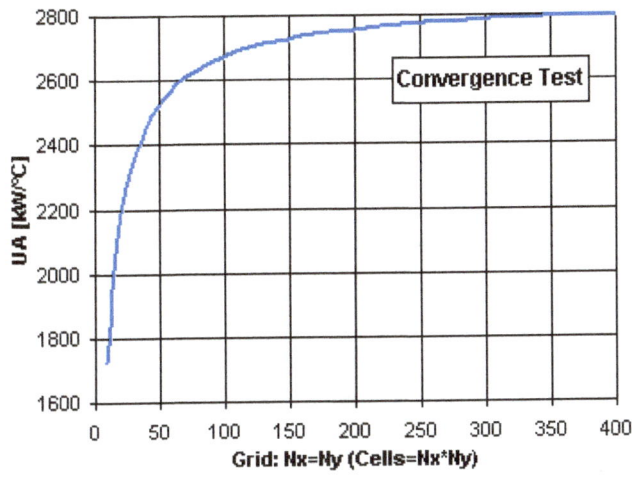

Explicit vs. Implicit Differences

Explicit temperature differences were used in this example up until this point. This over-estimates the thermal driving potential with a coarse grid. There are many possible corrections to the fully explicit difference that might be considered: average (arithmetic mean), log-mean, geometric mean, and harmonic mean.

Using only explicit temperature differences, it is possible to sweep through the grid in a single pass. This is not possible with implicit temperature differences, because the calculations for each cell depend on the results. This produces circular references. If you type =AVERAGE(G3-F15,G4-G15) into cell R4 and drag to fill the block, Excel® will display an error and blue arrows showing the circular references. You can get past this by enabling iterative calculations (see Tools/Options or File/Options/Formulas). This modification will slow the calculations considerably, but will significantly improve the accuracy of the results. The impact of this modification is shown in the next figure.

The 10x10 grid converges after 7 or 8 iterations. The initial hump at 2 or 3 iterations is overshoot and to be avoided. For grid sizes above about 15x15, this simple modification is not adequate, as it yields the same erroneous overshot solution. For our purposes here, a 10x10 grid with 8 implicit iterations of two-thirds/one-third weighting is adequate. Excel® handles the iterations automatically, as long as it doesn't get too far off, producing #VALUEs.

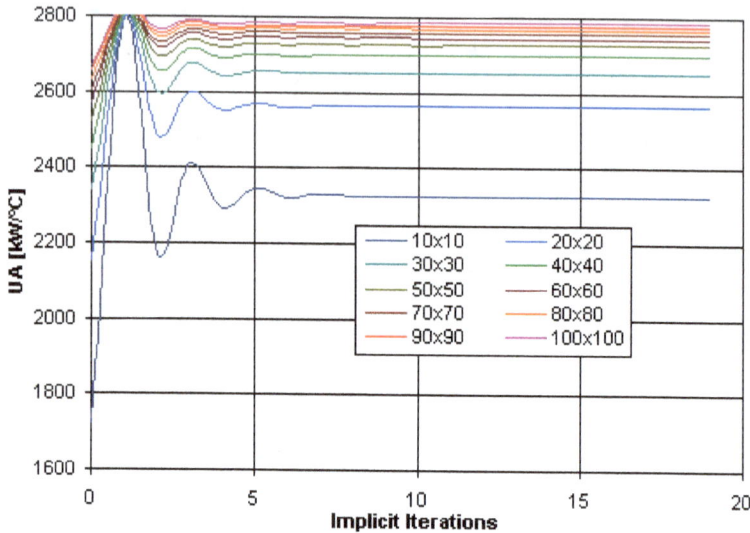

While this modification works well enough, it requires iterations and results in circular references in Excel®. The most efficient modification is to use the geometric mean, which has a closed-form solution. In terms of the cell indices, this becomes:

$$\Delta T_{i,j} = \sqrt{(\Delta T_A)(\Delta T_B)}$$
$$\Delta T_A = (T_H)_{i,j} - (T_C)_{i,j} \tag{11.5}$$
$$\Delta T_B = (T_H)_{i+1,j} - (T_C)_{i,j+1}$$

Substituting Equation 11.5 into 11.2 and solving 11.3 and 11.4 for the hot side exit temperature yields:

$$(T_H)_{i+1,j} = (T_H)_{i,j} + \frac{\Delta T_A \beta[(Ny + \alpha Nx)\beta - \sqrt{\gamma}]}{\delta Ny} \tag{11.6}$$

$$\alpha = \frac{\dot{m}_C C_{PC}}{\dot{m}_H C_{PH}} \tag{11.7}$$

$$\beta = \frac{UA}{\dot{m}_H C_{PH}} \tag{11.8}$$

$$\delta = \frac{2\alpha}{NxNy} \tag{11.9}$$

$$\gamma = (\alpha Nx^2 + Ny^2 + \delta)\beta^2 + \delta^2 \tag{11.10}$$

Not only does the geometric mean lead to a closed-form (non-iterative) solution, it supports a much coarser grid size than the arithmetic mean. Convergence with grid size is quite remarkable compared to the previous figure, which is why this adaptation is used whenever possible.

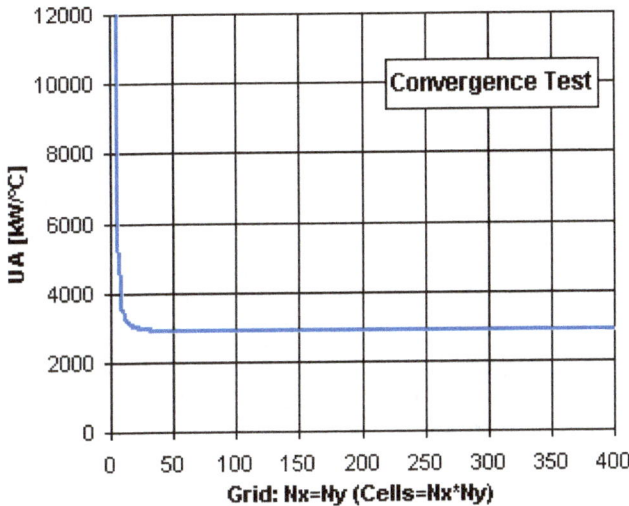

Chapter 12: Variable Properties

Fluid properties often vary only slightly throughout a heat exchanger and can be adequately approximated by average values; however, this is not always the case. The following extreme example of a simple double-pipe counterflow arrangement will be used to illustrate the potential difficulty of analyzing heat exchanger performance with significantly varying properties.

	A	B	C	D	E	F	G	H	I	J
1				Extreme Variable Property Example						
2	Inputs		Q	Th	Tc	dT	1/dT	CpH	CpC	dQ
3	mH	1000	0%	300.0	100.0	200.0	0.0050	6.031	0.904	14999
4	mC	1000	1%	302.5	116.6	185.9	0.0054	5.347	0.923	13942
5	UA	7500	2%	305.1	131.7	173.4	0.0058	4.820	0.941	13005
6	Q	6.26E+05	3%	307.8	145.5	162.3	0.0062	4.405	0.959	12171
7	LMTD method		4%	310.6	158.2	152.4	0.0066	4.071	0.976	11427
8	LMTD	200	5%	313.4	169.9	143.5	0.0070	3.798	0.993	10759
9	UA	3126	6%	316.2	180.7	135.5	0.0074	3.572	1.009	10159
10	error	-58.3%	7%	319.0	190.8	128.2	0.0078	3.381	1.025	9617
11	NTU-ε method		8%	321.9	200.2	121.7	0.0082	3.218	1.041	9127
12	CpH,avg	1.836	9%	324.7	209.0	115.8	0.0086	3.077	1.056	8682
13	CpC,avg	1.893	10%	327.5	217.2	110.4	0.0091	2.954	1.071	8277
14	mCh	1836	11%	330.3	224.9	105.4	0.0095	2.846	1.085	7907
15	mCc	1893	12%	333.1	232.2	100.9	0.0099	2.750	1.100	7569
16	R	0.970	13%	335.9	239.1	96.8	0.0103	2.665	1.114	7259
17	ε	0.667	14%	338.6	245.6	93.0	0.0108	2.587	1.128	6975
18	NTU	1.942	15%	341.3	251.8	89.5	0.0112	2.517	1.141	6713
19	UA	3566	16%	344.0	257.7	86.3	0.0116	2.454	1.155	6472
20	error	-52.4%	17%	346.6	263.3	83.3	0.0120	2.395	1.168	6249
21	Numerical method		18%	349.2	268.6	80.6	0.0124	2.341	1.181	6044
22	ΔTmean	76.7	19%	351.8	273.7	78.1	0.0128	2.292	1.194	5854
23	UA	8162	99%	684.5	497.1	187.4	0.0053	0.906	5.559	14054
24	error	8.8%	100%	700.0	499.6	200.4		0.889	6.257	

The units are immaterial and have been left blank intentionally. In this example, the specific heat of both fluids varies from inlet to exit by a factor of 7. This carefully selected variation in specific results in curvature of the T vs. Q curves, invalidating the assumptions made in developing both the LMTD and NTU-ε methods.

The T vs. Q curves are far apart at the inlet and exit (i.e., large ΔT) and close together in the middle (i.e., large ΔT). Simply averaging the specific heats over the length of the heat exchanger will not account for this variable temperature difference. In fact, there is no way to evaluate this process other than by numerically integrating or employing a finite difference or finite element technique, which would accomplish the same thing.

The LMTD and NTU-ε methods in this case have -58.3% and -50.7% errors, respectively. Even the numerical solution is off by 8.8% with 100 cells.

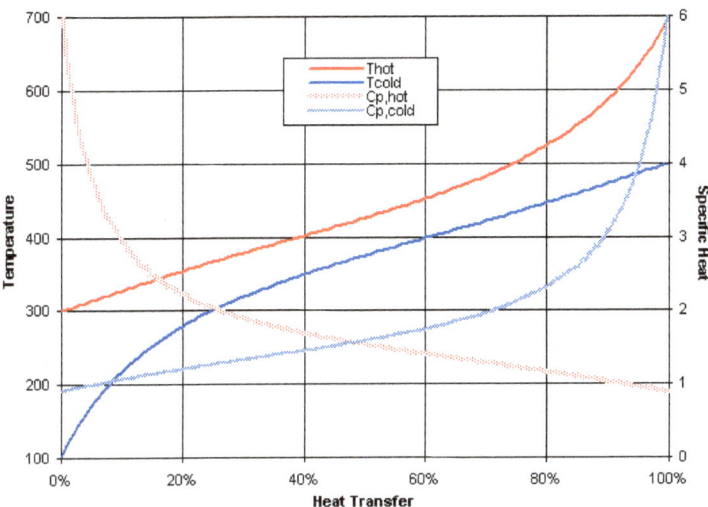

The geometric mean temperature difference is the only one that comes close to the correct solution with only a 3.2% error in overall conductance, **UA**. In this case the geometric mean is calculated by:

$$\Delta Tmean = \left[\prod_{i=1}^{n} \Delta T_i\right]^{\frac{1}{n}} \tag{12.1}$$

	A	B	C	D	E	F	G	H	I	J
1			Moderately Variable Property Example							
2	Inputs		Q	Th	Tc	dT	1/dT	CpH	CpC	dQ
3	mH	1000	0%	300.0	100.0	200.0	0.0050	1.96	1.24	22333
4	mC	1000	1%	311.4	118.0	193.4	0.0052	2.01	1.29	21594
5	UA	11167	2%	322.1	134.7	187.4	0.0053	2.06	1.34	20927
6	Q	1.22E+06	3%	332.3	150.3	182.0	0.0055	2.11	1.39	20320
7	LMTD method		4%	341.9	164.9	177.0	0.0056	2.15	1.44	19766
8	LMTD	129	5%	351.1	178.7	172.5	0.0058	2.20	1.49	19258
9	UA	9456	6%	359.9	191.6	168.3	0.0059	2.24	1.53	18789
10	error	-15.3%	7%	368.3	203.9	164.4	0.0061	2.28	1.57	18356
11	NTU-ε method		8%	376.4	215.6	160.8	0.0062	2.32	1.62	17953
12	CpH,avg	3.386	9%	384.1	226.7	157.4	0.0064	2.36	1.66	17578
13	CpC,avg	2.764	10%	391.5	237.3	154.3	0.0065	2.40	1.70	17227
14	mCh	3386	11%	398.7	247.4	151.3	0.0066	2.43	1.74	16898
15	mCc	2764	12%	405.7	257.1	148.6	0.0067	2.47	1.78	16589
16	R	0.816	13%	412.4	266.4	145.9	0.0069	2.51	1.82	16297
17	ε	0.817	14%	418.9	275.4	143.5	0.0070	2.54	1.85	16022
18	NTU	3.257	15%	425.2	284.1	141.2	0.0071	2.57	1.89	15762
19	UA	9002	16%	431.3	292.4	138.9	0.0072	2.61	1.93	15515
20	error	-19.4%	17%	437.3	300.4	136.8	0.0073	2.64	1.96	15280
21	Numerical method		18%	443.1	308.2	134.8	0.0074	2.67	2.00	15057
22	ΔTmean	105.3	19%	448.7	315.8	132.9	0.0075	2.70	2.03	14844
23	UA	11545	99%	698.0	621.2	76.9	0.0130	4.35	3.79	8582
24	error	3.4%	100%	700.0	623.4	76.6		4.36	3.80	

This second illustration has a more moderate variation of properties. The LMTD method is now only -15.3% off and the NTU-ε method is -19.4% off. The numerical solution is off by 3.4% with 100 cells. The T vs. Q curves are much closer to being typical, though still significantly curved.

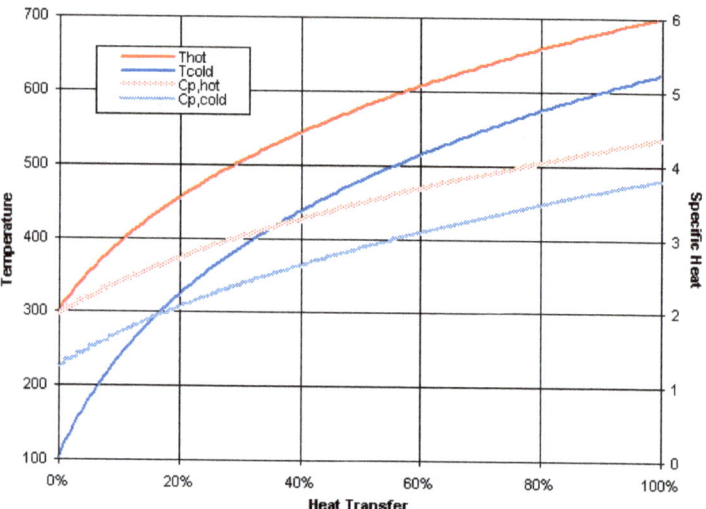

Any of these methods will work in the case of constant properties:

	A	B	C	D	E	F	G	H	I	J
1			Contstant Property Example							
2	Inputs		Q	Th	Tc	dT	1/dT	CpH	CpC	dQ
3	mH	1000	0%	300.0	100.0	200.0	0.0050	2.50	3.50	7928
4	mC	1000	1%	303.2	102.3	200.9	0.0050	2.50	3.50	7964
5	UA	3964	2%	306.4	104.5	201.8	0.0050	2.50	3.50	8000
6	Q	1.00E+06	3%	309.6	106.8	202.7	0.0049	2.50	3.50	8036
7	LMTD method		4%	312.8	109.1	203.6	0.0049	2.50	3.50	8072
8	LMTD	253	5%	316.0	111.4	204.6	0.0049	2.50	3.50	8109
9	UA	3955	6%	319.2	113.7	205.5	0.0049	2.50	3.50	8146
10	error	-0.2%	7%	322.5	116.1	206.4	0.0048	2.50	3.50	8182
11	NTU-ε method		8%	325.8	118.4	207.4	0.0048	2.50	3.50	8220
12	CpH,avg	2.500	9%	329.1	120.8	208.3	0.0048	2.50	3.50	8257
13	CpC,avg	3.500	10%	332.4	123.1	209.2	0.0048	2.50	3.50	8294
14	mCh	2500	11%	335.7	125.5	210.2	0.0048	2.50	3.50	8332
15	mCc	3500	12%	339.0	127.9	211.1	0.0047	2.50	3.50	8369
16	R	0.714	13%	342.4	130.3	212.1	0.0047	2.50	3.50	8407
17	ε	0.667	14%	345.7	132.7	213.1	0.0047	2.50	3.50	8445
18	NTU	1.582	15%	349.1	135.1	214.0	0.0047	2.50	3.50	8484
19	UA	3955	16%	352.5	137.5	215.0	0.0047	2.50	3.50	8522
20	error	-0.2%	17%	355.9	139.9	216.0	0.0046	2.50	3.50	8561
21	Numerical method		18%	359.3	142.4	217.0	0.0046	2.50	3.50	8600
22	ΔTmean	250.1	19%	362.8	144.8	217.9	0.0046	2.50	3.50	8639
23	UA	3998	99%	695.0	382.2	312.9	0.0032	2.50	3.50	12402
24	error	0.9%	100%	700.0	385.7	314.3		2.50	3.50	

The -0.2%, -0.2%, and 0.9% errors for the LMTD, NTU-ε, and numerical methods, respectively, are due to cumulative round-off and not reflective of any shortcomings in these analyses.

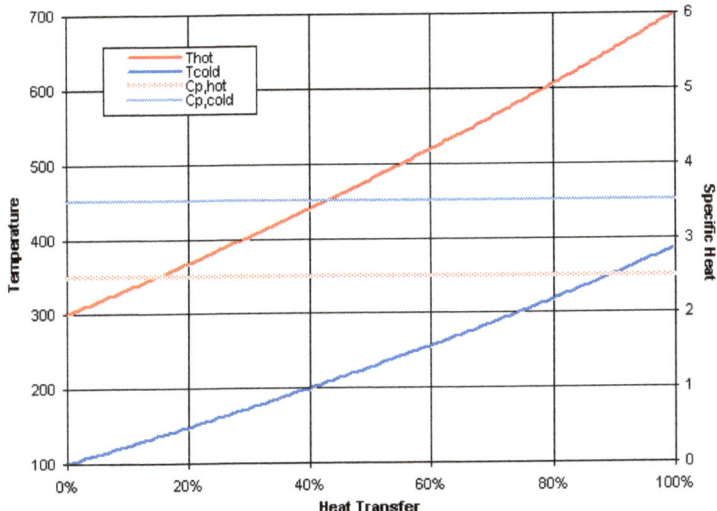

Chapter 13: Variable Conductance

Heat transfer coefficients may also vary within a heat exchanger leading to a variable overall conductance, UA_{local}. The impact of this is quite different than for variable properties because of how the governing differential equation was separated and integrated. In this first example, U varies is with ΔT^n, $n<0$.

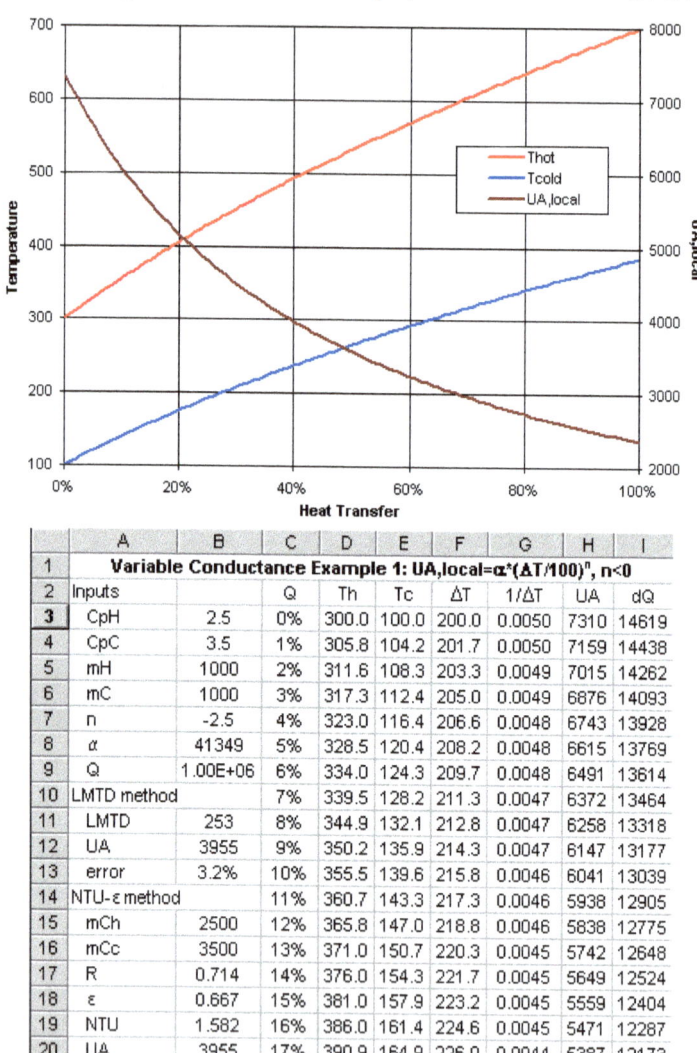

	A	B	C	D	E	F	G	H	I
1	Variable Conductance Example 1: UA,local=α*(ΔT/100)^n, n<0								
2	Inputs		Q	Th	Tc	ΔT	1/ΔT	UA	dQ
3	CpH	2.5	0%	300.0	100.0	200.0	0.0050	7310	14619
4	CpC	3.5	1%	305.8	104.2	201.7	0.0050	7159	14438
5	mH	1000	2%	311.6	108.3	203.3	0.0049	7015	14262
6	mC	1000	3%	317.3	112.4	205.0	0.0049	6876	14093
7	n	-2.5	4%	323.0	116.4	206.6	0.0048	6743	13928
8	α	41349	5%	328.5	120.4	208.2	0.0048	6615	13769
9	Q	1.00E+06	6%	334.0	124.3	209.7	0.0048	6491	13614
10	LMTD method		7%	339.5	128.2	211.3	0.0047	6372	13464
11	LMTD	253	8%	344.9	132.1	212.8	0.0047	6258	13318
12	UA	3955	9%	350.2	135.9	214.3	0.0047	6147	13177
13	error	3.2%	10%	355.5	139.6	215.8	0.0046	6041	13039
14	NTU-ε method		11%	360.7	143.3	217.3	0.0046	5938	12905
15	mCh	2500	12%	365.8	147.0	218.8	0.0046	5838	12775
16	mCc	3500	13%	371.0	150.7	220.3	0.0045	5742	12648
17	R	0.714	14%	376.0	154.3	221.7	0.0045	5649	12524
18	ε	0.667	15%	381.0	157.9	223.2	0.0045	5559	12404
19	NTU	1.582	16%	386.0	161.4	224.6	0.0045	5471	12287
20	UA	3955	17%	390.9	164.9	226.0	0.0044	5387	12173
21	error	3.2%	18%	395.8	168.4	227.4	0.0044	5305	12061
22	Numerical method		19%	400.6	171.9	228.7	0.0044	5225	11952
23	ΔTmean	260.8	20%	405.4	175.3	230.1	0.0043	5148	11846
24	UA,mean	3834	21%	410.1	178.7	231.5	0.0043	5073	11742

The temperature difference increases from left to right (i.e., the distance between the red and blue curves) and the local conductance decreases (i.e., the falling brown line). The LMTD and NTU-ε methods yield the same solution, off by only 3.2%. In this second example $n>0$.

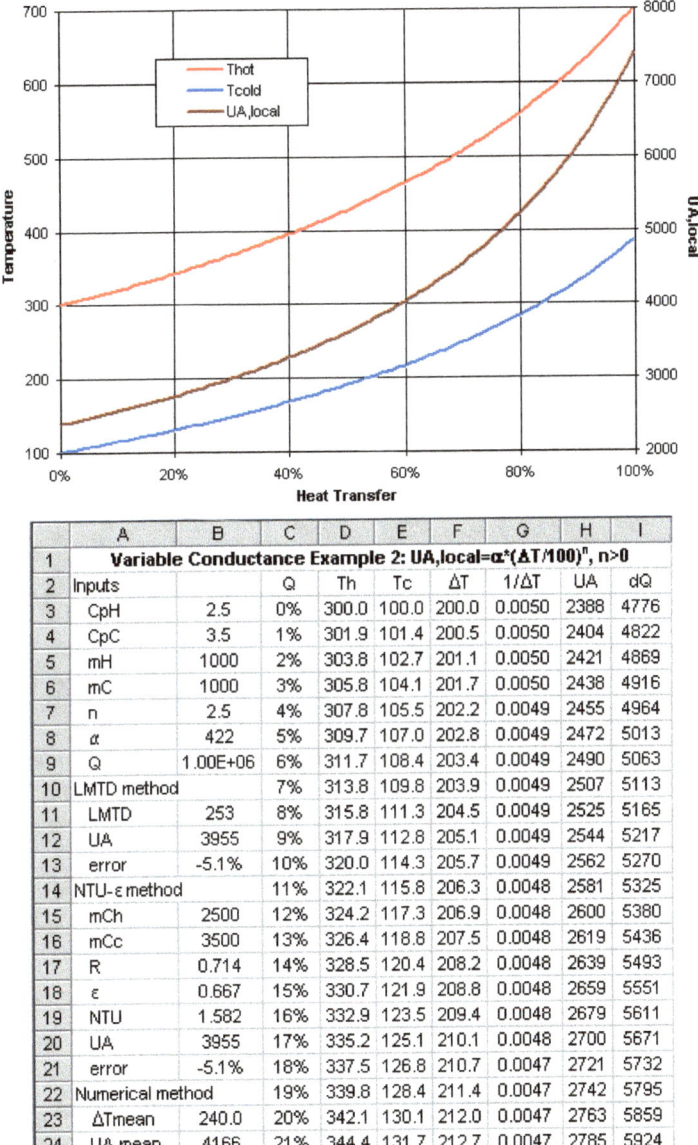

	A	B	C	D	E	F	G	H	I
1	Variable Conductance Example 2: UA,local=α*(ΔT/100)^n, n>0								
2	Inputs		Q	Th	Tc	ΔT	1/ΔT	UA	dQ
3	CpH	2.5	0%	300.0	100.0	200.0	0.0050	2388	4776
4	CpC	3.5	1%	301.9	101.4	200.5	0.0050	2404	4822
5	mH	1000	2%	303.8	102.7	201.1	0.0050	2421	4869
6	mC	1000	3%	305.8	104.1	201.7	0.0050	2438	4916
7	n	2.5	4%	307.8	105.5	202.2	0.0049	2455	4964
8	α	422	5%	309.7	107.0	202.8	0.0049	2472	5013
9	Q	1.00E+06	6%	311.7	108.4	203.4	0.0049	2490	5063
10	LMTD method		7%	313.8	109.8	203.9	0.0049	2507	5113
11	LMTD	253	8%	315.8	111.3	204.5	0.0049	2525	5165
12	UA	3955	9%	317.9	112.8	205.1	0.0049	2544	5217
13	error	-5.1%	10%	320.0	114.3	205.7	0.0049	2562	5270
14	NTU-ε method		11%	322.1	115.8	206.3	0.0048	2581	5325
15	mCh	2500	12%	324.2	117.3	206.9	0.0048	2600	5380
16	mCc	3500	13%	326.4	118.8	207.5	0.0048	2619	5436
17	R	0.714	14%	328.5	120.4	208.2	0.0048	2639	5493
18	ε	0.667	15%	330.7	121.9	208.8	0.0048	2659	5551
19	NTU	1.582	16%	332.9	123.5	209.4	0.0048	2679	5611
20	UA	3955	17%	335.2	125.1	210.1	0.0048	2700	5671
21	error	-5.1%	18%	337.5	126.8	210.7	0.0047	2721	5732
22	Numerical method		19%	339.8	128.4	211.4	0.0047	2742	5795
23	ΔTmean	240.0	20%	342.1	130.1	212.0	0.0047	2763	5859
24	UA,mean	4166	21%	344.4	131.7	212.7	0.0047	2785	5924

The LMTD and NTU-ε methods also yield the same solution in this case, off by only -5.1%. Variable *UA* has so much less dramatic impact on the analysis because the heat transfer is directly proportional it. While the heat transfer, Q, is also proportional to ΔT, ΔT decreases with increasing Q, so that ΔT is a function of itself.

Chapter 14. Two-Phase Flow Inside Tubes

The moisture separator/reheater presented in Chapter 9 is a real beast to analyze for the reasons already mentioned. In this chapter we will consider another reason: two-phase flow inside the tubes. There are several excellent texts on two-phase flow[10,11,12]. Collier in Section 2.5.2 describes Chisholm's Method[13] for pressure drop in horizontal pipes, which will be used here.

We begin with Lockart-Martinelli factor characterizing the two-phase flow.[14] This involves several parameters, including:

$$X^2 = \frac{\left(\frac{dp}{dx}\right)_f}{\left(\frac{dp}{dx}\right)_g} = \frac{\phi_g^2}{\phi_f^2} \qquad (14.1)$$

Here $dp/dx)_f$ and $dp/dx)_g$ are the pressure gradients due to friction for the liquid and vapor phases, respectively. The ratio of these two yields the square of empirical factor X. The two terms, φ_f and φ_g, are called the two-phase friction multipliers. It is furthermore assumed that these are related in the following form:

$$\phi_f^2 = 1 + \frac{C}{X} + \frac{1}{X^2} \qquad (14.2)$$

$$\phi_g^2 = 1 + CX + X^2 \qquad (14.3)$$

where C is a constant depending only on the combination of the two single-phase flow regimes given in the following table:

liquid	vapor	C
turbulent	turbulent	20
laminar	turbulent	12
turbulent	laminar	10
laminar	laminar	5

---

[10] Collier, J. G., *Convectivve Boiling and Condensation*, McGraw-Hill, 1972.
[11] Hsu, Y.-Y., and R. W. Graham, *Transport Processes in Boiling and Two-Phase Systems*, Hemisphere, 1976.
[12] Tong, L. S., *Boiling Heat Transfer and Two-Phase Flow*, Krieger, 1975.
[13] Chisholm, D., "The Influence of Mass Velocity on Friction Pressure Gradients During Steam-Water Flow," Paper 35, Thermodynamics and Fluid Mechanics Convention I Mech. Engrs. Bristol, 1968.
[14] Lockhart, R. W., and R. C. Martinelli, "Proposed Correlation of Data for Isothermal Two-Phase Two-Component Flow in Pipes," Chem. Eng. Prog., Vol. 45, No. 39, 1949.

The two-phase friction multipliers are shown in the following figure. The liquid multipliers start high and decrease with increasing X and the vapor ones do the opposite.

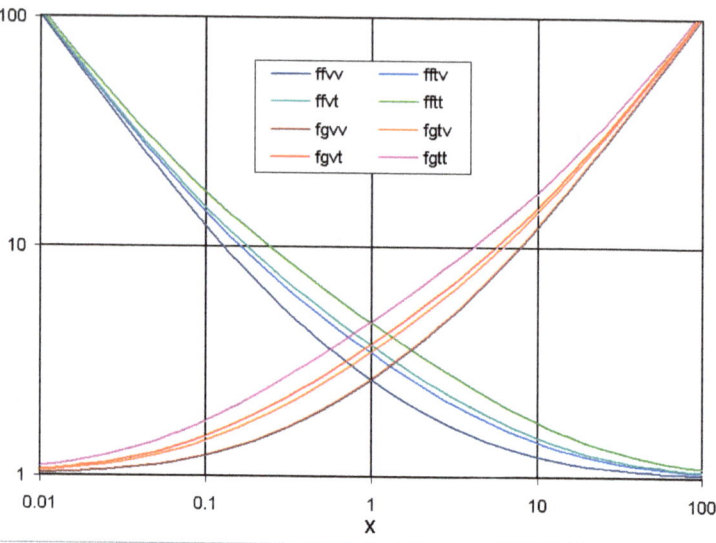

Colebrook-White[15] formula is used to calculate the single-phase friction factor for flow in pipes:

$$\frac{1}{\sqrt{f}} = -2\log_{10}\left(\frac{\varepsilon}{3.7D_h} + \frac{2.51}{\text{Re}\sqrt{f}}\right) \quad (14.4)$$

where D_h is the hydraulic diameter and Re is the Reynolds number. Chisholm modified this method, defining the coefficient, C, in the following manner:

$$C = \left[\lambda + (C_2 - \lambda)\left(\frac{v_{fg}}{V_g}\right)\right]\left[\sqrt{\frac{v_g}{v_f}} + \sqrt{\frac{v_f}{v_g}}\right] \quad (14.5)$$

$$\lambda = \frac{[2^{(2-n)} - 2]}{2} \quad (14.6)$$

$$C_2 = \frac{G*}{G} \quad (14.7)$$

---

[15] Colebrook, C. F. and White, C. M., "Experiments with Fluid Friction in Roughened Pipes". Proceedings of the Royal Society of London, Series A, Mathematical and Physical Sciences, Vol. 161, No. 906, pp. 367–381, 1937.

where v_f and v_g are the saturated liquid and vapor specific volumes, respectively, and v_{fg} is the difference between the two. For rough pipes $n=0$ and for smooth pipes $n=0.25$. The ratio, C_2, is reference mass flux, G^*, over the actual mass flux, G. For smooth tubes $G^*=2000$ kg/s/m² (1.47×10^6 lb/hr/ft²) and for rough tubes $G^*=1500$ kg/s/m² (1.1×10^6 lb/hr/ft²). Zivi's kinetic energy model[16] is used for the void ratio, α, is the ratio of the vapor to total areas in the pipe:

$$\alpha = \frac{A_g}{A} = \frac{1}{1+\left(\frac{1-x}{x}\right)\left(\frac{\rho_g}{\rho_f}\right)^{\frac{2}{3}}} \quad (14.8)$$

The slip ratio is the ratio of the vapor to the liquid velocities:

$$S = \frac{u_g}{u_f} = \left(\frac{\rho_g}{\rho_f}\right)^{\frac{1}{3}} \quad (14.9)$$

The two-phase flow regime is given by the model of Breber, Palen, and Taborek.[17] The conceptual regimes are illustrated in this first figure:

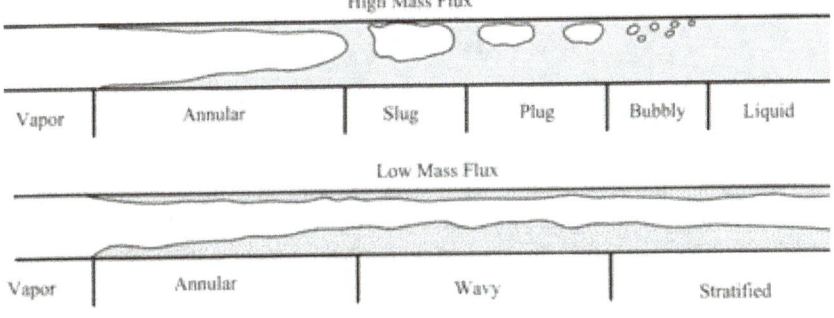

An excellent reference is the
Engineering Data Book III
by John R. Thome, published by
Wolverine Tube, Inc. and available on-line at:
https://pdfs.semanticscholar.org/9edc/9fe6557a5ecd15d7d773836a95fefa940c33.pdf

---

[16] Zivi, S.M., "Estimation of Steady State Steam Void Fraction by Means of the Principle of Minimum Entropy Production," ASME Journal of Heat Transfer, Vol. 86, pp. 247-252, 1964.

[17] Breber, G., J. W. Palen, and J. Taborek, "Prediction of Horizontal Tubeside Condensation of Pure Components Using Flow Regime Criteria," ASME Journal of Heat Transfer, Vol. 102, pp. 471-476, 1980.

The empirical relationship is illustrated in this second figure:

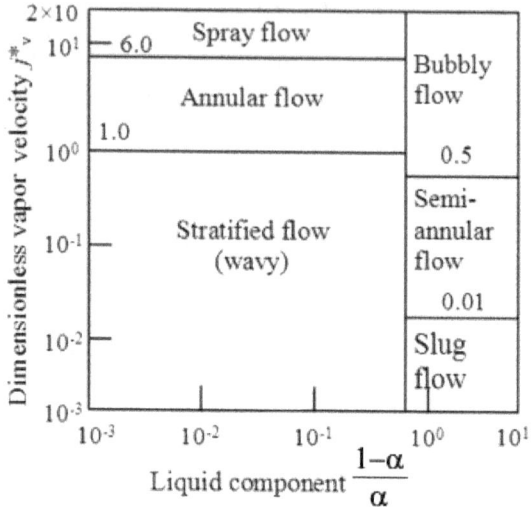

where the dimensionless vapor velocity, j^*_v, is given by:

$$j^*_v = \frac{x\dot{m}}{\sqrt{gD\rho_g(\rho_f - \rho_g)}} \qquad (14.10)$$

Once the flow regime is known, the heat transfer coefficient can be calculated using the Taitel-Dukler method[18]. The frictional pressure drop on the shell side is calculated using the method of Robinson and Briggs.[19] The frictional pressure drop on the shell side is calculated using the method of Briggs and Young.[20] It takes all of these steps just to calculate the pressure drop and heat transfer inside a single computational element along the pipe. This entire formulation has been written into a computer code that is listed in Appendix B.

Only select parts of the program output will be shown here. The complete output is included in the on-line archive. There are two banks of U-tubes: the high-pressure (higher temperature) at the top plus the low-pressure (lower

---

[18] Taitel and A.E. Dukler, "A Model for Predicting Flow Regime Transitions in Horizontal and Near Horizontal Gas-Liquid Flow", AIChE Journal, Vol. 22, No. 1, pp. 47–55, 1976.

[19] Robinson, K., K. and D. E. Briggs, "Pressure Drop of Air Flowing Across Triangular Pitch Banks of Finned Tubes," Eighth National Heat Transfer Conference, pp. 177-184, 1965.

[20] Briggs, D.E., and E.H. Young, "Convection Heat Transfer and Pressure Drop of Air Flowing Across Triangular Pitch Banks of Finned Tubes," Chem. Eng. Prog. Symp. Ser., Vol. 59, No. 41, pp. 1–10, 1963.

temperature) at the bottom. As these are U-tubes, the flow starts at the left, extends to the right, turns around, and flows back to the left. This first figure shows the flow patterns:

```
                         TWO-PHASE FLOW PATTERS INSIDE TUBES
 1H>  ENTR SUPR SUPR SUPR SUPR  SUPR SUPR SUPR SUPR  SUPR SUPR SUPR SUPR SUPR SUPR
 2H>  ENTR SUPR SUPR SUPR SUPR  SUPR SUPR SUPR SUPR  SUPR SUPR SUPR SUPR SUPR SUPR
 3H>  ENTR SUPR SUPR SUPR SUPR  SUPR SUPR SUPR SUPR  SUPR SUPR SUPR SUPR SUPR SUPR
 4H>  ENTR SUPR SUPR SUPR SUPR  SUPR SUPR SUPR SUPR  SUPR SUPR SUPR SUPR SUPR SUPR
 5H>  ENTR SUPR SUPR SUPR SUPR  SUPR SUPR SUPR SUPR  SUPR SUPR SUPR SUPR SUPR SUPR
 6H>  ENTR SUPR SUPR SUPR SUPR  SUPR SUPR SUPR SUPR  SUPR SUPR SUPR SUPR SUPR SUPR
 7H>  ENTR SUPR SUPR SUPR SUPR  SUPR SUPR SUPR SUPR  SUPR SUPR SUPR SUPR ANNU ANNU
 8H>  ENTR SUPR SUPR SUPR SUPR  SUPR SUPR SUPR SUPR  SUPR SUPR SUPR ANNU ANNU ANNU
 9H>  ENTR SUPR SUPR SUPR SUPR  SUPR SUPR SUPR SUPR  SUPR SUPR SUPR ANNU ANNU ANNU
10H>  ENTR SUPR SUPR SUPR SUPR  SUPR SUPR SUPR SUPR  SUPR SUPR ANNU ANNU ANNU ANNU
11H>  ENTR ANNU ANNU ANNU ANNU  ANNU ANNU ANNU ANNU  ANNU ANNU ANNU ANNU ANNU ANNU
12H>  ENTR ANNU ANNU ANNU ANNU  ANNU ANNU ANNU ANNU  ANNU ANNU ANNU ANNU ANNU ANNU
13H>  ENTR ANNU ANNU ANNU ANNU  ANNU ANNU ANNU ANNU  ANNU ANNU ANNU ANNU ANNU ANNU
14H>  ENTR ANNU ANNU ANNU ANNU  ANNU ANNU ANNU ANNU  ANNU ANNU ANNU ANNU ANNU ANNU
15H>  ENTR ANNU ANNU ANNU ANNU  ANNU ANNU ANNU ANNU  ANNU ANNU ANNU ANNU ANNU ANNU

16H<  ANNU ANNU ANNU ANNU ANNU  ANNU ANNU ANNU ANNU  ANNU ANNU ANNU ANNU ANNU BEND
17H<  ANNU ANNU ANNU ANNU ANNU  ANNU ANNU ANNU ANNU  ANNU ANNU ANNU ANNU ANNU BEND
18H<  ANNU ANNU ANNU ANNU ANNU  ANNU ANNU ANNU ANNU  ANNU ANNU ANNU ANNU ANNU BEND
19H<  ANNU ANNU ANNU ANNU ANNU  ANNU ANNU ANNU ANNU  ANNU ANNU ANNU ANNU ANNU BEND
20H<  ANNU ANNU ANNU ANNU ANNU  ANNU ANNU ANNU ANNU  ANNU ANNU ANNU ANNU ANNU BEND
21H<  ANNU ANNU ANNU ANNU ANNU  ANNU ANNU ANNU ANNU  ANNU ANNU ANNU ANNU ANNU BEND
22H<  ANNU ANNU ANNU ANNU ANNU  ANNU ANNU ANNU ANNU  ANNU ANNU ANNU ANNU ANNU BEND
23H<  ANNU ANNU ANNU ANNU ANNU  ANNU ANNU ANNU ANNU  ANNU ANNU ANNU ANNU ANNU BEND
24H<  ANNU ANNU ANNU ANNU ANNU  ANNU ANNU ANNU ANNU  ANNU ANNU ANNU ANNU ANNU BEND
25H<  ANNU ANNU ANNU ANNU ANNU  ANNU ANNU ANNU ANNU  ANNU ANNU ANNU ANNU ANNU BEND
26H<  ANNU ANNU ANNU ANNU ANNU  ANNU ANNU ANNU ANNU  ANNU ANNU ANNU ANNU ANNU BEND
27H<  ANNU ANNU ANNU ANNU ANNU  ANNU ANNU ANNU ANNU  ANNU ANNU ANNU ANNU ANNU BEND
28H<  ANNU ANNU ANNU ANNU ANNU  ANNU ANNU ANNU ANNU  ANNU ANNU ANNU ANNU ANNU BEND
29H<  ANNU ANNU ANNU ANNU ANNU  ANNU ANNU ANNU ANNU  ANNU ANNU ANNU ANNU ANNU BEND
30H<  SLUG SLUG ANNU ANNU ANNU  ANNU ANNU ANNU ANNU  ANNU ANNU ANNU ANNU ANNU BEND

31L<  ANNU ANNU ANNU ANNU ANNU  ANNU ANNU ANNU ANNU  ANNU ANNU ANNU ANNU ANNU ENTR
32L<  ANNU ANNU ANNU ANNU ANNU  ANNU ANNU ANNU ANNU  ANNU ANNU ANNU ANNU ANNU ENTR
33L<  ANNU ANNU ANNU ANNU ANNU  ANNU ANNU ANNU ANNU  ANNU ANNU ANNU ANNU ANNU ENTR
34L<  SUBC SUBC SLUG SLUG ANNU  ANNU ANNU ANNU ANNU  ANNU ANNU ANNU ANNU ANNU ENTR
35L<  SUBC SUBC SLUG SLUG ANNU  ANNU ANNU ANNU ANNU  ANNU ANNU ANNU ANNU ANNU ENTR
36L<  SUBC SUBC SLUG SLUG ANNU  ANNU ANNU ANNU ANNU  ANNU ANNU ANNU ANNU ANNU ENTR
37L<  SUBC SUBC SUBC SLUG SLUG  ANNU ANNU ANNU ANNU  ANNU ANNU ANNU ANNU ANNU ENTR
38L<  SUBC SUBC SUBC SLUG SLUG  WAVY ANNU ANNU ANNU  ANNU ANNU ANNU ANNU ANNU ENTR
39L<  SUBC SUBC SUBC SUBC SUBC  SLUG SLUG ANNU ANNU  ANNU ANNU ANNU ANNU ANNU ENTR
40L<  SUBC SUBC SUBC SUBC SUBC  SUBC SUBC SLUG WAVY  ANNU ANNU ANNU ANNU ANNU ENTR

41L>  BEND SUBC SUBC SUBC SUBC  SUBC SUBC SUBC SUBC  SUBC SUBC SUBC SUBC SUBC SUBC
42L>  BEND SUBC SUBC SUBC SUBC  SUBC SUBC SUBC SUBC  SUBC SUBC SUBC SUBC SUBC SUBC
43L>  BEND SUBC SUBC SUBC SUBC  SUBC SUBC SUBC SUBC  SUBC SUBC SUBC SUBC SUBC SUBC
44L>  BEND SUBC SUBC SUBC SUBC  SUBC SUBC SUBC SUBC  SUBC SUBC SUBC SUBC SUBC SUBC
45L>  BEND SUBC SUBC SUBC SUBC  SUBC SUBC SUBC SUBC  SUBC SUBC SUBC SUBC SUBC SUBC

46L<  SUBC SUBC SUBC SUBC SUBC  SUBC SUBC SUBC SUBC  SUBC SUBC SUBC SUBC SUBC ENTR
47L<  SUBC SUBC SUBC SUBC SUBC  SUBC SUBC SUBC SUBC  SUBC SUBC SUBC SUBC SUBC ENTR
48L<  SUBC SUBC SUBC SUBC SUBC  SUBC SUBC SUBC SUBC  SUBC SUBC SUBC SUBC SUBC ENTR
49L<  SUBC SUBC SUBC SUBC SUBC  SUBC SUBC SUBC SUBC  SUBC SUBC SUBC SUBC SUBC ENTR
50L<  SUBC SUBC SUBC SUBC SUBC  SUBC SUBC SUBC SUBC  SUBC SUBC SUBC SUBC SUBC ENTR

51L>  BEND SUBC SUBC SUBC SUBC  SUBC SUBC SUBC SUBC  SUBC SUBC SUBC SUBC SUBC SUBC
52L>  BEND SUBC SUBC SUBC SUBC  SUBC SUBC SUBC SUBC  SUBC SUBC SUBC SUBC SUBC SUBC
53L>  BEND SUBC SUBC SUBC SUBC  SUBC SUBC SUBC SUBC  SUBC SUBC SUBC SUBC SUBC SUBC
54L>  BEND SUBC SUBC SUBC SUBC  SUBC SUBC SUBC SUBC  SUBC SUBC SUBC SUBC SUBC SUBC
55L>  BEND SUBC SUBC SUBC SUBC  SUBC SUBC SUBC SUBC  SUBC SUBC SUBC SUBC SUBC SUBC
56L>  BEND SUBC SUBC SUBC SUBC  SUBC SUBC SUBC SUBC  SUBC SUBC SUBC SUBC SUBC SUBC
57L>  BEND SUBC SUBC SUBC SUBC  SUBC SUBC SUBC SUBC  SUBC SUBC SUBC SUBC SUBC SUBC
58L>  BEND ANNU SLUG SUBC SUBC  SUBC SUBC SUBC SUBC  SUBC SUBC SUBC SUBC SUBC SUBC
59L>  BEND ANNU ANNU ANNU SLUG  SUBC SUBC SUBC SUBC  SUBC SUBC SUBC SUBC SUBC SUBC
60L>  BEND ANNU ANNU ANNU ANNU  SLUG SLUG SUBC SUBC  SUBC SUBC SUBC SUBC SUBC SUBC
```

The fist column contains a number (tube row) plus a direction (< or >). Much of the top section is superheated (i.e., vapor only) and much of the bottom section is sub-cooled (i.e., liquid only). This is very undesirable, as the huge temperature differences along the tubes (superheated to sub-cooled) causes unequal thermal expansion and mechanical failure. The inputs in this case (including an orifice plate with differing diameter holes for each pipe) are based on actual operation, not hypothetical conditions. This next figure shows the temperature distribution throughout the tubes (blue=370°F to red=680°F):

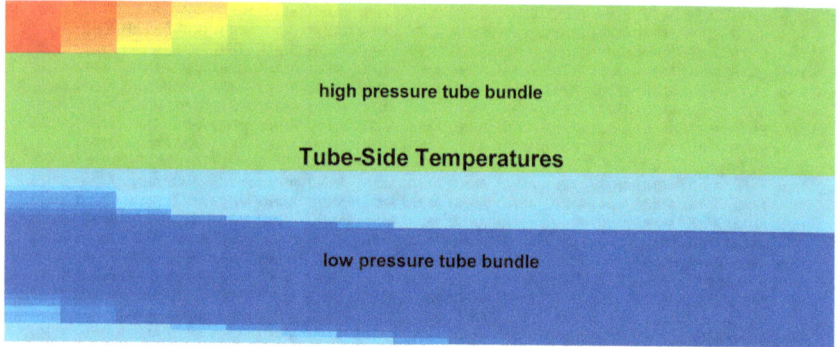

The distribution throughout the shell is similar:

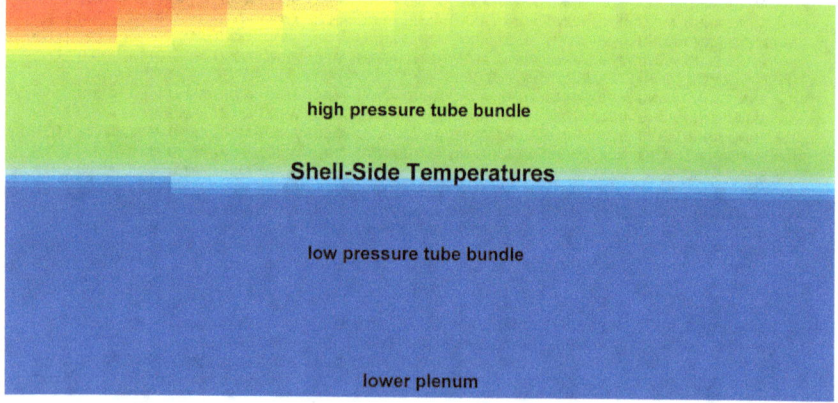

The void fraction is an important parameter in two-phase flow. A value of one indicates complete only vapor (saturated or superheated) and a value of zero indicates only liquid (saturated or sub-cooled). These correspond to the flow regimes shown in the previous figure, that is 1.000 corresponds to SUPR, 0.000 corresponds to SUBC, and everything in between corresponds to ANNUlar, WAVY, SLUG, etc. The predicted compressive stress in kilo-pounds per square inch is shown in the second figure.

VOID FRACTION (VAPOR AREA/TOTAL AREA)

1H>	0.000	1.000	1.000	1.000	1.000	1.000	1.000	1.000	1.000	1.000	1.000	1.000	1.000	1.000
2H>	0.000	1.000	1.000	1.000	1.000	1.000	1.000	1.000	1.000	1.000	1.000	1.000	1.000	1.000
3H>	0.000	1.000	1.000	1.000	1.000	1.000	1.000	1.000	1.000	1.000	1.000	1.000	1.000	1.000
4H>	0.000	1.000	1.000	1.000	1.000	1.000	1.000	1.000	1.000	1.000	1.000	1.000	1.000	1.000
5H>	0.000	1.000	1.000	1.000	1.000	1.000	1.000	1.000	1.000	1.000	1.000	1.000	1.000	1.000
6H>	0.000	1.000	1.000	1.000	1.000	1.000	1.000	1.000	1.000	1.000	1.000	1.000	1.000	1.000
7H>	0.000	1.000	1.000	1.000	1.000	1.000	1.000	1.000	1.000	1.000	1.000	1.000	1.000	0.997	0.997
8H>	0.000	1.000	1.000	1.000	1.000	1.000	1.000	1.000	1.000	1.000	1.000	1.000	0.991	0.974	0.974
9H>	0.000	1.000	1.000	1.000	1.000	1.000	1.000	1.000	1.000	1.000	1.000	1.000	0.977	0.957	0.957
10H>	0.000	1.000	1.000	1.000	1.000	1.000	1.000	1.000	1.000	1.000	1.000	1.000	0.968	0.936	0.911	0.911
11H>	0.000	0.409	0.407	0.405	0.404	0.403	0.401	0.400	0.399	0.398	0.398	0.397	0.396	0.396	0.396	0.396
12H>	0.000	0.408	0.406	0.403	0.402	0.400	0.398	0.397	0.396	0.395	0.394	0.393	0.392	0.392	0.392	0.392
13H>	0.000	0.407	0.404	0.401	0.399	0.396	0.394	0.392	0.391	0.389	0.388	0.387	0.387	0.386	0.386	0.386
14H>	0.000	0.406	0.401	0.397	0.394	0.391	0.388	0.386	0.384	0.382	0.380	0.379	0.379	0.378	0.378	0.378
15H>	0.000	0.404	0.398	0.393	0.388	0.384	0.380	0.377	0.374	0.372	0.370	0.369	0.368	0.368	0.368	0.368
16H<	0.373	0.373	0.374	0.374	0.374	0.375	0.375	0.375	0.376	0.376	0.376	0.377	0.377	0.377	0.378	0.000
17H<	0.383	0.383	0.384	0.384	0.385	0.385	0.385	0.386	0.386	0.386	0.387	0.387	0.387	0.388	0.388	0.000
18H<	0.391	0.391	0.391	0.392	0.392	0.393	0.393	0.393	0.394	0.394	0.394	0.395	0.395	0.395	0.396	0.000
19H<	0.396	0.396	0.397	0.397	0.398	0.398	0.398	0.399	0.399	0.399	0.400	0.400	0.401	0.401	0.401	0.000
20H<	0.400	0.400	0.401	0.401	0.402	0.402	0.402	0.403	0.403	0.403	0.404	0.404	0.404	0.405	0.405	0.000
21H<	0.499	0.499	0.522	0.544	0.569	0.594	0.622	0.651	0.682	0.713	0.746	0.779	0.814	0.848	0.882	0.000
22H<	0.472	0.472	0.495	0.518	0.542	0.567	0.595	0.624	0.659	0.695	0.734	0.776	0.819	0.865	0.911	0.000
23H<	0.407	0.407	0.431	0.454	0.479	0.506	0.535	0.566	0.601	0.639	0.684	0.734	0.789	0.847	0.910	0.000
24H<	0.333	0.333	0.356	0.379	0.405	0.432	0.462	0.496	0.534	0.576	0.624	0.678	0.745	0.820	0.905	0.000
25H<	0.259	0.259	0.281	0.303	0.328	0.354	0.384	0.418	0.457	0.501	0.552	0.612	0.684	0.775	0.885	0.000
26H<	0.189	0.189	0.208	0.228	0.251	0.275	0.304	0.336	0.375	0.419	0.471	0.535	0.613	0.711	0.848	0.000
27H<	0.131	0.131	0.147	0.165	0.185	0.207	0.233	0.263	0.299	0.342	0.394	0.459	0.542	0.651	0.804	0.000
28H<	0.080	0.080	0.094	0.108	0.126	0.144	0.167	0.194	0.226	0.265	0.314	0.377	0.460	0.575	0.746	0.000
29H<	0.041	0.041	0.052	0.064	0.078	0.094	0.113	0.136	0.164	0.198	0.243	0.301	0.381	0.498	0.683	0.000
30H<	0.008	0.008	0.017	0.028	0.039	0.052	0.068	0.087	0.111	0.141	0.180	0.234	0.310	0.427	0.629	0.000
31L<	0.050	0.050	0.058	0.067	0.076	0.085	0.096	0.109	0.126	0.144	0.168	0.198	0.237	0.287	0.355	0.000
32L<	0.028	0.028	0.035	0.042	0.050	0.057	0.067	0.079	0.094	0.111	0.133	0.162	0.201	0.253	0.331	0.000
33L<	0.011	0.011	0.017	0.023	0.029	0.036	0.044	0.054	0.067	0.082	0.103	0.130	0.167	0.221	0.305	0.000
34L<	0.000	0.000	0.001	0.006	0.011	0.017	0.023	0.032	0.043	0.056	0.074	0.098	0.132	0.185	0.274	0.000
35L<	0.000	0.000	0.000	0.004	0.007	0.014	0.020	0.028	0.039	0.052	0.069	0.092	0.125	0.176	0.264	0.000
36L<	0.000	0.000	0.001	0.004	0.008	0.014	0.021	0.029	0.039	0.052	0.069	0.092	0.125	0.176	0.265	0.000
37L<	0.000	0.000	0.000	0.002	0.006	0.012	0.019	0.027	0.037	0.050	0.067	0.089	0.122	0.173	0.261	0.000
38L<	0.000	0.000	0.000	0.000	0.004	0.008	0.015	0.022	0.032	0.044	0.060	0.082	0.114	0.163	0.252	0.000
39L<	0.000	0.000	0.000	0.000	0.000	0.001	0.005	0.013	0.021	0.032	0.047	0.067	0.097	0.144	0.233	0.000
40L<	0.000	0.000	0.000	0.000	0.000	0.000	0.000	0.002	0.013	0.025	0.042	0.067	0.109	0.193	0.000
41L>	0.000	0.000	0.000	0.000	0.000	0.000	0.000	0.000	0.000	0.000	0.000	0.000	0.000	0.000	0.000	0.000
42L>	0.000	0.000	0.000	0.000	0.000	0.000	0.000	0.000	0.000	0.000	0.000	0.000	0.000	0.000	0.000	0.000
43L>	0.000	0.000	0.000	0.000	0.000	0.000	0.000	0.000	0.000	0.000	0.000	0.000	0.000	0.000	0.000	0.000
44L>	0.000	0.000	0.000	0.000	0.000	0.000	0.000	0.000	0.000	0.000	0.000	0.000	0.000	0.000	0.000	0.000
45L>	0.000	0.000	0.000	0.000	0.000	0.000	0.000	0.000	0.000	0.000	0.000	0.000	0.000	0.000	0.000	0.000
46L<	0.000	0.000	0.000	0.000	0.000	0.000	0.000	0.000	0.000	0.000	0.000	0.000	0.000	0.000	0.000	0.000
47L<	0.000	0.000	0.000	0.000	0.000	0.000	0.000	0.000	0.000	0.000	0.000	0.000	0.000	0.000	0.000	0.000
48L<	0.000	0.000	0.000	0.000	0.000	0.000	0.000	0.000	0.000	0.000	0.000	0.000	0.000	0.000	0.000	0.000
49L<	0.000	0.000	0.000	0.000	0.000	0.000	0.000	0.000	0.000	0.000	0.000	0.000	0.000	0.000	0.000	0.000
50L<	0.000	0.000	0.000	0.000	0.000	0.000	0.000	0.000	0.000	0.000	0.000	0.000	0.000	0.000	0.000	0.000
51L>	0.000	0.000	0.000	0.000	0.000	0.000	0.000	0.000	0.000	0.000	0.000	0.000	0.000	0.000	0.000	0.000
52L>	0.000	0.000	0.000	0.000	0.000	0.000	0.000	0.000	0.000	0.000	0.000	0.000	0.000	0.000	0.000	0.000
53L>	0.000	0.000	0.000	0.000	0.000	0.000	0.000	0.000	0.000	0.000	0.000	0.000	0.000	0.000	0.000	0.000
54L>	0.000	0.000	0.000	0.000	0.000	0.000	0.000	0.000	0.000	0.000	0.000	0.000	0.000	0.000	0.000	0.000
55L>	0.000	0.000	0.000	0.000	0.000	0.000	0.000	0.000	0.000	0.000	0.000	0.000	0.000	0.000	0.000	0.000
56L>	0.000	0.000	0.000	0.000	0.000	0.000	0.000	0.000	0.000	0.000	0.000	0.000	0.000	0.000	0.000	0.000
57L>	0.000	0.000	0.000	0.000	0.000	0.000	0.000	0.000	0.000	0.000	0.000	0.000	0.000	0.000	0.000	0.000
58L>	0.000	0.009	0.003	0.000	0.000	0.000	0.000	0.000	0.000	0.000	0.000	0.000	0.000	0.000	0.000	0.000
59L>	0.000	0.024	0.016	0.009	0.003	0.000	0.000	0.000	0.000	0.000	0.000	0.000	0.000	0.000	0.000	0.000
60L>	0.000	0.042	0.030	0.020	0.013	0.006	0.002	0.000	0.000	0.000	0.000	0.000	0.000	0.000	0.000	0.000

COMPRESSIVE STRESS ON TUBES DUE TO THERMAL EXPANSION ASSUMING CLAMPED ENDS AND RIGID TUBE SHEETS

1H>	0	105.1	101.6	98.5	95.7	93.2	91.1	89.2	87.6	86.3	85.2	84.3	83.5	83.0	82.5	0
2H>	0	104.6	101.0	97.7	94.8	92.4	90.2	88.5	87.0	85.7	84.7	83.8	83.1	82.6	82.2	0
3H>	0	104.1	100.2	96.9	94.0	91.5	89.4	87.7	86.3	85.1	84.1	83.4	82.8	82.3	82.0	0
4H>	0	103.5	99.5	96.0	93.1	90.6	88.6	87.0	85.6	84.5	83.6	82.9	82.4	82.0	81.7	0
5H>	0	102.9	98.6	95.1	92.1	89.7	87.8	86.2	85.0	84.0	83.2	82.5	82.1	81.8	81.5	0
6H>	0	102.3	97.7	94.1	91.2	88.8	87.0	85.5	84.3	83.4	82.7	82.2	81.8	81.5	81.3	0
7H>	0	101.6	96.8	93.1	90.1	87.9	86.1	84.8	83.7	82.9	82.3	81.8	81.5	81.3	81.0	0
8H>	0	100.7	95.7	91.9	89.1	86.9	85.3	84.0	83.1	82.4	81.9	81.5	81.2	80.9	80.9	0
9H>	0	99.8	94.5	90.7	87.9	85.8	84.3	83.3	82.5	81.9	81.5	81.2	81.0	80.8	80.8	0
10H>	0	98.8	93.0	89.1	86.4	84.6	83.3	82.4	81.8	81.4	81.1	80.9	80.7	80.7	80.7	0
11H>	0	80.7	80.7	80.7	80.7	80.6	80.6	80.6	80.5	80.5	80.5	80.4	80.4	80.4	80.4	0
12H>	0	80.7	80.7	80.7	80.6	80.6	80.6	80.6	80.5	80.5	80.5	80.4	80.4	80.4	80.4	0
13H>	0	80.7	80.6	80.6	80.6	80.6	80.5	80.5	80.5	80.5	80.5	80.4	80.4	80.4	80.4	0
14H>	0	80.6	80.6	80.6	80.6	80.5	80.5	80.5	80.5	80.4	80.4	80.4	80.4	80.4	80.4	0
15H>	0	80.5	80.5	80.5	80.5	80.5	80.4	80.4	80.4	80.4	80.4	80.4	80.4	80.4	80.4	0
16H<	0	79.9	79.9	80.0	80.0	80.0	80.1	80.1	80.1	80.2	80.2	80.3	80.3	80.3	80.4	0
17H<	0	79.9	79.9	80.0	80.0	80.0	80.1	80.1	80.1	80.2	80.2	80.3	80.3	80.3	80.4	0
18H<	0	79.9	79.9	80.0	80.0	80.0	80.1	80.1	80.1	80.2	80.2	80.3	80.3	80.3	80.4	0
19H<	0	79.9	79.9	80.0	80.0	80.0	80.1	80.1	80.1	80.2	80.2	80.3	80.3	80.3	80.4	0
20H<	0	79.9	79.9	80.0	80.0	80.0	80.1	80.1	80.1	80.2	80.2	80.3	80.3	80.3	80.4	0
21H<	0	80.4	80.4	80.4	80.4	80.4	80.4	80.5	80.5	80.5	80.5	80.5	80.6	80.6	80.6	0
22H<	0	80.3	80.3	80.4	80.4	80.4	80.4	80.3	80.4	80.4	80.4	80.5	80.5	80.5	80.6	0
23H<	0	80.1	80.1	80.2	80.2	80.2	80.2	80.3	80.2	80.2	80.3	80.3	80.4	80.4	80.4	0
24H<	0	79.8	79.9	80.0	80.0	80.0	80.0	80.1	80.1	80.1	80.0	80.1	80.1	80.1	80.2	0
25H<	0	79.5	79.5	79.7	79.7	79.7	79.7	79.7	79.7	79.8	79.8	79.8	79.7	79.8	79.9	0
26H<	0	79.1	79.1	79.3	79.3	79.3	79.3	79.3	79.3	79.4	79.4	79.4	79.4	79.3	79.4	0
27H<	0	78.5	78.5	78.7	78.7	78.7	78.7	78.7	78.7	78.8	78.8	78.9	78.9	78.9	78.8	0
28H<	0	77.9	77.9	78.1	78.1	78.1	78.0	78.0	78.1	78.1	78.2	78.2	78.2	78.2	78.0	0
29H<	0	77.2	77.2	77.4	77.4	77.3	77.2	77.1	77.1	77.2	77.2	77.2	77.2	77.3	76.9	0
30H<	0	75.7	75.7	76.5	76.6	76.5	76.2	76.2	76.1	76.1	76.1	76.1	76.1	76.1	76.0	0
31L<	0	64.5	64.5	65.3	65.3	65.2	65.0	65.0	65.0	65.1	65.2	65.2	65.3	65.3	65.4	0
32L<	0	63.9	63.9	64.8	64.7	64.6	64.3	64.3	64.3	64.5	64.5	64.5	64.6	64.7	64.7	0
33L<	0	63.0	63.2	64.2	64.1	63.9	63.5	63.5	63.4	63.6	63.7	63.7	63.7	63.8	63.8	0
34L<	0	62.6	64.5	62.8	63.3	63.2	62.6	62.5	62.4	62.4	62.4	62.4	62.4	62.5	62.5	0
35L<	0	60.4	64.2	62.0	62.3	62.7	62.6	62.5	62.3	62.2	62.2	62.1	62.0	62.0	61.9	0
36L<	0	60.2	63.5	61.9	62.3	62.7	62.6	62.4	62.3	62.2	62.1	62.1	62.0	62.0	61.9	0
37L<	0	59.9	62.9	63.8	62.0	62.6	62.6	62.5	62.3	62.2	62.2	62.1	62.0	62.0	61.9	0
38L<	0	58.3	60.6	64.6	63.6	62.4	62.8	62.6	62.5	62.4	62.3	62.2	62.1	62.0	62.0	0
39L<	0	56.2	57.5	59.6	63.0	64.6	63.7	62.9	62.9	62.7	62.5	62.4	62.3	62.2	62.1	0
40L<	0	54.6	54.9	55.5	56.4	58.1	60.9	65.0	64.5	63.3	62.9	63.0	62.8	62.7	62.5	0
41L>	0	54.2	54.2	54.2	54.2	54.2	54.2	54.2	54.2	54.2	54.2	54.2	54.2	54.2	54.2	0
42L>	0	54.2	54.2	54.2	54.2	54.2	54.2	54.2	54.2	54.2	54.2	54.2	54.2	54.2	54.2	0
43L>	0	54.2	54.2	54.2	54.2	54.2	54.2	54.2	54.2	54.2	54.2	54.2	54.2	54.2	54.2	0
44L>	0	54.2	54.2	54.2	54.2	54.2	54.2	54.2	54.2	54.2	54.2	54.2	54.2	54.2	54.2	0
45L>	0	54.2	54.2	54.2	54.2	54.2	54.2	54.2	54.2	54.2	54.2	54.2	54.2	54.2	54.2	0
46L<	0	54.2	54.2	54.2	54.2	54.2	54.2	54.2	54.2	54.2	54.2	54.2	54.2	54.2	54.3	0
47L<	0	54.2	54.2	54.2	54.2	54.2	54.2	54.2	54.2	54.2	54.2	54.2	54.2	54.2	54.3	0
48L<	0	54.2	54.2	54.2	54.2	54.2	54.2	54.2	54.2	54.2	54.2	54.2	54.2	54.2	54.3	0
49L<	0	54.2	54.2	54.2	54.2	54.2	54.2	54.2	54.2	54.2	54.2	54.2	54.2	54.2	54.3	0
50L<	0	54.2	54.2	54.2	54.2	54.2	54.2	54.2	54.2	54.2	54.2	54.2	54.2	54.2	54.3	0
51L>	0	54.4	54.3	54.3	54.2	54.2	54.2	54.2	54.2	54.2	54.2	54.2	54.2	54.2	54.2	0
52L>	0	55.2	54.8	54.6	54.5	54.4	54.3	54.3	54.3	54.2	54.2	54.2	54.2	54.2	54.2	0
53L>	0	56.9	56.0	55.3	54.9	54.7	54.5	54.4	54.3	54.3	54.3	54.2	54.2	54.2	54.2	0
54L>	0	57.9	56.6	55.8	55.2	54.9	54.6	54.5	54.4	54.3	54.3	54.3	54.3	54.2	54.2	0
55L>	0	59.1	57.4	56.3	55.6	55.1	54.8	54.6	54.5	54.4	54.3	54.3	54.3	54.3	54.2	0
56L>	0	58.6	57.1	56.1	55.4	55.0	54.7	54.6	54.4	54.4	54.3	54.3	54.3	54.3	54.2	0
57L>	0	58.8	57.2	56.2	55.5	55.0	54.8	54.6	54.4	54.4	54.3	54.3	54.3	54.3	54.2	0
58L>	0	62.5	63.8	61.8	59.1	57.4	56.2	55.5	55.1	54.8	54.6	54.5	54.4	54.3	54.3	0
59L>	0	62.7	62.9	62.6	63.8	62.2	59.3	57.5	56.3	55.5	55.1	54.8	54.6	54.4	54.4	0
60L>	0	62.5	62.6	62.8	62.8	62.1	63.4	60.2	58.0	56.6	55.7	55.2	54.8	54.6	54.5	0

Chapter 15. Condensation in Crossflow

In this chapter we will consider the condensation of steam on the outside of tubes in a crossflow when the velocity is non-trivial. Such is definitely the case when steam exiting a low-pressure turbine may reach sonic velocity, turns downward in a hood, and impinges on the top row of tubes in a large water-cooled condenser. By the time the steam reaches the bottom row of tubes (that isn't flooded), the kinetic energy is spent and the velocity is negligible. This interesting process occurs because, at typical condenser operating conditions (1.5 in.HgA), the density of water in the liquid state is 25,000 times that of the corresponding vapor.

The original experimental work on this type of condensation and associated formula for calculating the heat transfer coefficient is due to Nusselt. This formula may be found in any heat transfer text as well as readily on-line. Shekriladze and Gomelauri modified Nusselt's correlation[21] to account for non-trivial vapor velocity.

$$Nu = 0.728 \text{Re}^{\frac{1}{2}} = \frac{hd}{k} \qquad (15.1)$$

where the film Reynolds number is given by:

$$\text{Re} = \left[\frac{\rho_f (\rho_f - \rho_g) h_{fg} g d^3}{k\mu \Delta T} \right]^{\frac{1}{2}} \qquad (15.2)$$

Thermal resistance to heat transfer in the condensation of steam on a cooled surface arises from the thin film of liquid that forms and must run off the surface as condensation proceeds. A crossflow in the vapor produces a shear stress, which strips away the liquid, thinning the film and reducing the resistance. This is the basis for Shekriladze and Gomelauri modification.[22] The modified Nusselt number is given by:

$$Nu = \frac{0.9 + 0.728 F^{\frac{1}{2}}}{\left(1 + 3.44 F^{\frac{1}{2}} + F^2\right)^{\frac{1}{4}}} \text{Re}^{\frac{1}{2}} \qquad (15.3)$$

F is equal to the Prandtl number divided by Froude and Jacob number, or:

---

[21] Nusselt, W., "Die Oberflächenkondensation des Wasserdampfes [The Surface Condensation of Water Vapor]," VDI [Association of German Engineers], 1916.
[22] Shekriladze, I. G. and Gomelauri, V. I., "Theoretical Study of Laminar Film Condensation of Flowing Vapor," International Journal of Heat and Mass Transfer, Vol. 9, pp. 581–591, 1966.

$$F = \left(\frac{\mu h_{fg}}{uk\Delta T}\right)\sqrt{\frac{g}{d}} \qquad (15.4)$$

This formulation can be applied to several types of heat exchangers, including a water-cooled steam surface condenser such as would be found in a power plant. The inputs to such a model include geometry and operating conditions. Thermal and transport properties are also needed. These are provided on the *properties* tab and are accessed via an interpolation macro, also included in the spreadsheet condensation_crossflow.xls.

	A	B	C	D
1	**Condensation on Horizontal Tubes in Crossflow**			
2	INPUTS	symbol	units	value
3	Steam Turbine Exhaust			
4	turbine annulus area	Aan	ft²	72
5	number of ends	Nan	-	4
6	Condenser Geometry			
7	tube length	L	ft	40.0
8	number of tubes	N	-	12,000
9	tube outside dia.	do	in	1.125
10	tube gauge	ga	-	22
11	tube wall conduct.	kw	BTU/hr/ft/°F	9.32
12	tube pitch	pitch	in	1.125
13	Steam			
14	steam flow	stm	lbm/hr	2,500,000
15	steam quality	x	-	91%
16	Cooling Water			
17	flow	Qccw	gpm	240,000
18	inlet temperature	Tccw,in	°F	60
19	SOLUTION			
20	backpressure	Psat	in.HgA	1.84

The solution is the operating pressure (or turbine backpressure). This is solved iteratively using a bisection search when you push the button provided. The cooling water inlet temperature is an input and the outlet temperature is calculated from the duty (i.e., the steam condensed).

21	PROPERTIES	symbol	units	value
22	Cooling Water			
23	density	ρ	lbm/ft³	62.37
24	specific heat	Cp	BTU/lbm/°F	1.0004
25	viscosity	μ	lbm/ft/hr	2.712
26	thermal cond.	k	BTU/hr/ft/°F	0.3423
27	Prandtl Number	Pr	-	7.93
28	Steam			
29	liquid density	ρf	lbm/ft³	62.01
30	vapor density	ρg	lbm/ft³	0.0027
31	latent heat	hfg	BTU/lbm	1037.6
32	liquid viscosity	μf	lbm/ft/hr	1.678
33	liquid thermal cond.	kf	BTU/hr/ft/°F	0.3611
34	CALCULATIONS	symbol	units	value
35	Condenser Geometry			
36	tube wall thick.	wt	in	0.028
37	tube inside dia.	di	in	1.069
38	water flow area	Af	ft²	74.79
39	surface area	As	ft²	141,372
40	steam flow area	Ap	ft²	410.79

The film Reynolds number depends weakly on the temperature difference across the condensing film. A (spatially) constant value is used here, equal to the overall log-mean temperature difference.

41	Cooling Water			
42	duty	duty	BTU/hr	2.358E+09
43	mass flow rate	mccw	lbm/hr	1.201E+08
44	outlet temperature	Tccw,out	°F	79.63
45	velocity inside tubes	ut	ft/sec	7.15
46	tube side Reynolds	Re,t	-	52,729
47	tube side Nusselt	Nu,t	-	315
48	tube side ht. tr. coef.	ht	BTU/hr/ft/°F	1212
49	tube wall ht. tr. coef.	hw	BTU/hr/ft/°F	3894
50	Steam			
51	operating pressure	Psat	psia	0.9046
52	saturation temp.	Tsat	°F	98.38
53	exit specific volume	1/ρan	ft³/lbm	332.5
54	annulus velocity	Van	ft/sec	802
55	log mean temp. diff.	LMTD	°F	27.4
56	required conductance	U	BTU/hr/°F	608.9
57	film Reynolds number	Re	-	2.874E+05

The heat transfer for each row of tubes (in 20 groups) is calculated sequentially so that the local steam velocity can be used to calculate the impact on condensing heat transfer coefficient.

			Local Heat Transfer Calculations					
vertical position	ht. tr. BTU/hr	quality -	sp.vol. ft³/lbm	velocity ft/sec	F -	Nu -	hc BTU/hr/ft²/°F	U BTU/hr/ft²/°F
top	1.21E+08	86.23%	315.4	533.1	6.12	553.2	2130	627.2
2	1.21E+08	81.55%	298.3	504.2	6.47	550.4	2120	626.3
3	1.21E+08	76.88%	281.2	475.4	6.86	547.5	2108	625.3
4	1.21E+08	72.22%	264.2	446.6	7.30	544.3	2096	624.2
5	1.21E+08	67.57%	247.1	417.8	7.80	540.8	2083	623.0
6	1.20E+08	62.93%	230.2	389.1	8.38	537.1	2069	621.8
7	1.20E+08	58.30%	213.2	360.5	9.05	533.1	2053	620.3
8	1.20E+08	53.68%	196.3	331.9	9.82	528.8	2036	618.8
9	1.20E+08	49.07%	179.5	303.4	10.75	524.1	2018	617.1
10	1.19E+08	44.48%	162.7	275.0	11.85	518.9	1999	615.3
11	1.19E+08	39.90%	146.0	246.7	13.21	513.4	1977	613.2
12	1.18E+08	35.34%	129.3	218.5	14.92	507.2	1954	610.9
13	1.18E+08	30.80%	112.7	190.5	17.12	500.5	1928	608.4
14	1.17E+08	26.28%	96.1	162.5	20.06	493.0	1899	605.5
15	1.17E+08	21.79%	79.7	134.7	24.20	484.7	1867	602.2
16	1.16E+08	17.32%	63.4	107.1	30.44	475.2	1830	598.3
17	1.15E+08	12.89%	47.1	79.7	40.91	464.1	1788	593.7
18	1.14E+08	8.50%	31.1	52.6	62.02	450.7	1736	587.9
19	1.12E+08	4.17%	15.3	25.8	126.34	432.9	1667	579.8
bottom	1.08E+08	0.00%	0.0	0.0	∞	390.3	1503	558.6

The variation in Nusselt number is illustrated in the following figure:

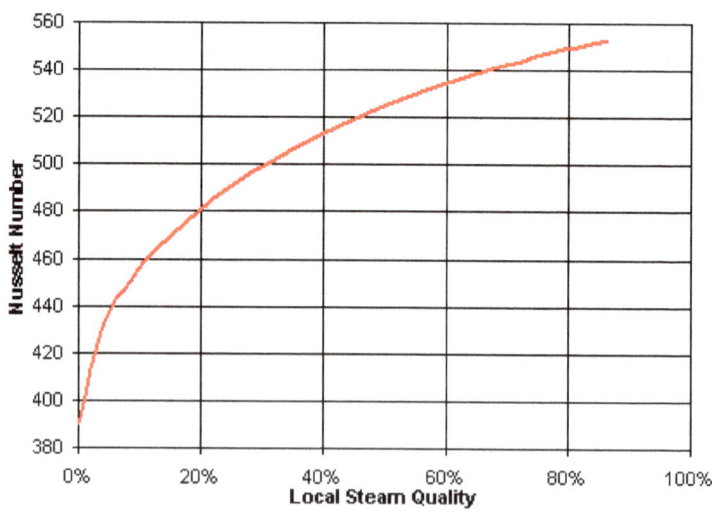

There is also a button and macro to vary the cooling water inlet temperature and produce a table of backpressures:

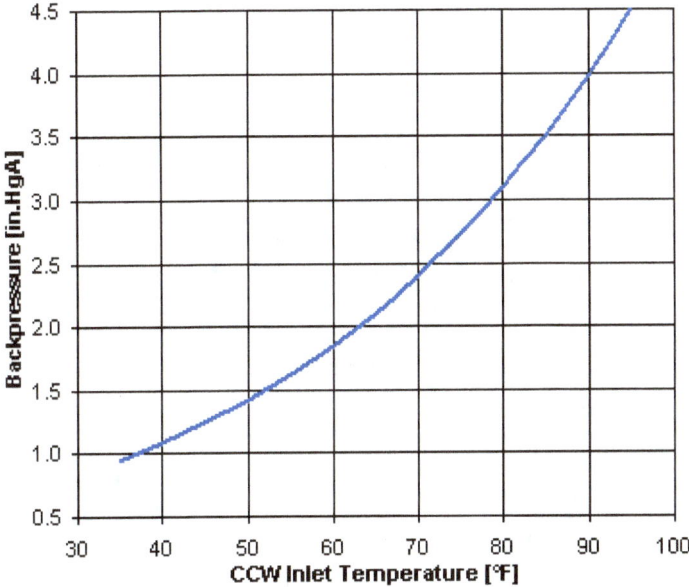

Chapter 16. Operational Data

It is often assumed that all data exhibit normally distributed variability; however, this is often not the case. We will first consider some 7966 hourly averages representing the performance of an actual high-pressure feedwater heater. The total heat transfer and overall conductance (following the analysis in Chapter 7) are shown in this first figure as it varies throughout the year.

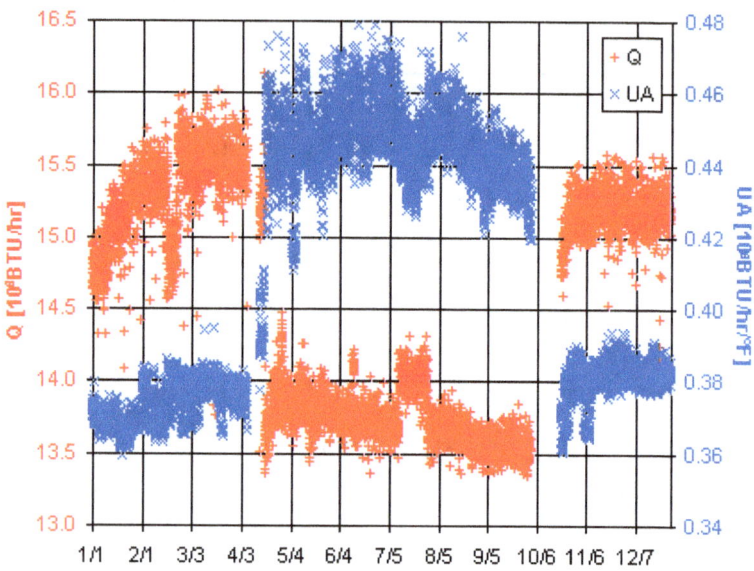

There are clearly two modes of operation: one in the colder months and another in the warmer months. This arises from a change in boiler operation to meet seasonal emissions targets. The probability of occurrence of this same data is shown in the second figure.

The two modes of operation produce two peaks in the probability. Note that the area under the red (Q) and blue (UA) curves both sum to unity (i.e., 100%) so that the horizontal scale is immaterial. The split between these two modes of operation is so close that the area under each peak sums to one-half (i.e., 50%). The difference is only 0.25%. Such distinctions don't occur by chance. This is a response to a purposefully implemented standard operating procedure (SOP).

This data can be presented in statistical form by counting the number of occurrences in each interval over the range of observed values. The resulting probability distributions are fairly normal, as is often the case with actual operating data of this sort. The variability or spread of the data so plotted is an indication of the uncertainty of the measurements as well as the operation in addition to the two distinct modes. All of the data and graphs are in the spreadsheet HP_FWH_operational.xls, which is included in the on-line archive.

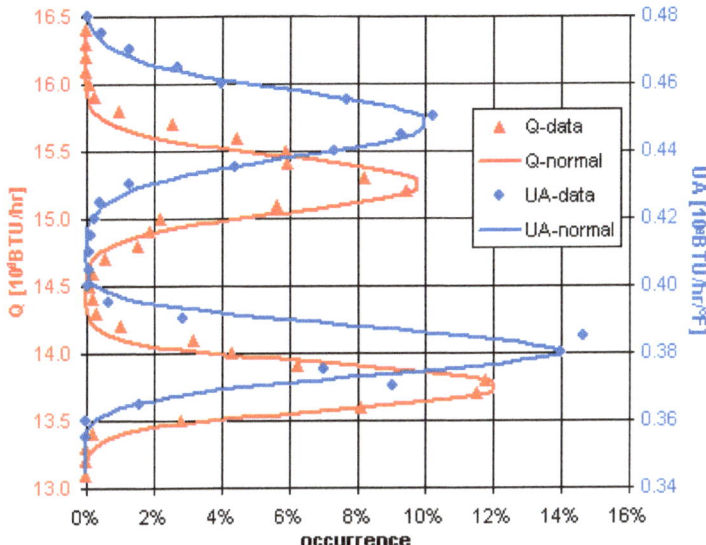

The heat transfer, **Q**, is well approximated by two combined normal distributions: 13.75±0.16 and 15.25±0.20 million BTU/hr. The conductance, **UA**, is also well approximated by two combined normal distributions: 0.3803±0.0071 and 0.4480±0.0098 million BTU/hr/°F.

We will next consider similar operational data, this time from a condenser. As the water flow in this case is based on pump curves, rather than an accurate measurement, this cannot be considered a code-level test in accordance with PTC-12.5; nevertheless, this is often the best that can be done under the circumstances. At this plant, the condenser cooling water (CCW) flows through buried concrete pipe with no accessible straight run in support of a pitot traverse, the preferred method of measurement. Dye dilution was also rejected for lack of an acceptable injection point. The dimensions of the condenser are:

DESIGN INFORMATION		
description	units	value
number of shells (HP/LP)	-	2
area (per shell)	ft²	195,000
nominal steam flow (per shell)	lbm/hr	1,665,000
nominal duty (per shell)	BTU/hr	1.20E+09
number of CCW pumps	-	3
nominal CCW flow	gpm	111,267
TESTING INFORMATION		
condensate nozzle diameter	in	7.500
condensate nozzle Cd	-	0.9021

This is the beginning of the operational data:

	A	B	C	D	E	F	G	H	I	J
1						MEASURED DATA				
2	Description	CCW inlet temp.	CCW exit temp.	CCW pump head	LPTA exh. pres. gov. end	LPTA exh. pres. gen. end	LPTB exh. pres. gov. end	LPTB exh. pres. gen. end	hotwell temp.	condensate flow nozzle delta-P
3	Units	Deg F	Deg F	in.H2O	in.HgA	in.HgA	in.HgA	in.HgA	Deg F	in.H2O
4	2/26/14 9:27	80.29	93.54	186.39	1.701	1.718	2.105	2.077	100.95	408.23
5	2/26/14 9:27	80.28	93.53	188.35	1.707	1.722	2.100	2.074	100.94	408.79
6	2/26/14 9:28	80.28	93.53	185.39	1.703	1.718	2.099	2.072	100.94	407.92
7	2/26/14 9:28	80.26	93.51	192.77	1.703	1.721	2.096	2.069	100.91	408.72
8	2/26/14 9:29	80.25	93.49	187.39	1.709	1.726	2.104	2.076	100.89	408.67
9	2/26/14 9:29	80.23	93.48	187.86	1.711	1.726	2.106	2.077	100.89	405.63
10	2/26/14 9:30	80.23	93.47	186.17	1.713	1.730	2.112	2.084	100.90	402.02
11	2/26/14 9:30	80.24	93.48	189.36	1.716	1.733	2.116	2.087	100.95	401.51
12	2/26/14 9:31	80.22	93.46	184.60	1.715	1.728	2.113	2.086	101.00	407.29
13	2/26/14 9:31	80.22	93.46	186.40	1.716	1.735	2.111	2.085	101.03	415.64
14	2/26/14 9:32	80.22	93.46	186.64	1.720	1.735	2.111	2.085	101.04	413.42
15	2/26/14 9:32	80.22	93.46	187.58	1.725	1.740	2.120	2.094	101.05	404.64
16	2/26/14 9:33	80.21	93.44	182.51	1.727	1.742	2.127	2.100	101.11	401.63
17	2/26/14 9:33	80.21	93.45	184.38	1.732	1.749	2.137	2.109	101.14	402.46
18	2/26/14 9:34	80.20	93.43	186.15	1.736	1.751	2.138	2.107	101.22	405.96
19	2/26/14 9:34	80.19	93.43	185.74	1.736	1.754	2.142	2.115	101.29	409.87
20	2/26/14 9:35	80.20	93.43	182.34	1.742	1.755	2.143	2.116	101.37	409.32
21	2/26/14 9:35	80.20	93.43	185.42	1.740	1.759	2.146	2.118	101.44	404.79

This is the calculation section:

K	L	M	N	O	P	Q	R	S	T	U	V	W	X
						CALCULATIONS							
CCW flow	duty	cond. density	cond. flow	LPTA exh. pres.	LPTB exh. pres.	LPTA exh. temp.	LPTB exh. temp.	LPTA exhaust quality	LPTB exhaust quality	shell A LMTD	shell B LMTD	shell A U	shell B U
gpm	BTU/hr	lbm/ft³	lbm/hr	in.HgA	in.HgA	Deg F	Deg F	-	-	Deg F	Deg F	BTU/hr/ft²/°F	
457,219	3.018E+09	61.980	3,372,124	1.710	2.091	95.88	102.54	0.8612	0.8644	11.98	12.02	645.9	644.0
457,129	3.017E+09	61.980	3,374,446	1.715	2.087	95.96	102.48	0.8604	0.8635	12.08	11.95	640.3	647.1
457,268	3.018E+09	61.980	3,370,850	1.711	2.086	95.90	102.46	0.8615	0.8646	12.01	11.94	644.2	648.1
457,140	3.016E+09	61.980	3,374,185	1.712	2.082	95.93	102.41	0.8603	0.8634	12.06	11.91	641.5	649.6
457,148	3.016E+09	61.980	3,373,994	1.717	2.090	96.04	102.53	0.8604	0.8635	12.18	12.04	635.1	642.1
457,635	3.019E+09	61.981	3,361,398	1.718	2.091	96.05	102.55	0.8645	0.8676	12.21	12.08	633.9	640.6
458,209	3.023E+09	61.980	3,346,434	1.722	2.098	96.11	102.66	0.8694	0.8725	12.28	12.20	631.3	635.4
458,291	3.023E+09	61.980	3,344,267	1.725	2.101	96.17	102.71	0.8700	0.8732	12.32	12.24	629.0	633.1
457,369	3.017E+09	61.979	3,368,202	1.722	2.099	96.11	102.68	0.8620	0.8651	12.28	12.23	629.7	632.4
456,025	3.007E+09	61.979	3,402,518	1.725	2.098	96.19	102.66	0.8508	0.8538	12.37	12.21	623.6	631.6
456,384	3.010E+09	61.979	3,393,416	1.728	2.098	96.22	102.66	0.8536	0.8567	12.40	12.21	622.3	632.1
457,793	3.019E+09	61.979	3,357,203	1.733	2.107	96.32	102.80	0.8655	0.8686	12.50	12.36	619.2	626.3
458,271	3.021E+09	61.978	3,344,665	1.734	2.113	96.36	102.90	0.8696	0.8727	12.55	12.48	617.3	620.9

The same '67 steam property functions used in the spreadsheet. Perhaps the most critical measurement is the turbine exhaust pressure. In this case there are two low-pressure turbines, each with two ends: one on the governor end and the other on the generator end. The average pressures are shown in this next figure:

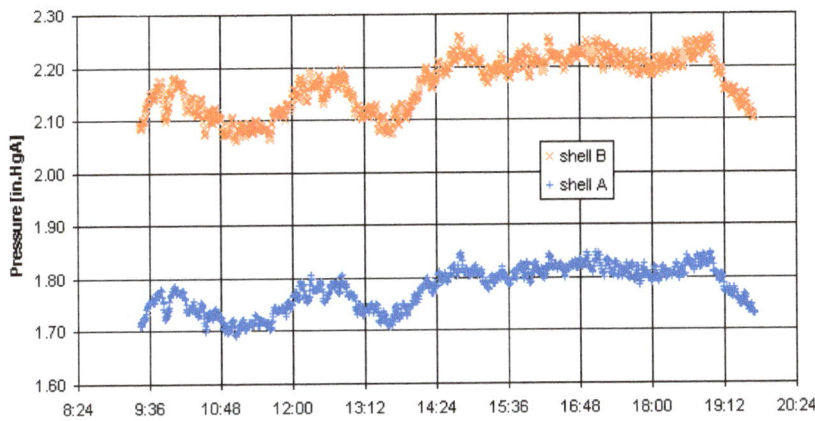

The saturation temperature associated with all four pressures is consistent with the hotwell temperature, something of great concern in these tests. If the hotwell temperature is higher, then the pressures cannot possibly be correct. If the hotwell temperature is more than 1.8°F/1.0°C lower (i.e., *sub-cooling*), the pressures are suspect. Here, this is not an issue.

Condenser cooling water flow was determined by measuring the head across the pumps and the curves. This is rarely an accurate determination of flow. If it were, there would be no point in testing pumps. Such curves are often based on scale models and may easily be off by 15%. In this case, the indicated flow was remarkably close to the design.

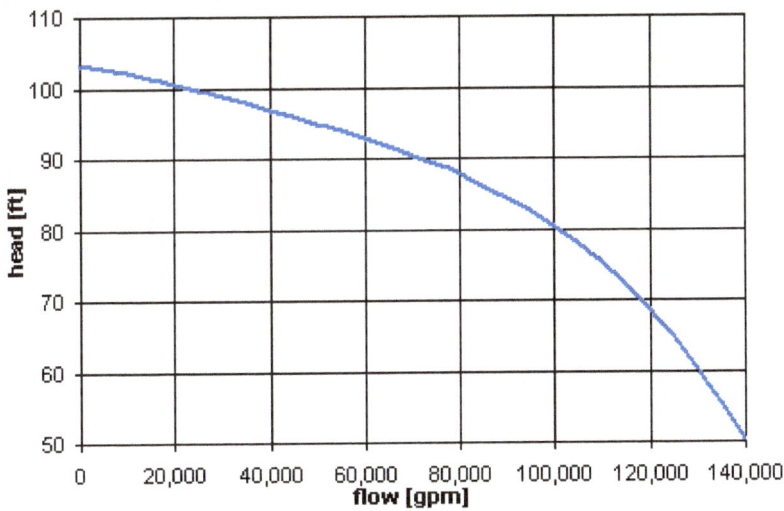

The LMTDs are well behaved, stable, and also close to the design values:

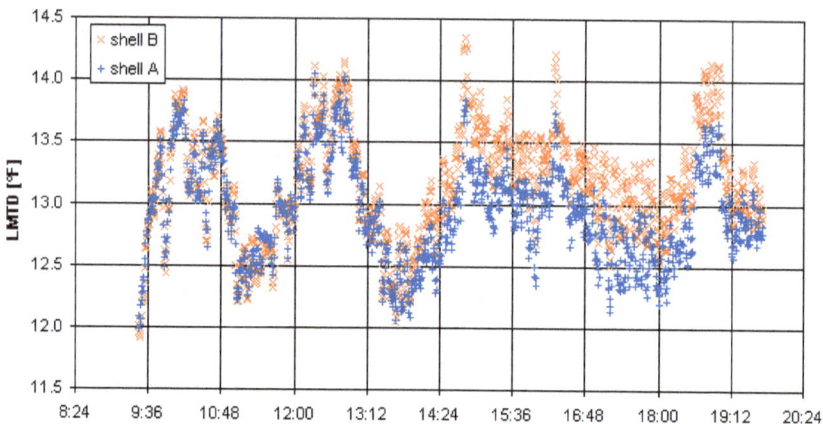

The two sections of the condenser are side-by-side. Although there were two fittings where temperature probes were inserted, there is no reason to presume that these two points accurately represent the average temperature between the two sections. The duty of the two sections was assumed equal in these calculations for lack of any better option. The overall conductance, U, of each shell is shown in this next figure.

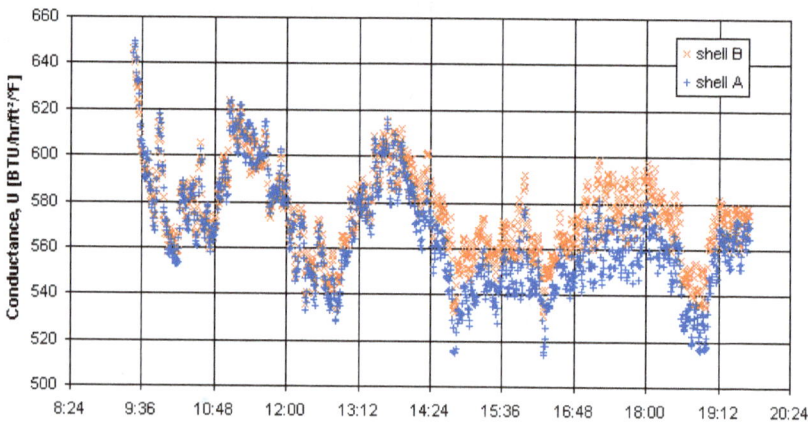

This data indicates a single mode of operation and all of the variables exhibited a fairly normal distribution of values. Only the U is shown here, as this was the result of greatest interest and a direct indication of the cleanliness of the condenser.

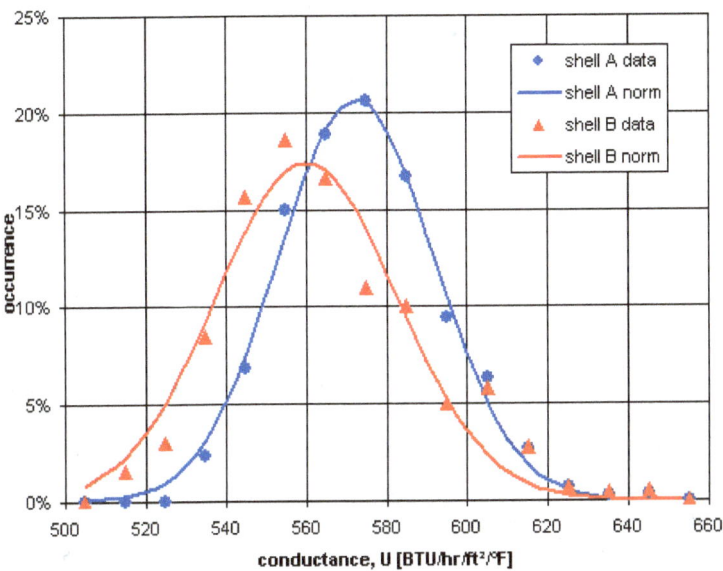

The A shell (blue dots) has a mean of 572.6 with a standard deviation of 19.28 BTU/hr/ft²/°F and the B shell (red triangles) has a mean of 560.2 with a standard deviation of 22.3 BTU/hr/ft²/°F.

Chapter 17. Monte Carlo Methods

A model or computational method utilizing random values is often called Monte Carlo after the famous casino. Monte Carlo models have several uses in heat exchanger analysis, including the estimate of uncertainty in calculated results (e.g., heat transfer, mean temperature difference, and overall conductance) based on the uncertainty of the inputs (e.g., flows, temperatures, and properties). In this chapter we will explore this method of arriving at uncertainty.

At the core of any Monte Carlo method is a function to generate normally distributed random numbers. Most random number generators (e.g., the rand() function in Excel® and it's equivalent in VBA® rnd()), return uniformly distributed random numbers. What this means is that, if you create a large list of such numbers, sort these, separate them into groups, and count the number in each group, you will get a nearly flat curve. That's not what we want. For a Monte Carlo model, we must have numbers that form a bell-shaped curve.

The simplest way of generating normally distributed random numbers from uniformly distributed ones is to take the average of several of the latter. Experience has shown that 12 are sufficient to accomplish the desired result. Sample code is provided in Appendix C.

The following spreadsheet illustrates the Monte Carlo Method for estimating heat exchanger uncertainty:

	A	B	C	D	E	F	G	H	I
1	Monte Carlo Simulation (10,000 cases)								
2	name	units	mean	uncer					
3	mH	kg/s	4.00	0.40	push to update				
4	CpH	kJ/kg/°C	3.50	0.05					
5	mC	kg/s	9.00	0.90					
6	CpC	kJ/kg/°C	2.50	0.04					
7	THi	°C	500	0.67					
8	THo	°C	300	0.67					
9	TCi	°C	250	0.50					
10	TCo	°C	375	0.50					
11	A	m²	100	N/A					
12									
13	case	mH	CpH	mC	CpC	THi	THo	TCi	TCo
14	0	4.000	3.500	9.000	2.500	500.0	300.0	250.0	375.0
15	1	4.025	3.496	9.275	2.503	499.9	300.3	250.1	374.9
16	2	3.901	3.494	8.911	2.493	500.1	300.1	250.0	374.8
17	3	3.937	3.487	9.114	2.498	499.6	300.0	250.0	374.9
18	4	3.895	3.505	8.636	2.485	500.2	299.9	250.0	375.0
19	5	4.135	3.502	9.077	2.504	499.8	300.1	250.1	374.8
20	6	3.886	3.508	8.889	2.508	500.0	299.7	250.1	375.0

While the randomized values could be entered into the spreadsheet with function calls, doing so results in a prohibitive recalculation time (>1 hour). Instead, a button is provided that fills in the cells with the same values that would be obtained by using function calls. As both flows and all four temperatures are measured, this results in an ambiguity (or discrepancy) between the hot and cold side heat transfer and overall conductance. The heat transfer for both sides is illustrated in this first figure, ΔT in the second, and U in the third:

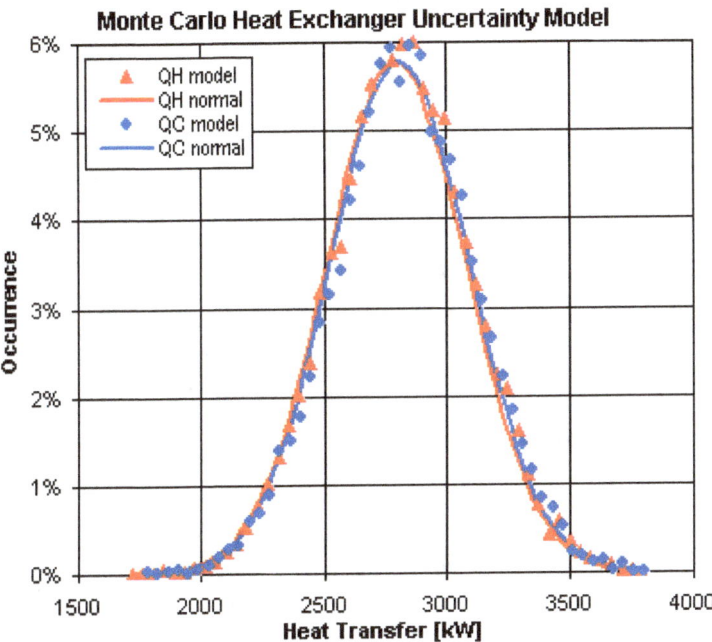

It should be no surprise that the results fall out on a bell-shaped curve. This is to be expected, considering the individual points were generated with normally distributed random numbers. The Monte Carlo Method quickly translates uncertainty in the inputs to uncertainty in the outputs. While it is possible to obtain this result analytically, the algebra is quite cumbersome. We might consider the following equation:

$$(\lambda_{Qh} \pm \sigma_{Qh}) = (\lambda_{\dot{m}h} \pm \sigma_{\dot{m}h})(\lambda_{Cph} \pm \sigma_{Cph})[(\lambda_{Thi} \pm \sigma_{Thi}) - (\lambda_{Tho} \pm \sigma_{Tho})] \quad (17.1)$$

where λ_X represents the mean of X and σ_X represents the uncertainty or variation of X. As there are four occurrences of $\pm$ on the right side of Equation 17.1, there are eight possible values, even with all of the parameters held constant. A similar situation exists for the LMTD:

$$(\lambda_{LMTD} \pm \sigma_{LMTD}) = \frac{(\lambda_{\Delta T1} \pm \sigma_{\Delta T1}) - (\lambda_{\Delta T2} \pm \sigma_{\Delta T2})}{\log\left(\dfrac{\lambda_{\Delta T1} \pm \sigma_{\Delta T1}}{\lambda_{\Delta T2} \pm \sigma_{\Delta T2}}\right)} \quad (17.2)$$

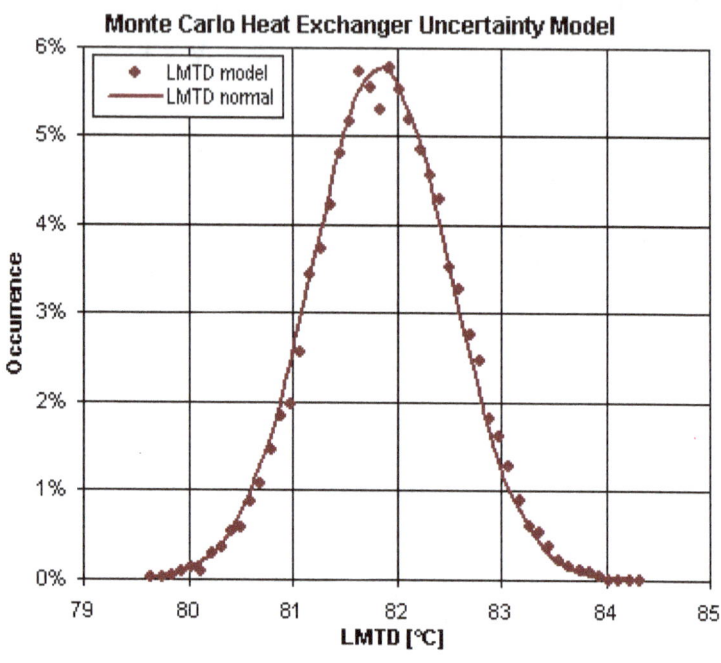
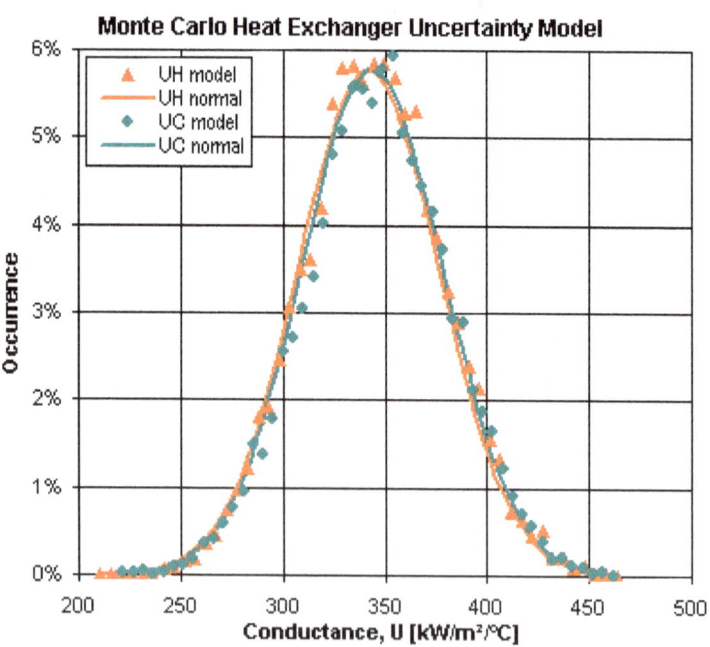

This simulation with 10,000 cases is adequate to achieve occurrence levels that well approximate bell-shaped curves, as illustrated in the previous three figures. Not surprisingly, the Monte Carlo simulation is much closer to the ideal than actual data in the previous chapter. Ten thousand is about the upper limit for Excel®. Real, compiled code can easily handle one million cases, which yields smooth bell-shaped curves, as shown in these next figures. The code is listed in Appendix C.

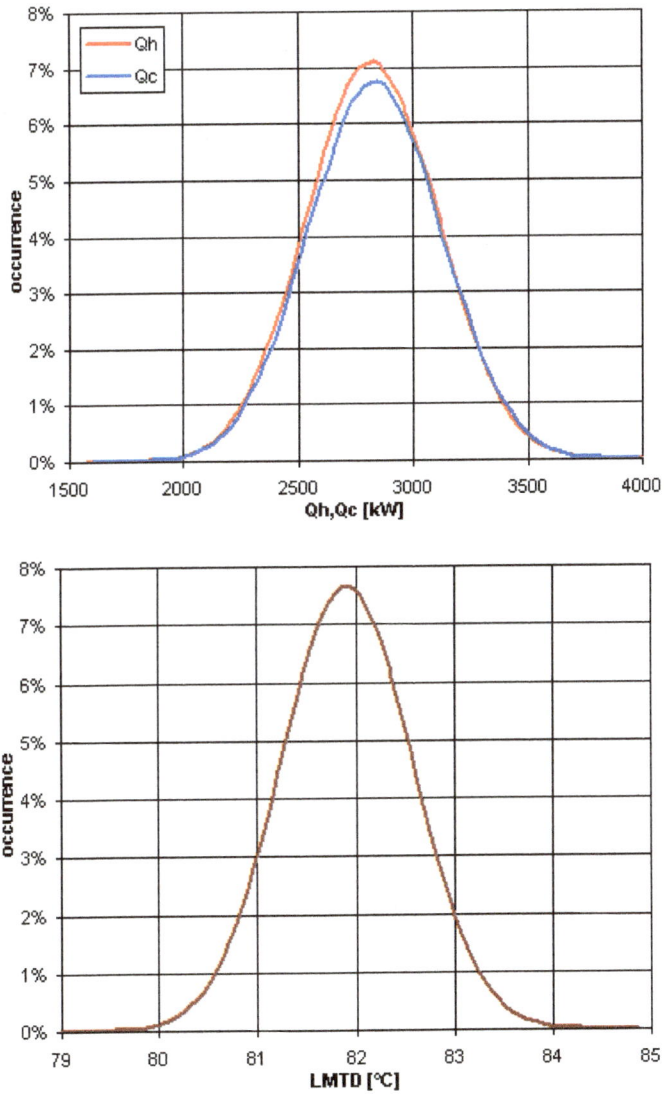

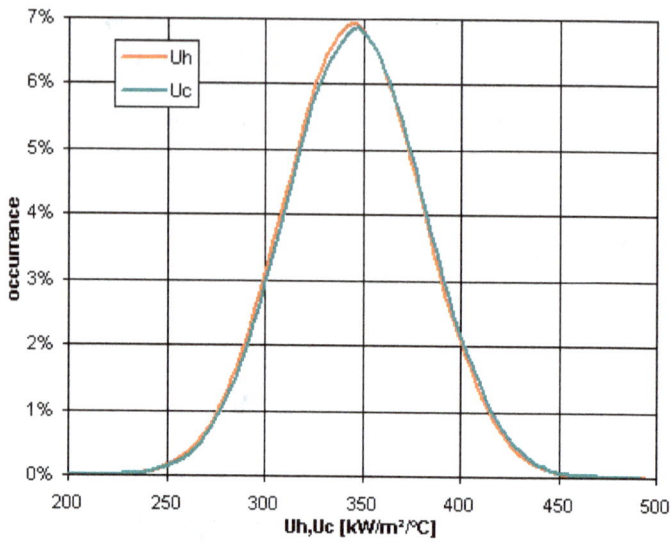

Chapter 18. Heat Recovery Steam Generators

In this text we will consider various aspects of design and testing heat recovery steam generators (HRSGs). Steam properties are covered in Appendix D and exhaust gas properties in Appendix E. Combustion, in so far as it applies to HRSGs, is covered in Appendix F. Design is divided into two sections: 1) individual components and 2) splitting and placement along the gas stream. Performance calculations and testing are covered after the design sections.

Orientation

The elements (components) of a HRSG (i.e., economizers, evaporators, and superheaters) are crossflow tube-in-shell heat exchangers, having steam (liquid or vapor) on the inside and exhaust gas on the outside. While some components may be designated as *counterflow* by some manufacturers (e.g., Nooter-Eriksen), this is merely conceptual or for the purposes of calculation. Nobody puts horizontal tube banks in a HRSG running counter-current to the gas flow. You might choose to perform heat transfer calculations as if such a component were counterflow, but be aware that this is not physically the case with any.

Phase Change

HRSGs are somewhat unusual in that they are designed specifically to separate the phase-change and non-phase-change sections. It is assumed (and forced in GateCycle™) that there be no evaporation in an economizer or superheater. That is, liquid enters and leaves any and all economizers, while dry (saturated or superheated) vapor enters and leaves any and all superheaters. All of the phase change occurs in the evaporators. GateCycle™ allows subcooled liquid at the inlet of an evaporator, but only saturated vapor at the exit. This means that, except at the inlet of the evaporator, there is no phase-change anywhere except in the evaporators and there is only phase change in the evaporators (apart from the initial heating, which may or may not be in physically separate elements). All of the manufacturers that I have worked with follow this same pattern. Even what are called *once-through* boilers typically have three sections (economizer, evaporator, and superheater). Because of this peculiarity, heat transfer calculations within HRSG components are more specific than for heat exchangers in general.

Chapter 19. Analytical vs. Numerical

As mentioned in Chapter 18, HRSG elements are crossflow tube-in-shell heat exchangers; so we will first consider the available methods for calculating heat transfer in such devices. There are three primary methods: 1) NTU-effectiveness, 2) F-LMTD, and 3) numerical. All three methods can be found in most heat transfer texts, for example the one by Lindon C. Thomas.[23] The most common and easily implemented numerical method is finite difference, which can be found in Appendix P of Thomas' professional version textbook. We will compare these three methods for several cases.

The NTU-Effectiveness Method

This method is similar to the LMTD method, but takes a slightly different approach to solving the same differential equation. The effectiveness, P, is defined as the ratio of the actual to the maximum heat transfer:

$$P = \frac{Q}{Q_{MAX}} \qquad (19.1)$$

The maximum possible heat transfer would be equal to the difference in the two inlet temperatures times the minimum product of the mass flow and specific heat or:

$$Q_{MAX} = (\dot{m}C_P)_{MIN}(T_{H,in} - T_{C,in}) \qquad (19.2)$$

where the minimum could be the hot or cold side. The ratio of the minimum to maximum product of mass and specific heat is given the symbol, R:

$$R = \frac{(\dot{m}C_P)_{MIN}}{(\dot{m}C_P)_{MAX}} \qquad (19.3)$$

The number of transfer units, NTU, is defined as:

$$NTU = \frac{UA}{(\dot{m}C_P)_{MIN}} \qquad (19.4)$$

The formula relating P(NTU,R) may be found in Appendix N-1 of the previous reference and in the spreadsheet crossflow1.xls.

$$P = \frac{1 - e^{-\Gamma R}}{R} \qquad (19.5)$$

$$\Gamma = 1 - e^{-NTU} \qquad (19.6)$$

---

[23] https://www.amazon.com/Lindon-C.-Thomas/e/B001HOVPI4

The F-LMTD Method

The heat transfer (Q) is the product of the overall conductance (UA), the log-mean temperature difference (LMTD), and the fudge factor (F):

$$Q = UA \times F \times LMTD \qquad (19.7)$$

The factor F for one side mixed (outside the tubes) and one side unmixed (inside the tubes) in any heat transfer text. Most often this is given graphically, as in the following figure:

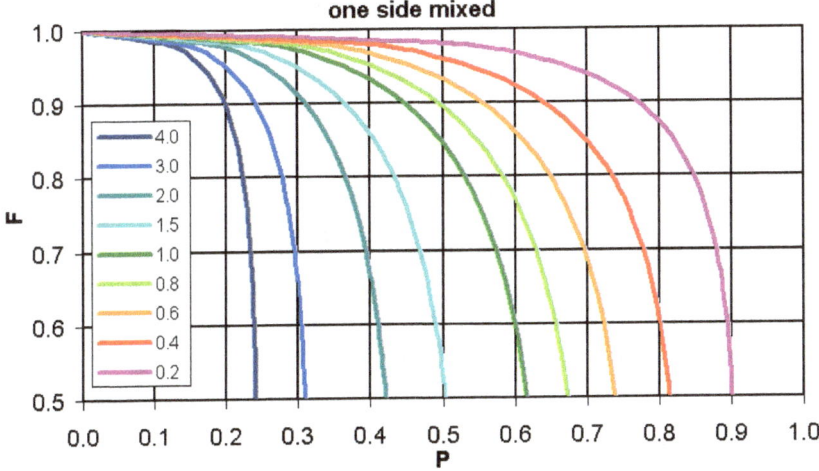

This figure appears in Section 11-8-1 of Thomas' text and was originally published by Bowman, Mueller, and Nagle.[24] The figure along with the formulas can be found in the online archive in folder examples\crossflow in spreadsheet crossflow1.xls. The factor F is given by the following relationship with Γ as before from Equation 19.6:

$$F = \frac{\ln\left(\dfrac{R\,e^{-\Gamma R}}{R-1+e^{-\Gamma R}}\right)}{NTU(1-R)} \qquad (19.8)$$

If R=0, this reduces to:

$$F = \frac{1-e^{-\Gamma}}{NTU\,e^{-\Gamma}} \qquad (19.9)$$

---

[24] Bowman, R. A., Mueller, A. C., and Nagel, W. M., "Mean Temperature Difference in Heat Exchanger Design," Transactions of the ASME, No. 62, p. 283, 1940.

The Finite Difference Numerical Method

The first step of implementing the Finite Difference Method (FDM) is to break the domain up into smaller cells (or elements) and to consider the heat transfer within one of these cells. The physical orientation is immaterial.

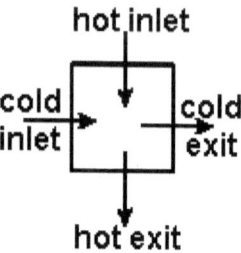

The fully explicit method (often called Euler's) considers only the inlet temperatures; thus the heat transfer within this cell is given by:

$$\Delta Q = \Delta UA(T_{HOT,IN} - T_{COLD,IN}) \qquad (19.10)$$

If the domain is broken up into a uniform rectangular grid of nx*ny cells and the overall heat transfer coefficient, U, is presumed to be uniform over the heat transfer surface, $\Delta UA = UA/nx/ny$. The conservation of energy (1st Law of Thermodynamics) for this cell is:

$$\Delta Q = \Delta m_{HOT} C_{P,HOT} (T_{HOT,IN} - T_{HOT,OUT}) \qquad (19.11)$$

$$\Delta Q = \Delta m_{COLD} C_{P,COLD} (T_{COLD,OUT} - T_{COLD,IN}) \qquad (19.12)$$

For now the specific heats, C_P, will be considered constant. We will next consider variable properties. The cell mass flows are presumed equally distributed: $\Delta m_{HOT} = m_{HOT}/nx$, $\Delta m_{COLD} = m_{COLD}/ny$.

Example 2.1

All three calculations are implemented in the spreadsheet crossflow1.xls, which is based on Example 11-21 in Thomas' text. Hot water enters from the top at 65°C and a rate of 5.25 kg/s. Cold air enters from the left side at 15°C and a rate of 2.39 kg/s. The specific heats are 4.19 and 1.01 kJ/kg/°C for the hot and cold sides, respectively. The overall heat transfer coefficient, U, is 205 W/m²/°C and the total surface area is 5.98 m². The P-NTU method can be calculated based on either the hot or cold side and so it is implemented both ways to illustrate that the same end result is obtained. The F-LMTD has only one variant, which produces the same result. The calculations are exactly as indicated by Equations 19.1 through 19.9. Inputs and outputs are shown in the following figure, which is taken directly from the spreadsheet:

	A	B	C	D
1	crossflow example 11-211			
2	symbol	units	hot	cold
3	U	W/m²/°C	205	
4	Cp	kJ/kg/°C	4.19	1.01
5	m	kg/s	5.25	2.39
6	Tin	°C	65.0	15.0
7	Tout	°C	62.9	34.5
8	q	kW	47.1	
9	LMTD	°C	38.52	
10	P	-	0.043	0.390
11	R	-	9.113	0.110
12	symbol	units	one mixed	
13	F-LMTD method			
14	F	-	0.995	
15	UA	kW/°C	1.23	
16	A	m²	5.98	
17	P-NTU method			
18	NTU	-	0.0558	0.509
19	UA	kW/°C	1.23	1.23
20	A	m²	5.98	5.98
21	user inputs in blue			
22	calculations in orange			
23	linked cells in green			

The hot and cold temperatures are shown in this next figure:

Q	R	S	T	U	V	W	X	Y	Z	AA	AB	AC
crossflow (hot mixed/cold unmixed) - explicit - 10x10 grid												
	1.23					hot side inlet						
		65.0	65.0	65.0	65.0	65.0	65.0	65.0	65.0	65.0	65.0	
		64.8	64.8	64.8	64.8	64.8	64.8	64.8	64.8	64.8	64.8	
		64.6	64.6	64.6	64.6	64.6	64.6	64.6	64.6	64.6	64.6	
		64.3	64.3	64.3	64.3	64.3	64.3	64.3	64.3	64.3	64.3	
		64.1	64.1	64.1	64.1	64.1	64.1	64.1	64.1	64.1	64.1	
		63.9	63.9	63.9	63.9	63.9	63.9	63.9	63.9	63.9	63.9	
		63.7	63.7	63.7	63.7	63.7	63.7	63.7	63.7	63.7	63.7	
		63.5	63.5	63.5	63.5	63.5	63.5	63.5	63.5	63.5	63.5	
		63.2	63.2	63.2	63.2	63.2	63.2	63.2	63.2	63.2	63.2	
	Texit	63.0	63.0	63.0	63.0	63.0	63.0	63.0	63.0	63.0	63.0	
	62.8	62.8	62.8	62.8	62.8	62.8	62.8	62.8	62.8	62.8	62.8	
						hot side outlet						
	15.0	17.5	20.0	22.2	24.4	26.5	28.4	30.3	32.1	33.7	35.3	
cold side inlet	15.0	17.5	19.9	22.2	24.4	26.4	28.4	30.2	32.0	33.7	35.2	cold side outlet
	15.0	17.5	19.9	22.2	24.3	26.4	28.3	30.2	31.9	33.6	35.2	
	15.0	17.5	19.9	22.2	24.3	26.3	28.3	30.1	31.8	33.5	35.1	
	15.0	17.5	19.9	22.1	24.3	26.3	28.2	30.0	31.8	33.4	35.0	
	15.0	17.5	19.8	22.1	24.2	26.2	28.1	30.0	31.7	33.3	34.9	
	15.0	17.5	19.8	22.1	24.2	26.2	28.1	29.9	31.6	33.2	34.8	
	15.0	17.5	19.8	22.0	24.1	26.1	28.0	29.8	31.5	33.2	34.7	
	15.0	17.5	19.8	22.0	24.1	26.1	28.0	29.8	31.5	33.1	34.6	
	15.0	17.4	19.8	22.0	24.1	26.0	27.9	29.7	31.4	33.0	34.5	
										Texit	34.9	

Notice that the hot temperatures are mixed (i.e., the same horizontally) and the cold temperatures are not mixed (i.e., vary horizontally and vertically) The temperature differences (fully explicit) and heat transfer for each cell are shown in this next figure:

AD	AE	AF	AG	AH	AI	AJ	AK	AL	AM
\multicolumn{10}{c}{crossflow - explicit - 10x10 grid}									
\multicolumn{10}{c}{ΔT in each cell}									
\multicolumn{10}{c}{use max(0,dT) to prevent overshoot}									
50.0	47.5	45.0	42.8	40.6	38.5	36.6	34.7	32.9	31.3
49.8	47.2	44.8	42.6	40.4	38.3	36.4	34.5	32.8	31.1
49.6	47.0	44.6	42.4	40.2	38.2	36.2	34.4	32.6	31.0
49.3	46.8	44.4	42.2	40.0	38.0	36.1	34.2	32.5	30.8
49.1	46.6	44.2	42.0	39.9	37.8	35.9	34.1	32.3	30.7
48.9	46.4	44.0	41.8	39.7	37.7	35.7	33.9	32.2	30.6
48.7	46.2	43.9	41.6	39.5	37.5	35.6	33.8	32.1	30.4
48.5	46.0	43.7	41.4	39.3	37.3	35.4	33.6	31.9	30.3
48.2	45.8	43.5	41.2	39.2	37.2	35.3	33.5	31.8	30.2
48.0	45.6	43.3	41.1	39.0	37.0	35.1	33.3	31.6	30.0
\multicolumn{10}{c}{ΔQ in each cell}									
0.61	0.58	0.55	0.52	0.50	0.47	0.45	0.43	0.40	0.38
0.61	0.58	0.55	0.52	0.50	0.47	0.45	0.42	0.40	0.38
0.61	0.58	0.55	0.52	0.49	0.47	0.44	0.42	0.40	0.38
0.61	0.57	0.55	0.52	0.49	0.47	0.44	0.42	0.40	0.38
0.60	0.57	0.54	0.52	0.49	0.46	0.44	0.42	0.40	0.38
0.60	0.57	0.54	0.51	0.49	0.46	0.44	0.42	0.40	0.38
0.60	0.57	0.54	0.51	0.49	0.46	0.44	0.41	0.39	0.37
0.59	0.56	0.54	0.51	0.48	0.46	0.43	0.41	0.39	0.37
0.59	0.56	0.53	0.51	0.48	0.46	0.43	0.41	0.39	0.37
0.59	0.56	0.53	0.50	0.48	0.45	0.43	0.41	0.39	0.37
								q	48.1

The calculated exit hot temperature (bold number below Texit on the left side of the second figure, half-way down) is 62.8°C, close, but not exactly equal, to the analytical result, 62.9°C (cell C7 in the first figure). The calculated exit cold temperature (bold number beside Texit at the bottom right of the second figure) is 34.9°C, close, but not exactly equal, to the analytical result, 34.9°C (cell D7 in the first figure). The calculated total heat transfer (bold number beside q in the bottom right corner of the third figure) is 48.1 kW is close, but not exactly equal, to the analytical result, 47.1 kW (cell C8 in the first figure).

Discretization

While we could use more cells, reducing the size of each one and increasing the discretization of the FDM calculations in the Excel® spreadsheet, this would be both inefficient and unnecessary. Instead, we will implement the entire calculation in C. The source code (crossflow1.x) as well as a little batch file to compile it (_compile.bat) can be found in this same folder. The analytical functions are very similar to the VBA code in the spreadsheet:

```
double F1mix(double R,double NTU)
```

```
{
  double G;
  G=1.-exp(-NTU);
  if(fabs(R-1.)<0.01)
    return((1.-exp(-G))/exp(-G)/NTU);
  return(log(R*exp(-G*R)/(R-1.+exp(-G*R)))/NTU/(1.-R));
}

double NTUofF(double R,double F)
{
  int iter;
  double N1,N2,NTU;
  N1=0.01;
  N2=100.;
  for(iter=0;iter<32;iter++)
  {
    NTU=(N1+N2)/2.;
    if(F1mix(R,NTU)>F)
      N1=NTU;
    else
      N2=NTU;
  }
  return(NTU);
}

double P1mix(double R,double NTU)
{
  double G;
  G=1.-exp(-NTU);
  return((1.-exp(-G*R))/R);
}

double NTU1mix(double R,double P)
{
  int iter;
  double N1,N2,NTU;
  N1=0.01;
  N2=100.;
  for(iter=0;iter<32;iter++)
  {
    NTU=(N1+N2)/2.;
    if(P1mix(R,NTU)<P)
      N1=NTU;
    else
      N2=NTU;
  }
  return(NTU);
}

double fLMTD(double dT1,double dT2)
```

```
{
if(dT1<=0.||dT2<=0.)
  return(0.);
if(fabs(dT1-dT2)<0.01)
  return(sqrt(dT1*dT2));
return((dT1-dT2)/log(dT1/dT2));
}
```

The analytical results are duplicated, as shown below:

```
crossflow example 1
62.86 hot side outlet temperature [°C]
34.51 cold side outlet temperature [°C]
0.0428 hot side P []
9.1128 hot side R []
0.0558 hot side NTU []
1.228 hot side UA [kW/°C]
5.989 hot side A [m²]
0.3902 cold side P []
0.1097 cold side R []
0.5089 cold side NTU []
1.228 cold side UA [kW/°C]
5.992 cold side A [m²]
38.15 LMTD [°C]
0.9953 F []
1.240 UA [kW/°C]
6.050 A [m²]
```

The FDM results are implemented in a function that accepts grid size as an input so that we can investigate the convergence:

```
void FDM(int nx,int ny,int list)
  {
  int i,j;
  double dQ,dT,Qh,*Tc,*Th;
  Tc=calloc(ny*(nx+1),sizeof(double));
  Th=calloc(ny+1,sizeof(double));
  Th[0]=Thi;
  for(i=0;i<ny;i++)
    Tc[(nx+1)*i]=Tci;
  for(Tco=i=0;i<ny;i++)
    {
    for(Qh=j=0;j<nx;j++)
      {
      dT=fmax(0.,Th[i]-Tc[(nx+1)*i+j]);
      dQ=UA*dT/nx/ny;
      Qh+=dQ;
      Tc[(nx+1)*i+j+1]=Tc[(nx+1)*i+j]+dQ/(mC/ny)/CpC;
      }
    Tco+=Tc[(nx+1)*i+nx];
    Th[i+1]=Th[i]-Qh/mH/CpH;
    }
  Tco/=ny;
```

values that are independent of the number of cells). This may seem like an extreme result, but is not uncommon for explicit methods, which is why implicit methods are often employed. Implicit methods use the exiting as well as entering temperatures. As these are not known from the outset, the process is implicit and requires iterative calculation. The only alternative to iterative calculations arises when Equations 19.11, 19.12, and 19.13 can be solved simultaneously.

$$\Delta T = \frac{(T_{HOT,IN} - T_{COLD,IN}) + (T_{HOT,OUT} - T_{COLD,OUT})}{2} \quad (19.13)$$

This cannot be entirely implemented in this case, as the hot side is presumed to be mixed, but can be partially implemented. The approximate solution is:

$$Q = \frac{2m_C C_{PC} m_H C_{PH} (T_{HI} - T_{CI}) UA}{(m_C C_{PC} + m_H C_{PH}) UA + 2m_C C_{PC} m_H C_{PH}} \quad (19.14)$$

The implicit variant is also implemented in the same code (crossflow1.c). The output for 10x10 is listed below:

```
implicit FDM
<---------------------Tcold--------------------->Thot
17.5,19.9,22.1,24.3,26.3,28.2,30.1,31.8,33.5,35.0,64.8
17.5,19.8,22.1,24.2,26.3,28.2,30.0,31.7,33.4,35.0,64.6
17.5,19.8,22.1,24.2,26.2,28.1,29.9,31.7,33.3,34.9,64.3
17.5,19.8,22.0,24.1,26.2,28.1,29.9,31.6,33.2,34.8,64.1
17.5,19.8,22.0,24.1,26.1,28.0,29.8,31.5,33.2,34.7,63.9
17.4,19.8,22.0,24.1,26.1,27.9,29.7,31.5,33.1,34.6,63.7
17.4,19.7,21.9,24.0,26.0,27.9,29.7,31.4,33.0,34.5,63.5
17.4,19.7,21.9,24.0,26.0,27.8,29.6,31.3,32.9,34.4,63.3
17.4,19.7,21.9,23.9,25.9,27.8,29.5,31.2,32.8,34.4,63.1
17.4,19.7,21.8,23.9,25.9,27.7,29.5,31.2,32.8,34.3,62.8
nx=10, ny=10, Q=47.45, Tho=62.84, Tco=34.66
```

The final values for 100x100 cells are:

nx=100, ny=100, Q=47.44, Tho=62.84, Tco=34.65

```
Tho=Th[ny];
Q=(mH*CpH*(Thi-Tho)+mC*CpC*(Tco-Tci))/2.;
}
```

Notice that it's only necessary to allocate Th[ny+1] and Tc[ny*(nx+1)]. extra row (ny+1) or column (nx+1) in these arrays provides the cold and inlet conditions, respectively. The same number of cells is present in spreadsheet. Because the hot side is mixed, it is not necessary to allocate a dimensional array, only a one-dimensional. The output for 10x10 cells is:

```
<----------------------Tcold-------------------->Tho
17.6,20.0,22.3,24.5,26.6,28.6,30.4,32.2,33.9,35.5,64.
17.6,20.0,22.3,24.5,26.5,28.5,30.4,32.1,33.8,35.4,64
17.5,20.0,22.3,24.4,26.5,28.4,30.3,32.1,33.7,35.3,64
17.5,19.9,22.2,24.4,26.4,28.4,30.2,32.0,33.6,35.2,64
17.5,19.9,22.2,24.3,26.4,28.3,30.2,31.9,33.6,35.1,63
17.5,19.9,22.2,24.3,26.3,28.3,30.1,31.8,33.5,35.0,63
17.5,19.9,22.1,24.3,26.3,28.2,30.0,31.8,33.4,34.9,6
17.5,19.9,22.1,24.2,26.2,28.1,30.0,31.7,33.3,34.9,6
17.5,19.8,22.1,24.2,26.2,28.1,29.9,31.6,33.2,34.8,6
17.5,19.8,22.0,24.1,26.1,28.0,29.8,31.5,33.1,34.7,6
nx=10, ny=10, Q=48.48, Tho=62.80, Tco=35.09
```

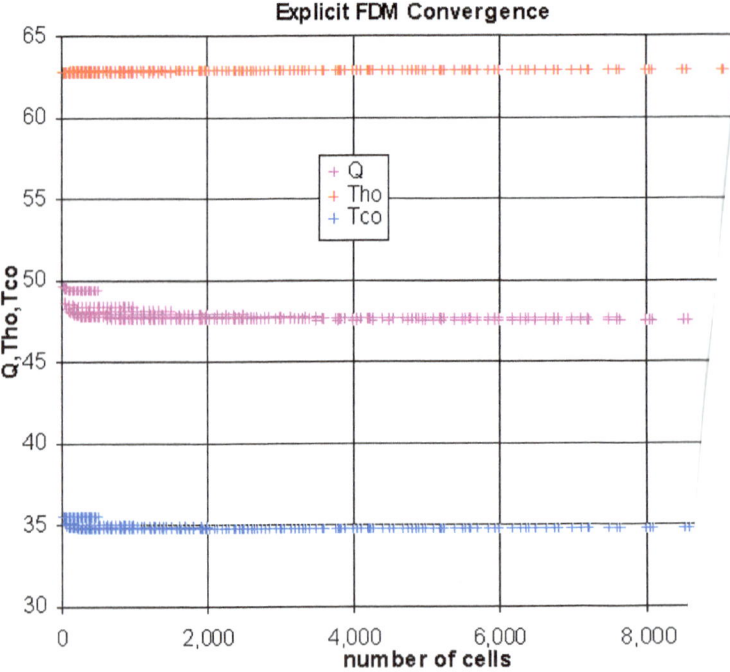

The explicit FDM convergence (to Q, Tho, and Tco) is ill preceding figure, which is also included in the spreadsheet along of the program. It takes at least 2,000 cells (≈45x45) to converge

The following figure shows why implicit methods are so often used:

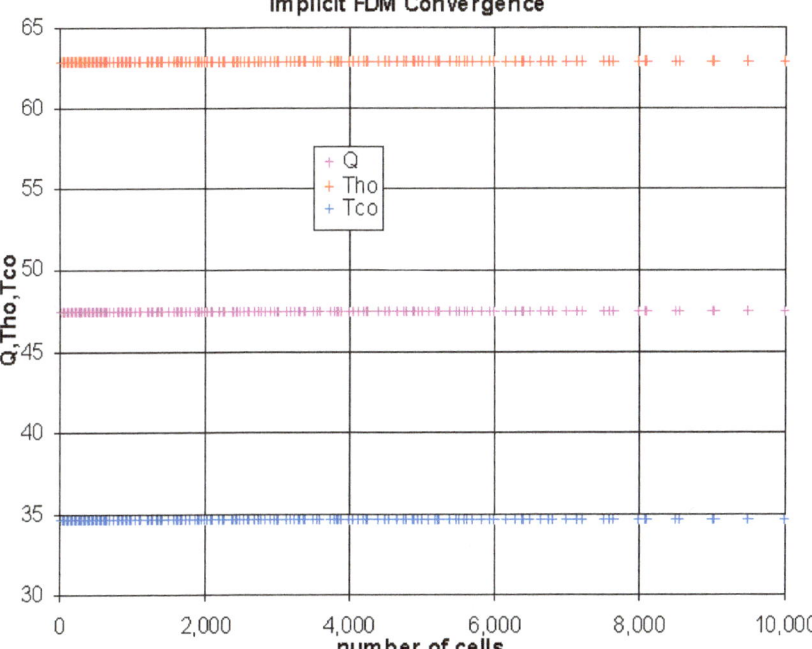

The implicit FDM converges even with 5x5 cells, an order of magnitude fewer cells and a similar reduction in calculation time. Notice that the final converged values for the FDM method (explicit or implicit) do not exactly match the analytical values for either the P-NTU or F-LMTD methods. The final values of Q are 47.05 and 47.44 kW for the analytical and numerical methods, respectively. The final values of Tho are 62.86 and 62.84°C for the analytical and numerical methods, respectively. The final values of Tco are 34.49 and 34.65°C for the analytical and numerical methods, respectively. Some of these differences are due to round off in the input parameters and the example in the textbook.

Variable Properties and Heat Transfer Coefficient

The FDM is naturally suited to variable properties and heat transfer coefficients; whereas, the analytical methods are not. If these change throughout the heat exchanger, one might achieve greater accuracy by using average values, in general, this only works when the quantities vary linearly *and* simultaneously. If C_P rises while U falls or vice versa, this is not the same as if they both rise or both fall together. Averages do not account for variation in quantities that combine multiplicatively or by division, let alone the sort of relationship in Equation 19.14. We will first consider the case of C_{PH} varying positively with temperature from 0.5 to 1.5 times the mean, such that the average is 1.0. We will also generate contours of the cold side temperatures in order to facilitate comparisons. The code (crossflow2.c) in the same folder is setup to do this conveniently. The cold side temperature contours are:

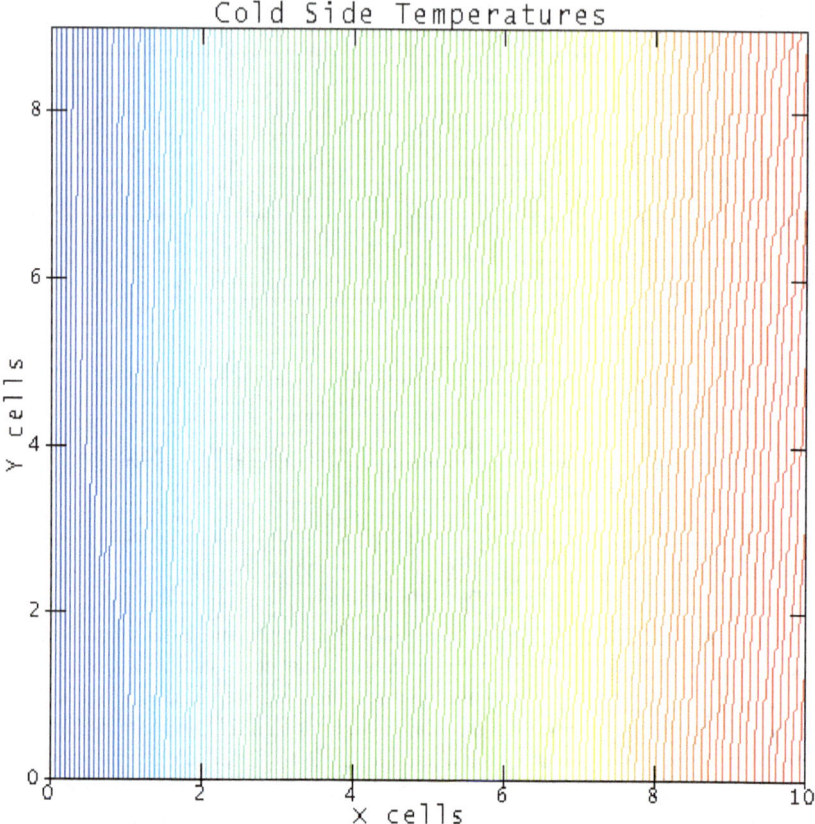

Having variable specific heat will distort the temperature profile and also change the overall result. The specific heat is:

```
double CpH(double T)/*hot side specific heat [kJ/kg°C]*/
    {
    return(4.19*(0.5+(T-62.84)/(65.-62.84)));
    }
```

The results are:
```
implicit FDM
    <----------------------Tcold---------------------->Thot
    17.5,19.9,22.1,24.3,26.3,28.2,30.1,31.8,33.5,35.1,64.9
    17.5,19.9,22.1,24.2,26.3,28.2,30.0,31.8,33.4,35.0,64.7
    17.5,19.8,22.1,24.2,26.2,28.2,30.0,31.7,33.4,34.9,64.5
    17.5,19.8,22.1,24.2,26.2,28.1,29.9,31.7,33.3,34.9,64.4
    17.5,19.8,22.0,24.2,26.2,28.1,29.9,31.6,33.2,34.8,64.2
```

```
17.5,19.8,22.0,24.1,26.1,28.0,29.8,31.5,33.2,34.7,64.0
17.4,19.8,22.0,24.1,26.1,28.0,29.8,31.5,33.1,34.6,63.8
17.4,19.8,22.0,24.0,26.0,27.9,29.7,31.4,33.0,34.6,63.6
17.4,19.7,21.9,24.0,26.0,27.8,29.6,31.3,32.9,34.5,63.3
17.4,19.7,21.9,23.9,25.9,27.8,29.5,31.2,32.8,34.4,63.0
variable: nx=10, ny=10, Q=46.59, Tho=63.01, Tco=34.74
constant: nx=10, ny=10, Q=47.45, Tho=62.84, Tco=34.66
```

The difference contours are shown below:

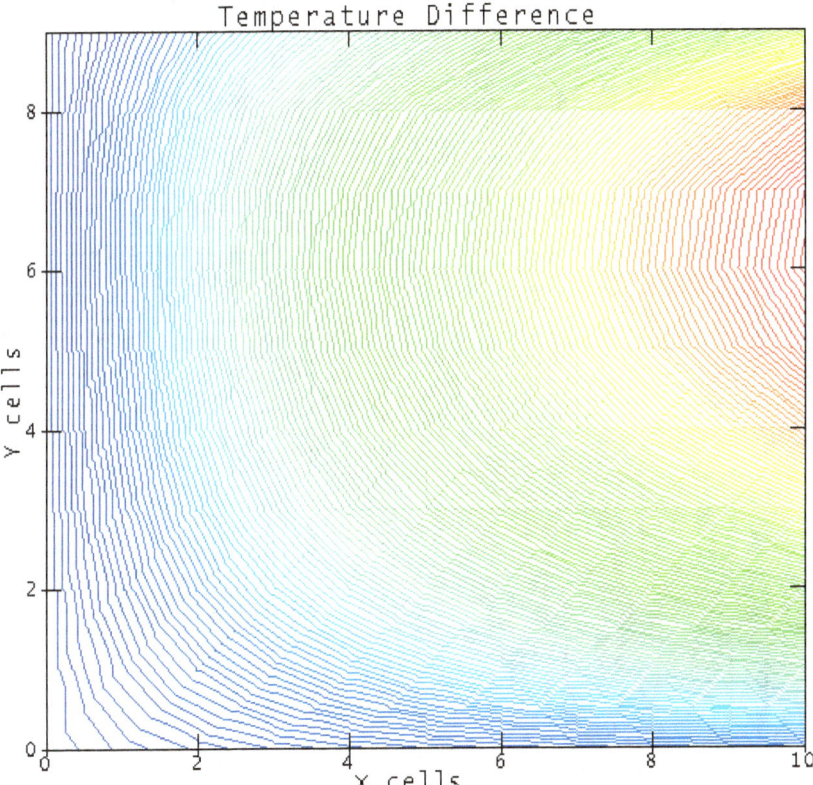

While the differences are very small (<0.123°C), they are not uniformly distributed. This example illustrates two things: 1) even changing Cp from 0.5 to 1.5 about the mean doesn't produce extreme differences and 2) the differences vary over the heat exchanger. We will now do the same thing with the cold side specific heat by modifying the same code. The specific heat function is very similar:

```
double CpC(double T)/*cold side specific ht [kJ/kg/°C]*/
  {
  return(1.01*(0.5+(T-15.)/(34.66-15.)));
  }
```

The output is as follows:

```
implicit FDM
<---------------------Tcold--------------------->Thot
19.9,22.9,25.2,27.1,28.8,30.4,31.7,32.9,34.1,35.1,64.8
19.9,22.8,25.2,27.1,28.8,30.3,31.7,32.9,34.0,35.1,64.6
19.8,22.8,25.1,27.1,28.8,30.3,31.6,32.8,34.0,35.0,64.4
19.8,22.8,25.1,27.0,28.7,30.2,31.6,32.8,33.9,35.0,64.2
19.8,22.7,25.1,27.0,28.7,30.2,31.5,32.7,33.9,34.9,64.0
19.8,22.7,25.0,26.9,28.6,30.1,31.4,32.7,33.8,34.8,63.8
19.8,22.7,25.0,26.9,28.6,30.1,31.4,32.6,33.7,34.8,63.6
19.7,22.7,25.0,26.9,28.5,30.0,31.3,32.6,33.7,34.7,63.4
19.7,22.6,24.9,26.8,28.5,30.0,31.3,32.5,33.6,34.7,63.2
19.7,22.6,24.9,26.8,28.4,29.9,31.2,32.5,33.6,34.6,63.0
variable: nx=10, ny=10, Q=46.65, Tho=62.95, Tco=34.87
constant: nx=10, ny=10, Q=47.45, Tho=62.84, Tco=34.66
```

The difference contours are shown below:

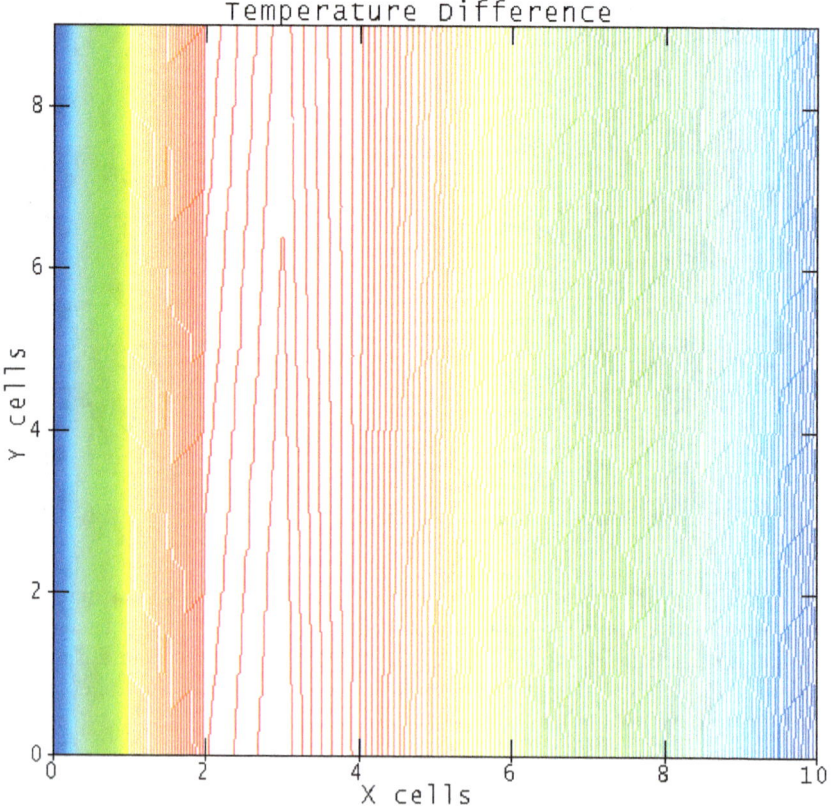

The differences resulting from variable cold side specific heat are 25 times as large as with varying the hot side specific heat by the same ratio (≤3.071°C compared to ≤0.123°C). An entirely different pattern arises also. If nothing else, this demonstrates that simply using average properties and hoping for the best is not a wise path, especially if you might face liquidated damages for inadequate performance. The temperatures themselves (rather than the differences) are very similar to the previous Cold Side Temperatures graph. We next vary both specific heats in the same way as before. The temperature differences are:

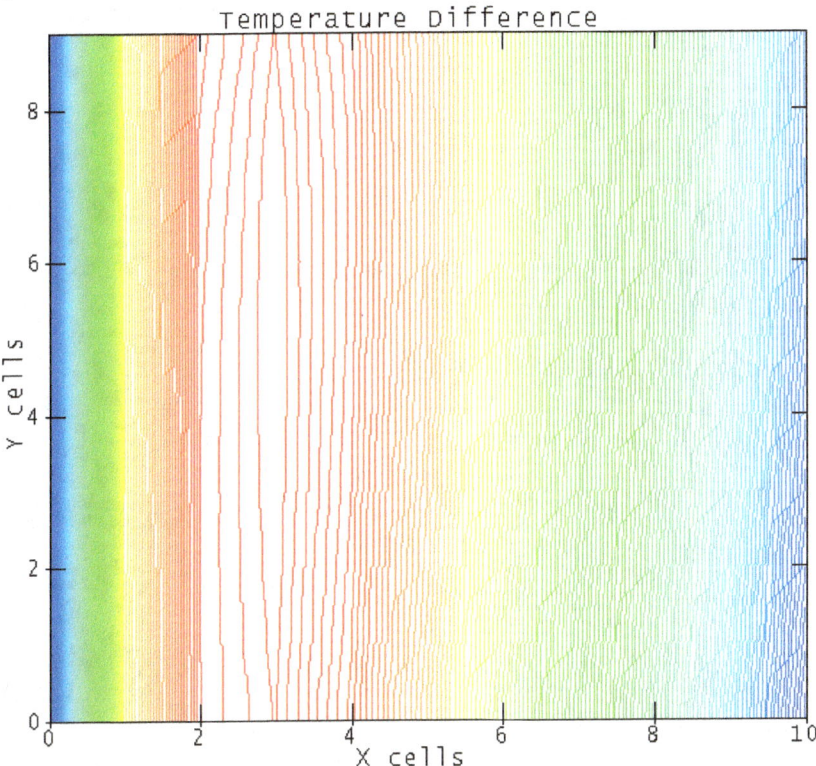

The output is:

```
implicit FDM
<---------------------Tcold--------------------->Thot
19.9,22.9,25.2,27.2,28.9,30.4,31.7,33.0,34.1,35.2,64.9
19.9,22.8,25.2,27.1,28.8,30.3,31.7,32.9,34.1,35.1,64.7
19.8,22.8,25.2,27.1,28.8,30.3,31.6,32.9,34.0,35.1,64.6
19.8,22.8,25.1,27.1,28.8,30.3,31.6,32.8,34.0,35.0,64.4
19.8,22.8,25.1,27.0,28.7,30.2,31.6,32.8,33.9,35.0,64.2
19.8,22.8,25.1,27.0,28.7,30.2,31.5,32.7,33.9,34.9,64.1
19.8,22.7,25.0,27.0,28.6,30.1,31.5,32.7,33.8,34.9,63.9
19.8,22.7,25.0,26.9,28.6,30.1,31.4,32.6,33.8,34.8,63.7
19.7,22.7,25.0,26.9,28.6,30.0,31.4,32.6,33.7,34.7,63.4
19.7,22.6,24.9,26.8,28.5,30.0,31.3,32.5,33.6,34.7,63.2
variable: nx=10, ny=10, Q=45.98, Tho=63.16, Tco=34.94
constant: nx=10, ny=10, Q=47.45, Tho=62.84, Tco=34.66
```

The maximum difference is ≤3.099°C. The pattern is similar. We note that that $m_C C_{PC}$ is much less than $m_H C_{PH}$, so that the cold side specific heat and mass flow dominate this problem. You can modify this program (crossflow2.c) to demonstrate a reflexive pattern when the hot side specific heat and mass flow dominate. We will now vary the conductance by a factor of 0.5 to 1.5 over the hot side, using constant cold and hot side Cp. The temperature difference contours are (<0.2206°C):

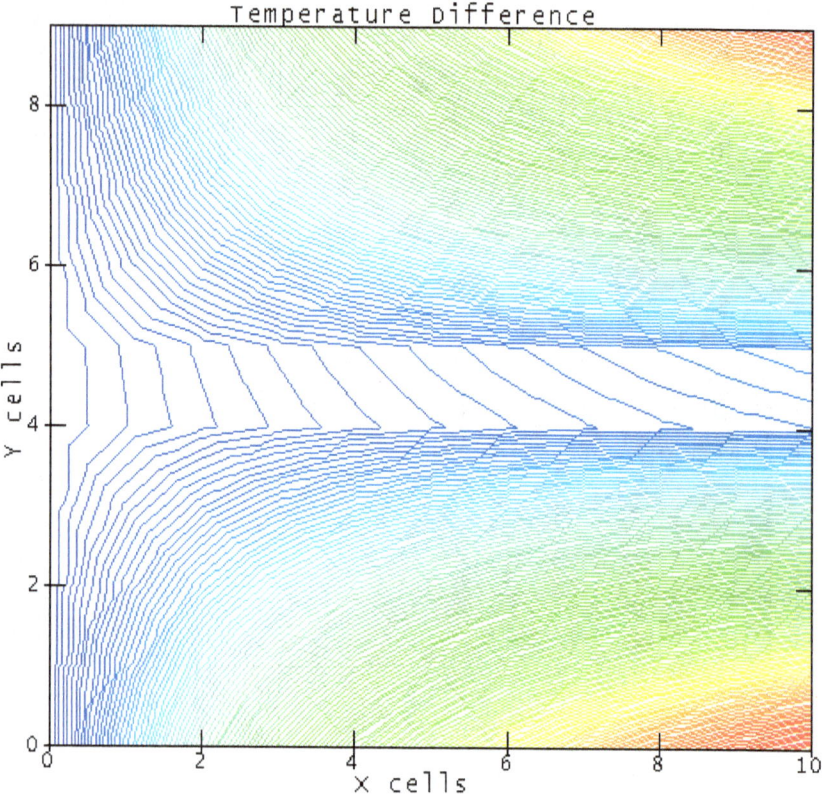

The results are:

```
implicit FDM
<----------------------Tcold-------------------->Thot
17.5,19.9,22.2,24.4,26.4,28.4,30.3,32.0,33.7,35.3,64.8
17.5,19.9,22.2,24.3,26.4,28.3,30.1,31.9,33.6,35.1,64.6
17.5,19.9,22.1,24.3,26.3,28.2,30.0,31.8,33.4,35.0,64.3
17.5,19.8,22.1,24.2,26.2,28.1,29.9,31.7,33.3,34.9,64.1
17.5,19.8,22.0,24.1,26.1,28.0,29.8,31.5,33.2,34.7,63.9
17.4,19.8,22.0,24.1,26.0,27.9,29.7,31.4,33.0,34.6,63.7
17.4,19.7,21.9,24.0,26.0,27.8,29.6,31.3,32.9,34.5,63.5
17.4,19.7,21.9,23.9,25.9,27.7,29.5,31.2,32.8,34.3,63.3
17.4,19.7,21.8,23.9,25.8,27.7,29.4,31.1,32.7,34.2,63.1
```

Summary

These illustrations reveal the importance of accounting for variations in properties and heat transfer coefficients and also that the analytical methods are not able to handle this. Specifically, averaging the variable parameters and using these in the analytical solutions, hoping to achieve more accuracy is not a rigorous approach because all averaging schemes necessarily presume some distribution throughout the domain and the distributions have been shown (by the preceding contour graphs of temperature differences) to vary from case-to-case, invalidating such presumptions. The numerical approach (implicit FDM) has been shown to rapidly converge and is able to handle all such variations throughout the domain; thus, it will be used when variations are expected.

Chapter 20. Expected Variations

The properties that vary throughout a heat exchanger are either 1) thermodynamic or 2) transport. The thermodynamic property of interest is the constant pressure specific heat, C_P, which appears in the heat transfer equations presented in the previous chapter. Transport properties include: density, viscosity, and thermal conductivity. These three impact the heat transfer coefficient, U, which includes convective and/or phase change (in this case boiling) on the inside and outside of the tubes. The tube material and fouling factors complete the calculation of U.

Specific Heat

The following figure shows the enthalpy of steam (SF-95, see Appendix D) vs. temperature for a series of pressures spanning the range found in a HRSG (atmospheric to critical).

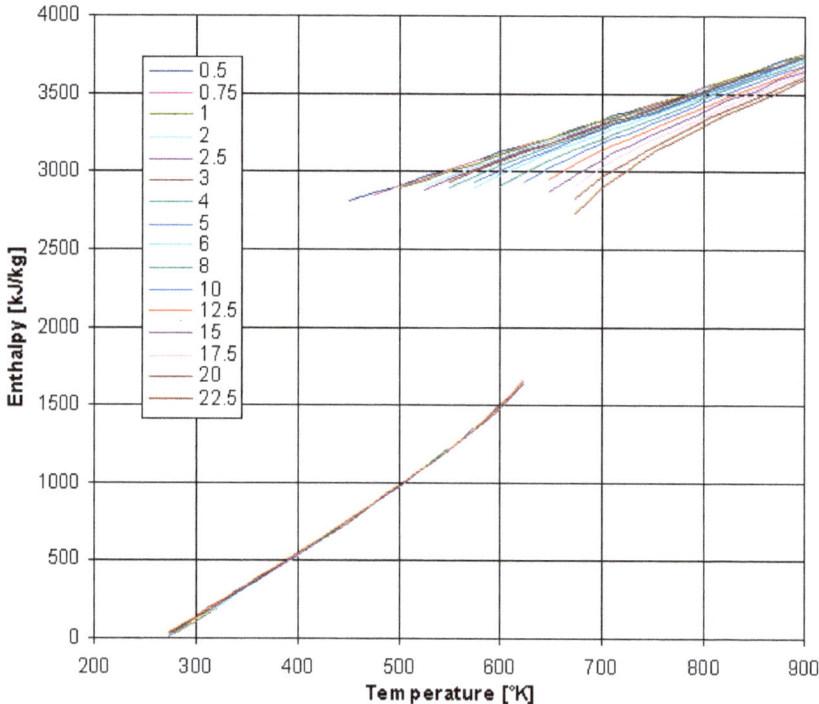

The subcooled liquids are on the lower left and the superheated vapors are on the upper right. All of the subcooled liquids are essentially on top of each other and form an almost straight line, indicating a nearly constant specific heat. This means that using the average specific heat is adequate for any economizer. The superheated vapors do fan out and have different slopes, but very little curvature. This means that using the average specific heat may also be adequate for most any superheater.

We must also consider the gas specific heat, for which we use NASA Glenn (see Appendix E). Typical mole fractions are unfired: 69.90% N2, 11.10% O2, 3.80% CO2, 14.30% H2O, 0.90% Ar and fired: 69.54% N2, 9.59% O2, 4.51% CO2, 15.46% H2O, 0.90% Ar. The mass-weighted specific heats are shown below:

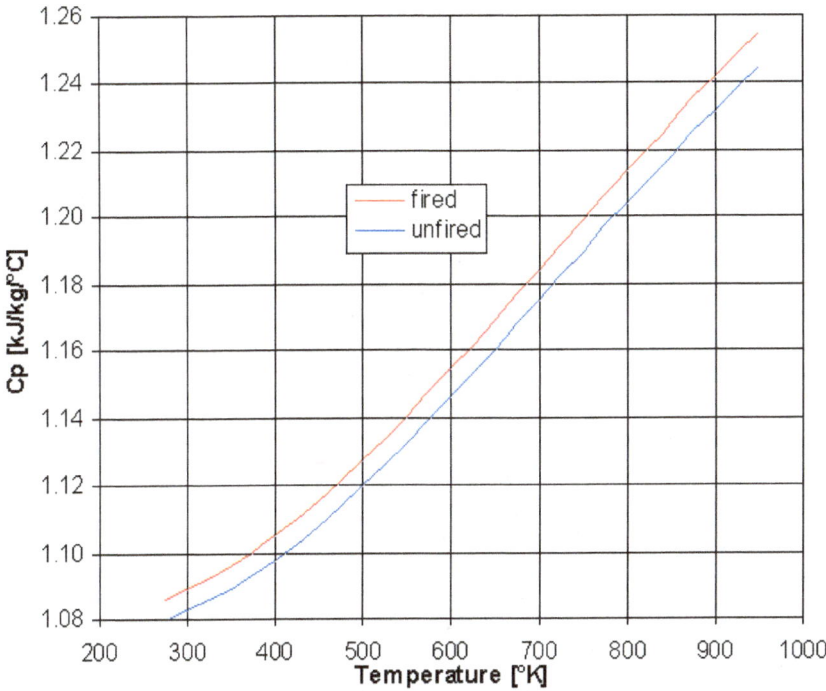

These vary little (<15%) over the range of expected temperatures and do so almost linearly; thus, using the average value will probably be adequate.

Heat Transfer Coefficient

The overall heat transfer coefficient, U, in a typical HRSG will vary at most from 10 to 50 W/m²/°C (2 to 8 BTU/hr/ft²/°F). The lower value might be seen in a significantly fouled or corroded unit, while the higher value would be difficult to achieve in a pristine one. The default value for all economizers, evaporators, and superheaters in GateCycle™ is 8 BTU/hr/ft²/°F. A more reasonable span would be 15 to 25 W/m²/°C (2.5 to 4.4 BTU/hr/ft²/°F). I would be skeptical of reported values outside this range.

The overall heat transfer coefficient is a conductance and accumulates by a series of reciprocals, as in the following equation:

$$U = \cfrac{1}{\cfrac{1}{h_{FI}} + \cfrac{1}{h_{CI}} + \cfrac{1}{h_W} + \cfrac{1}{h_{CO}} + \cfrac{1}{h_{FO}}} \quad (20.1)$$

where h_{FI} is the fouling conductance on the inside of the tubes, h_{CI} is the convection (or evaporation) conductance on the inside of the tubes, h_W is the tube wall conductance, h_{CO} is the convection on the outside of the tubes, and h_{FO} is the fouling conductance on the outside of the tubes. The fouling contributions are most often presented as resistances rather than conductances, that is:

$$r_{FI} = \frac{1}{h_{FI}} \quad (20.2)$$

Combining the reciprocals in this way results in the smallest value controlling the overall result. The tube wall resistance is quite low (or the conductance is quite high) compared to either convective or evaporative contributions. The convective coefficient is at least an order of magnitude less than the evaporative, making them (inside and/or outside) controlling. As water (and steam) have much higher thermal conductivity and density than exhaust gas, the outside convective coefficient is significantly smaller than the inside, even with fins. See Appendix H for more on heat transfer coefficients.

These considerations lead to the observation that the overall conductance for this type of heat exchanger will most likely be within the range of 15 to 25 W/m²/°C (2.5 to 4.4 BTU/hr/ft²/°F), but not necessarily vary by this much inside a single HRSG component. The biggest differences are specific to the particular design, rather than the location along one tube or another. Entrance and exit effects, as well as headers and manifolds, impact the local conductance, but considerable effort is invested in minimizing such impacts.

Chapter 21. Actual Designs

Reputable HRSG manufacturers know what they're doing. Equipment of this type is very expensive and comes with various guarantees, which if not met, result in significant monetary penalties. The most experienced manufacturers have devoted years to developing and fine-tuning their calculations so as to meet expectations as efficiently as possible. Not surprisingly, they are very protective of this information.

HRSG Design Data

In the folder examples\hrsg of the online archive, you will find a data file (cases.h) and spreadsheet (cases.xls) containing 421 cases from 4 different manufacturers, sufficiently "scrubbed" to obfuscate the source. These cases contain only the thermal performance data and not the proprietary specifics (e.g., tube size, tube length, tube spacing, fin size, fin shape, fin spacing, bending angles, manifold and plenum geometries, etc.). I was personally involved with testing each of these units and know that they performed as expected.

The crossflow program (hrsg.c) has been modified to process this data, displaying the results from all three analysis methods (P-NTU, F-LMTD, and implicit FDM). The calculations are performed using English units, as this is the form in which the design data was supplied. The IF-97 steam properties are used; however, you can easily change this to one of the others (IF-67, KKHM-69, NBS/NRC-84, or SF-95) by modifying the batch file (_compile.bat). The spreadsheet lists the results for all but the FDM and also includes the calculations. Below is a sample:

	A	B	C	D	E	F	G	H	I	J
1				typical HRSG design data						
2	case	area	mC	mH	Pci	Pco	Tci	Tco	Thi	Tho
3	name	ft²	lb/hr	lb/hr	psia	psia	°F	°F	°F	°F
4	1RHTR3	67,711	430,190	3,307,390	309	298	953	1044	1093	1070
5	1HPSH2	127,004	376,950	3,307,390	1277	1232	866	1052	1070	1024
6	1RHTR2	67,711	430,190	3,307,390	311	309	823	953	1024	992
7	1HPSH1	69,351	376,950	3,307,390	1295	1277	577	866	992	893
8	1RHTR1	51,772	430,190	3,307,390	318	311	683	823	893	858
9	1IPSH2	12,859	71,220	1,103,820	325	318	480	555	601	590
10	1LPSH2	25,718	63,710	2,203,570	56	49	479	581	601	595
11	1HPEC2	264,571	376,950	3,307,390	1304	1295	429	569	593	521
12	1LPSH1	25,718	63,710	2,205,260	57	56	289	479	521	510
13	1IPSH1	12,859	71,220	1,102,130	326	326	425	480	521	512
14	1HPEC1	260,865	376,950	2,691,720	1317	1304	292	429	444	366
15	1IPEC	59,288	108,730	615,670	346	326	290	420	444	350
16	1LPEC2	240,115	549,390	3,307,390	70	57	184	284	390	242
17	1LPEC1	156,811	549,390	3,307,390	78	70	81	184	242	174

Figure 4.1 Typical HRSG Design Data

	K	L	M	N	O
1	calculations				
2	Qc	Qh	Q	CpC	CpH
3	MBTU/hr	MBTU/hr	error	BTU/lb/°F	BTU/lb/°F
4	20.9	24.5	16%	0.534	0.322
5	42.4	48.6	14%	0.605	0.319
6	29.5	33.5	13%	0.528	0.317
7	89.4	102.4	14%	0.821	0.313
8	32.0	35.7	11%	0.532	0.308
9	3.3	3.5	7%	0.611	0.289
10	3.2	3.8	17%	0.494	0.289
11	62.7	68.2	8%	1.188	0.287
12	64.7	6.9	162%	5.341	0.283
13	2.7	2.8	3%	0.695	0.284
14	54.2	57.8	6%	1.049	0.275
15	14.9	15.9	7%	1.053	0.275
16	55.6	131.4	81%	1.011	0.268
17	56.5	58.4	3%	0.999	0.260

Figure 4.2 Basic Calculation Results

	P	Q	R	S	T	U
1	P-NTU method					
2	Rc	Rh	Pc	Ph	NTUc	NTUh
3	-	-	-	-	-	-
4	0.22	4.62	0.650	0.164	1.21	0.37
5	0.22	4.63	0.912	0.225	100.00	100.00
6	0.22	4.62	0.647	0.159	1.19	0.34
7	0.30	3.34	0.696	0.239	1.52	0.65
8	0.22	4.46	0.667	0.167	1.28	0.36
9	0.14	7.34	0.620	0.091	1.04	0.16
10	0.05	20.26	0.836	0.049	1.92	0.32
11	0.47	2.12	0.854	0.439	100.00	100.00
12	0.54	1.84	0.819	0.047	100.00	0.05
13	0.16	6.31	0.573	0.094	0.92	0.15
14	0.53	1.87	0.901	0.513	100.00	100.00
15	0.68	1.48	0.844	0.610	100.00	100.00
16	0.63	1.60	0.485	0.718	0.86	100.00
17	0.64	1.57	0.640	0.422	1.73	1.18

Figure 4.3 P-NTU Calculations Part A

	V	W	X	Y	Z
1			P-NTU method		
2	UAc	UAh	Uc	Ua	U
3	MBTU/hr/°F	MBTU/hr/°F	MBTU/hr/ft²/°F	MBTU/hr/ft²/°F	error
4	0.277	0.392	4.09	5.79	34%
5	22.792	105.638	179.46	831.77	129%
6	0.271	0.356	4.00	5.25	27%
7	0.470	0.672	6.77	9.70	36%
8	0.293	0.371	5.65	7.17	24%
9	0.045	0.052	3.53	4.04	13%
10	0.061	0.206	2.35	8.02	109%
11	44.769	94.758	169.21	358.16	72%
12	34.030	0.032	1323.18	1.24	200%
13	0.045	0.048	3.53	3.72	5%
14	39.539	74.081	151.57	283.98	61%
15	11.449	16.907	193.11	285.16	38%
16	0.480	88.756	2.00	369.64	198%
17	0.948	1.011	6.05	6.45	6%

Figure 4.4 P-NTU Calculations Part B

	AA	AB	AC	AD	AE
1			F-LMTD method		
2	F	LMTD	UA	U	U
3	-	°F	MBTU/hr/°F	MBTU/hr/ft²/°F	error
4	0.942	78.1	0.308	4.55	8%
5	0.012	64.5	58.386	459.71	9%
6	0.946	113.0	0.295	4.35	6%
7	0.885	206.6	0.524	7.56	8%
8	0.939	114.6	0.315	6.08	5%
9	0.976	73.4	0.047	3.68	3%
10	0.928	54.6	0.069	2.70	48%
11	0.014	50.6	91.432	345.59	31%
12	0.999	107.8	0.332	12.91	98%
13	0.979	61.1	0.046	3.59	1%
14	0.015	37.0	103.950	398.48	83%
15	0.015	39.3	25.318	427.03	79%
16	0.935	79.6	1.256	5.23	97%
17	0.798	74.1	0.972	6.20	1%

Figure 4.5 F-LMTD Calculations

Several things should stand out from even a cursory inspection of the spreadsheet. The first of these is the heat transfer error (column M), that is, the difference between the cold (steam) and hot (gas) side values. Occasionally this arises from a typographic error, but most of the time it is simply round-off in the temperatures. All four of these manufacturers utilize software to perform the calculations, so there shouldn't be a significant internal error other than round-off. There is at least one manufacturer that still performs all calculations by *hand*, that is, without the aid of a computer program—truly hard to believe in the 21st century![25] The maximum error in heat transfer for these 421 cases is 198%, the minimum is 1%, and the average is 34%. I know from extensive testing experience that most of this discrepancy is in the gas side temperatures. Some, but not all, manufacturers assume a certain heat loss from each component (e.g., 5%), which would explain a positive bias in the percent error, but nothing like what you see in the spreadsheet.

P-NTU Results

The next items to notice in the spreadsheet are the differences between the P-NTU results (i.e., U) for the cold side and hot side. If there were no discrepancy between the hot and cold side heat transfer, these two values should be equal. One of the advantages of having this data in a spreadsheet is that it facilitates sorting on any of the columns. Considering only those cases with no more than 10% error in the heat balance, this leaves only 141 cases (33% of the total). Among these, the discrepancy in U is as high as 200%, with an average of 24%.

F-LMTD Results

The discrepancies in U for the F-LMTD method are also large (up to 100%), with an average of 41%. Again, eliminating all those with a heat transfer discrepancy of more than 10%, reduces the average error to 24%. In this one respect, the two methods (P-NTU and F-LMTD) are similar. This is not surprising, considering the expected variations described in the previous chapter.

Implicit FDM Results

The implicit FDM results are calculated by hrsg.c and written out to iFDM.csv, which has been appended to cases.xls as AF:AG. For the full set of cases, these values of U exhibit discrepancies of up to 100% with an average of 41%. Eliminating all but the cases with no more than 10% heat balance error reduces the average discrepancy to 22%.

Adjusted Data

An interesting exercise would be to use the Excel® Solver() to adjust each of the temperatures (or flows) so as to minimize the heat transfer error. The most likely choice of variable is the hot side inlet temperature, as any of the other temperatures may result in a temperature difference violation of the 2nd Law of Thermodynamics (see spreadsheet cases_adjusted.xls). This is also implemented in a code (adjust.c), which creates an output file (adjusted.csv), which has also been converted to an include file (adjusted.h). This include file can be used with

---

[25] I offered to write software for them that would faithfully implement their existing process, but they balked at the suggestion.

the program (hrsg.c) to create an adjusted version of iFDM.csv, which has been appended to adjusted.xls.

Even after adjusting the either the hot side inlet or hot side exit temperatures to minimize the discrepancy in cold side and hot side heat transfer, the average discrepancy in U is still 10%, 5%, and 10% for the P-NTU, F-LMTD, and iFDM methods, respectively.

<u>Reasonableness</u>

Not only must the cold and hot side heat transfer agree, but also the result (U) must be reasonable. The most extreme criterion ($1 \leq U \leq 10$) is met by only 346 of the 421 cases (82% of the total) P-NTU method, 359 for the F-LMTD and 397 for the iFDM. A much more reasonable expectation would be $2 \leq U \leq 8$, which leaves 254, 246, and 278 for the three methods, respectively. The following figure shows all of the U values for all of the cases:

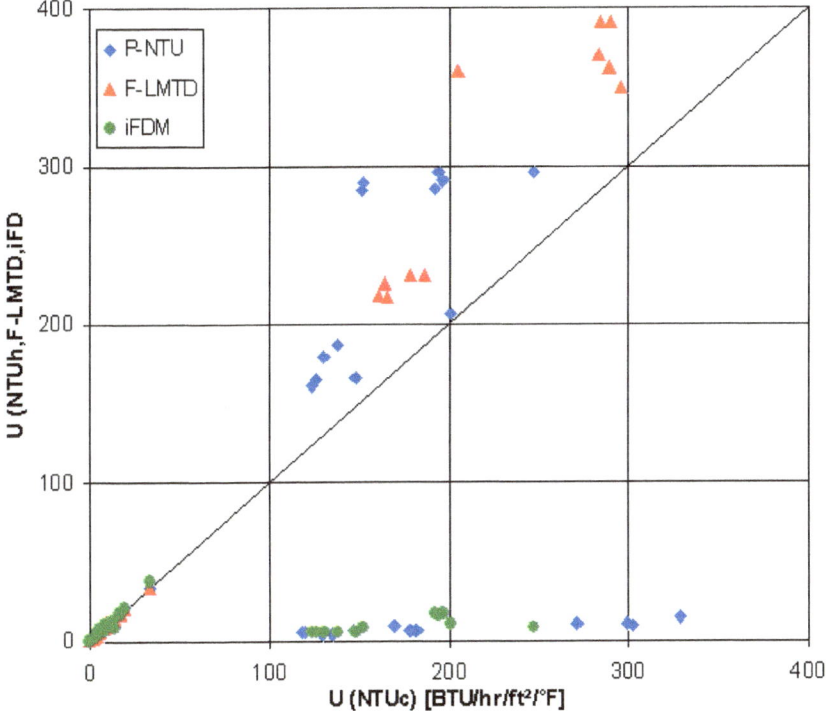

Clearly, many of these are outliers, in spite of having adjusted the gas temperatures to close the heat transfer discrepancies. The bottom left corner of this same figure is shown below:

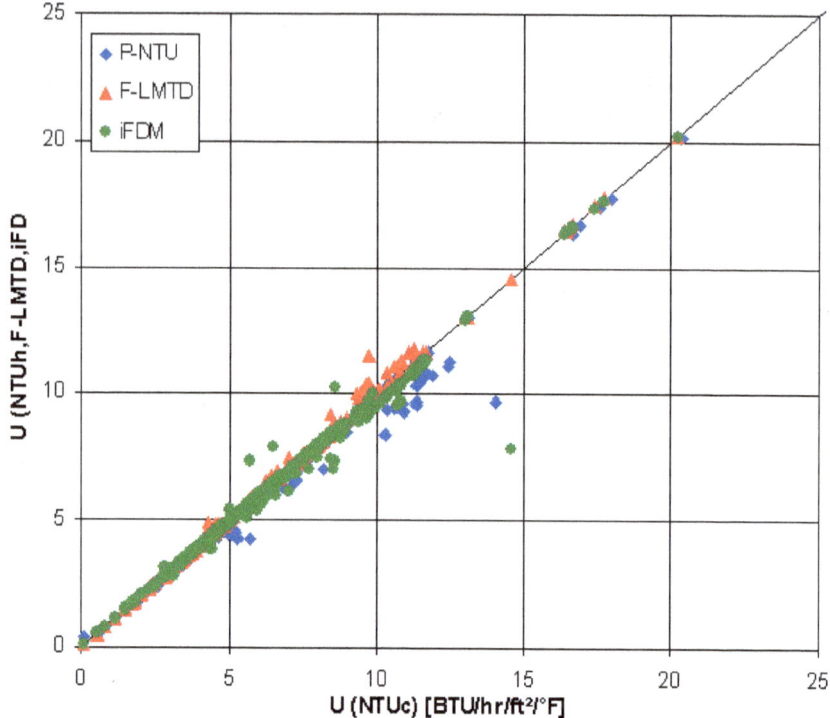

For the somewhat reasonable values ($1 \leq U \leq 12$), all three methods are essentially the same (i.e., the blue diamonds, red triangles, and green circles are practically on top of each other). What this means is that, for this particular type of heat exchanger, it doesn't matter which method you use, as long as the input values are reasonable.

Chapter 22. Heat Release Diagram

Perhaps the most revealing presentation of a HRSG thermal design is the heat release diagram. This is a plot of gas (hot side) and steam (cold side) temperatures vs. the cumulative heat transfer. GateCycle™ produces these graphs for you, but they are not complicated to generate. You will find a spreadsheet (HRSG1.xls) in the online archive in folder examples\release that solves a single-pressure HRSG and generates the associated heat release diagram. There is an English and an SI tab so that it works with both units. Typical results are as follows:

A	B	C	D	E	F	G	H	I
			Single Pressure HRSG					
case name	mC lb/hr	mH lb/hr	Pci psia	Pco psia	Tci °F	Tco °F	Thi °F	Tho °F
SH	300,000	3,000,000	1800	1750	621	1050	1100	986
EV	300,000	3,000,000	1800	1800	621	621	986	824
EC	300,000	3,000,000	1950	1800	120	616	824	641

J	K	L	M	N	O	P	Q
			Single Pressure HRSG				
Q MBTU/hr	CpC BTU/lb/°F	CpH BTU/lb/°F	Rc -	Rh -	Pc -	Ph -	NTUc -
108.86	0.846	0.319	0.27	3.77	0.896	0.237	####
150.73	∞	0.310	∞	0.00	0.000	0.443	0.59
164.13	1.103	0.299	0.37	2.71	0.704	0.260	1.69

R	S	T	U	V	W	X
NTUh -	LMTD °F	F -	UA MBTU/hr/°F	ΣQ MBTU/hr	Th °F	Tc °F
0.91	158.50	0.78	0.69	0.00	1100	1050
0.59	276.46	1.00	0.55	108.86	986	621
0.60	341.30	0.90	0.48	259.58	824	616
		total	1.71	423.71	641	120

user inputs in blue
calculations in orange

The calculations are updated automatically when you change the values in bold blue text.

This next figure is the heat release diagram:

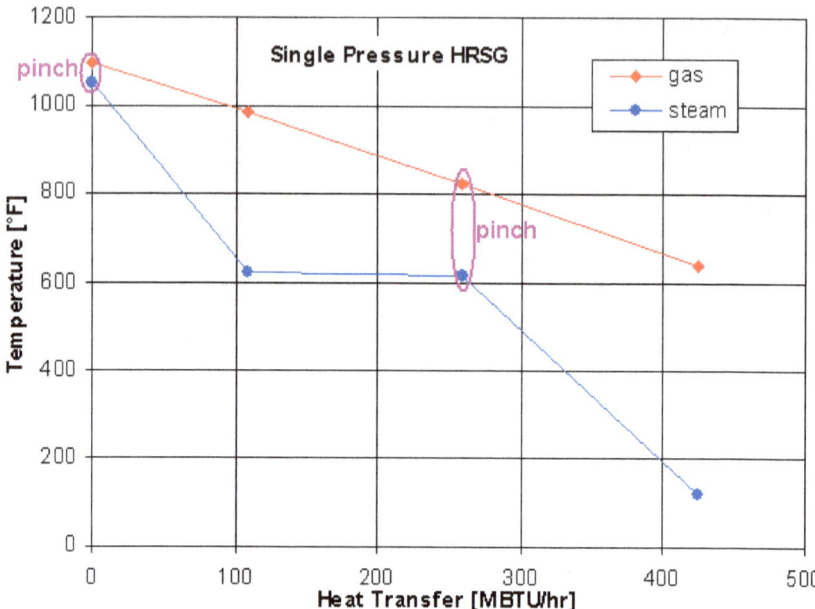

The slope of the lines is equal to minus the reciprocal of the mass flow rate times the specific heat (i.e., $-1/m_C/C_{PC}$ and $-1/m_H/C_{PH}$). There is almost no curvature in the red (hot gas side) line because the specific heat is almost constant over this temperature range. The three sections of the blue line (from left to right) correspond to the superheater, evaporator, and economizer, respectively. The first and third sections of the blue line has negligible curvature, because the specific heat of the vapor and liquid, respectively, is almost constant over the temperature range. The flat portion of the blue line arises from the evaporation process, which is isothermal. The specific heat would be equal to the latent heat of vaporization divided by the temperature change, which is zero, yielding an infinite result. The reciprocal results in a zero slope for this section. While there are some differences from one design to another and single to double to triple pressure systems, the important features of the shape are the same.

Pinch Points

The two locations along the process in the preceding figure where the red and blue lines are closest together are called *pinch* points. The left pinch point occurs at the exit of the superheater. The right pinch point occurs at the exit of the economizer. Pinch points control the operation of a HRSG and are the predominant factor in a thermal design. The required surface area is inversely proportional to the thickness (in degrees) of the pinch points. The tighter the pinch points, the more surface is needed. More surface means more material and more expense.

This maxim is worthy of repeating...
Heat exchangers are sold by the pound (or kg).

To obtain a cost-effective design (and win the bid), the engineer must effectively manage pinch points in a HRSG. The following does not occur:

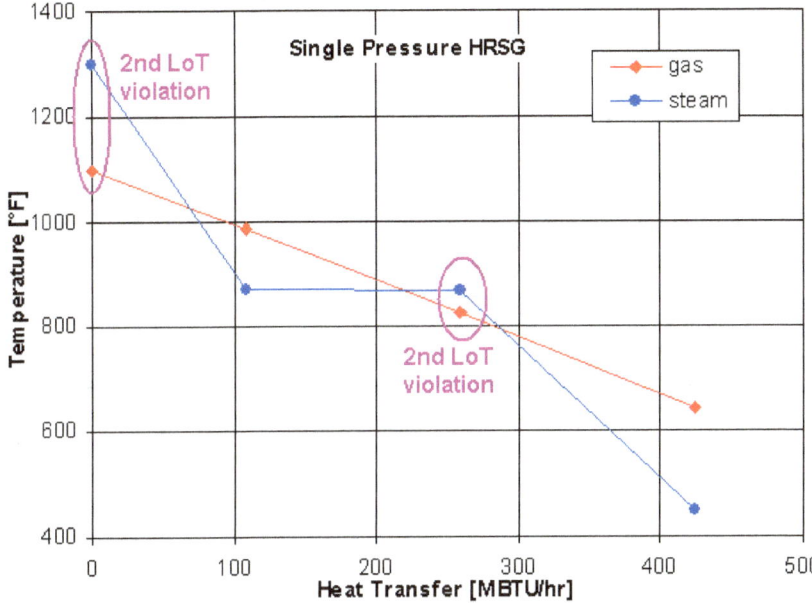

as this would be a violation of the 2nd Law of Thermodynamics (2LoT). Even an infinitely large heat exchanger (A=∞) with infinite conductance (U=∞, or zero thermal resistance) could not achieve this contradiction.[26]

---

[26] Just in case you're interested... Yes, I've seen it on drawings. It's usually a mistake that can be resolved. Twice in 20 years of experience it was a financial disaster (i.e., a lawsuit).

Impact of Steam Flow

There is a button in the spreadsheet to vary the steam flow and produce a modified heat release diagram for three different steam flow rates. There is hardly any impact on the hot side (gas) temperature line except for the end point. The terminal point of the cold side (steam) remains the same also. The flat portion on the steam line remains at the same level, only shifting left to right and growing in length. The right side of the flat zone is closest to the gas line, so steam flow directly impacts the pinch point; thus, the pinch point determines the steam flow.

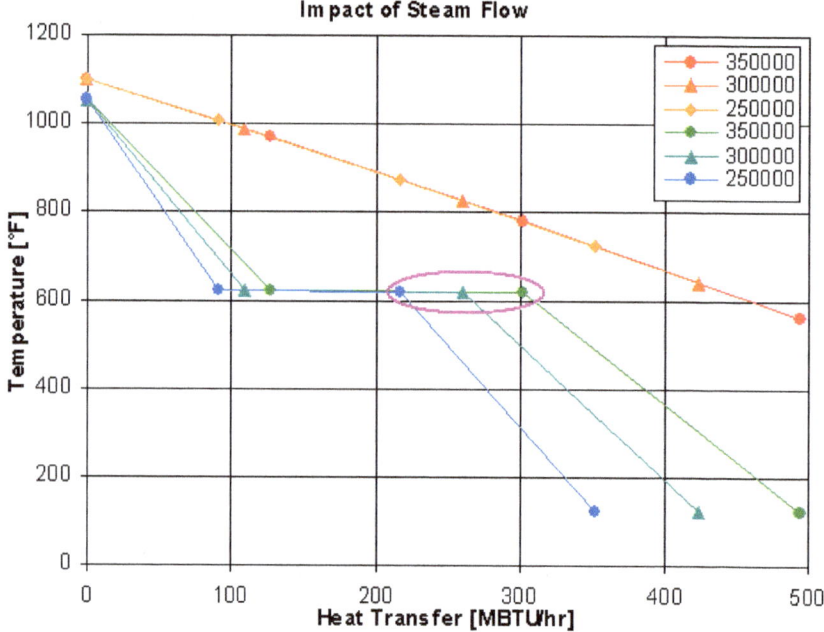

Impact of Steam Pressure

There is a second button that varies the steam pressure. Not only does changing the pressure impact the vertical position (i.e., saturation temperature), but also the length of the flat portion, which changes the position of the end point and reduces the pinch at constant steam flow.

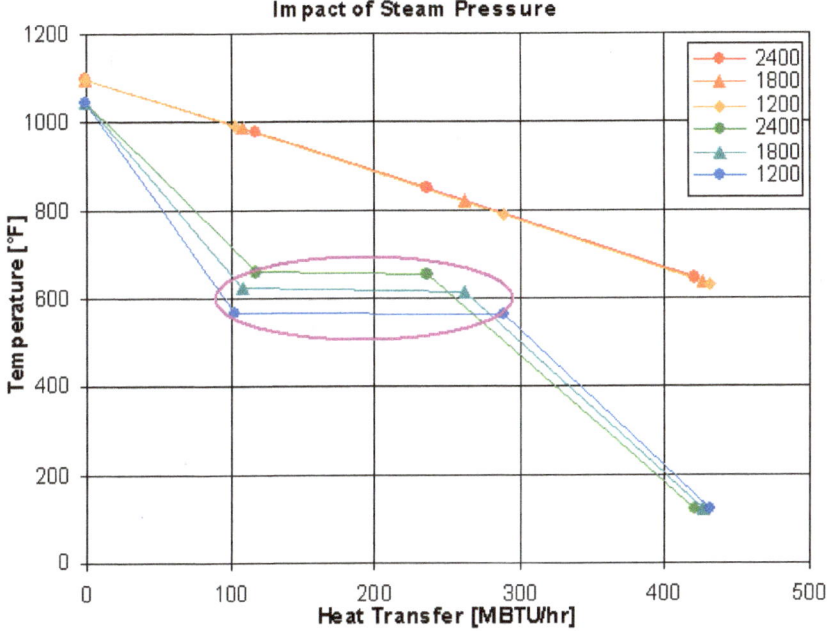

Impact of Gas Flow

The third button varies the gas flow. The slope of the lines (except in the evaporator) is proportional to the inverse of the product of the mass flow and specific heat; so this will change the slope of the hot side (gas) line and also impact the pinch point.

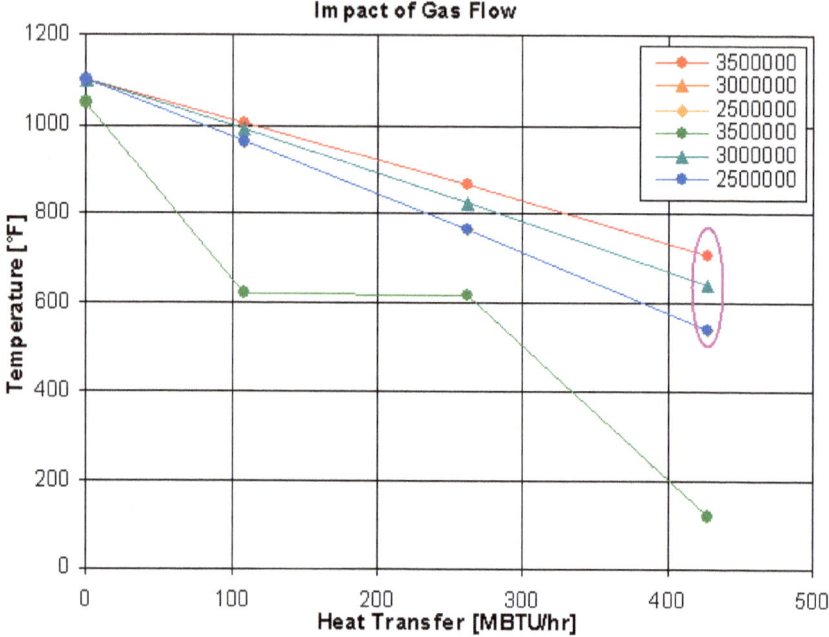

Impact of Gas Temperature

The fourth button varies the gas temperature. The slope of the hot (gas) line doesn't change, but the whole line shifts up and down; so this will impact the pinch point.

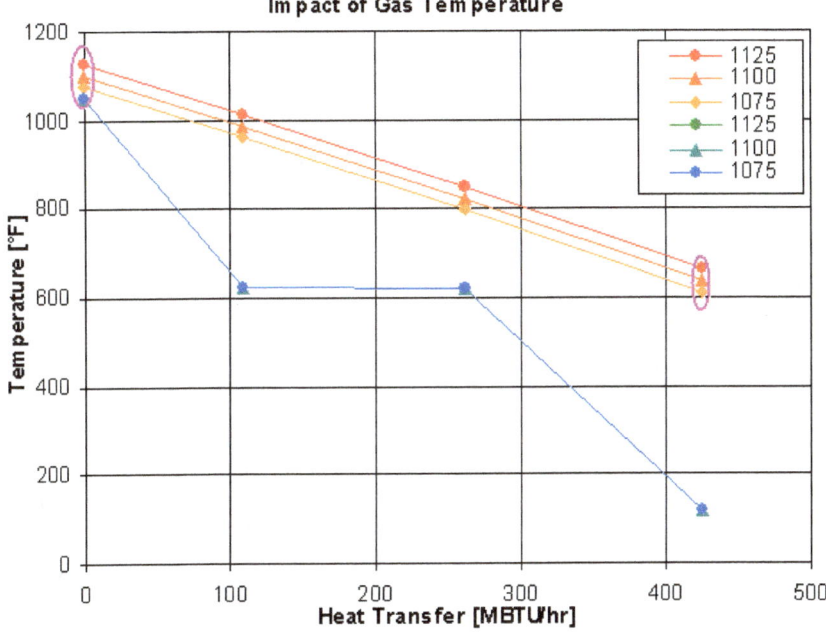

Three-Pressure HRSG

The most common arrangement is a three-pressure HRSG. The low pressure (LP) is mostly for control, venting, and purification. The intermediate pressure (IP) is added to the cold reheat and the high pressure (HP) forms the main steam. The reheat is necessary because continued expansion of the steam without reheating would result in far too much moisture and damage of the steam turbine. This is readily apparent from a Mollier Diagram. A typical GateCycle™ heat release diagram for a three-pressure system is shown in this next figure:

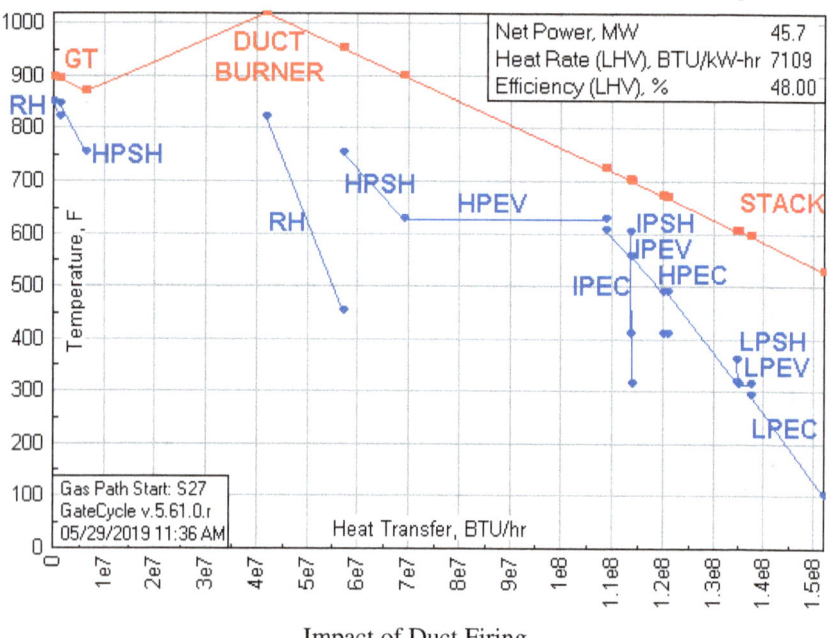

Impact of Duct Firing

Notice in the above figure how the duct burner pushes the red (hot gas) side temperature up to the right and also moves the right-hand-side farther away from the blue (cold steam) line, relieving the pinch. This is why duct firing is so prevalent in combined cycle power plants (CCPPs). While it's beyond the scope of this book, the incremental heat rate for duct firing is not what you might expect. The overall plant test code, ASME PTC46[27], there is a correction for duct firing, known as ω_7/Δ_7. The change in heat input (from design or guarantee point) is ω_7 and the associated change in net power output is Δ_7. While the correction is applied downward (more heat input from the duct burner requires an adjustment downward in net power output to arrive at the corrected value), the graph is most often drawn with an upward slope, as illustrated below:

---

[27] *Performance Test Code on Overall Plant Performance*, American Society of Mechanical Engineers, 2003.

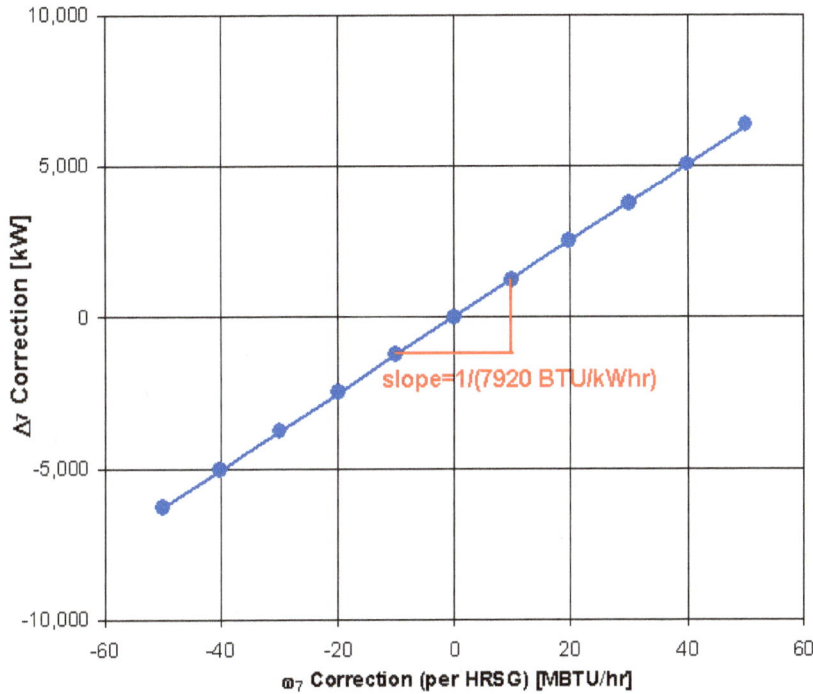

Two things about this figure are quite interesting: 1) it's a straight line and 2) the slope is less (worse) than the net plant heat rate, which in this case is 6650 BTU/kWhr. Duct firing is used to produce more power, but it isn't *cheap* power, because it increases the heat rate (decreases the efficiency). I have built well over 100 GateCycle™ models. More than 50 of these have been complete, including correction curves. In all of my experience, I have never seen a case in which the slope of the ω_7/Δ_7 curve was equal to the reciprocal of the base net plant heat rate, nor have I ever seen a case with significant curvature to this line.

Typical Three-Pressure CCPP

We will discuss the placement of elements in the HRSG in a subsequent chapter. For now, we will work with a typical design and see how to build a spreadsheet to perform the calculations in the absence of commercial software. You will find a an example along with the source code (CCPP4.c) in the online archive that accompanies my text, *Thermodynamic Cycles*. We will not dwell on how to solve the system or attach the properties, as we are only considering the thermal design of HRSGs in this present text. The current example system is illustrated in the following figure:

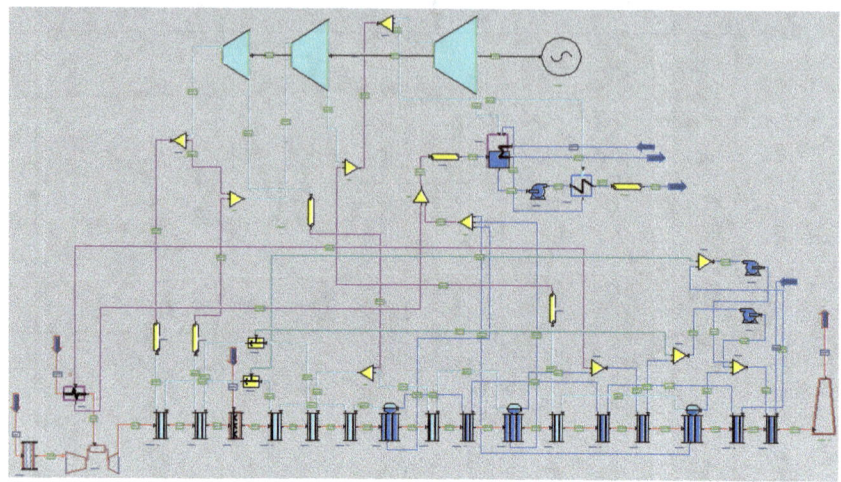

The corresponding spreadsheet (CCPP3.xls) is in folder examples\release, which only contains the HRSG. The entire GateCycle™ model can be found in the folder examples\CCPP. The heat release diagram for this system is shown below. GateCycle™ displays all of the steam paths in the same color and as disjointed segments, which is understandable but may be confusing. The Excel® spreadsheet allows us to draw all three steam paths distinctively.

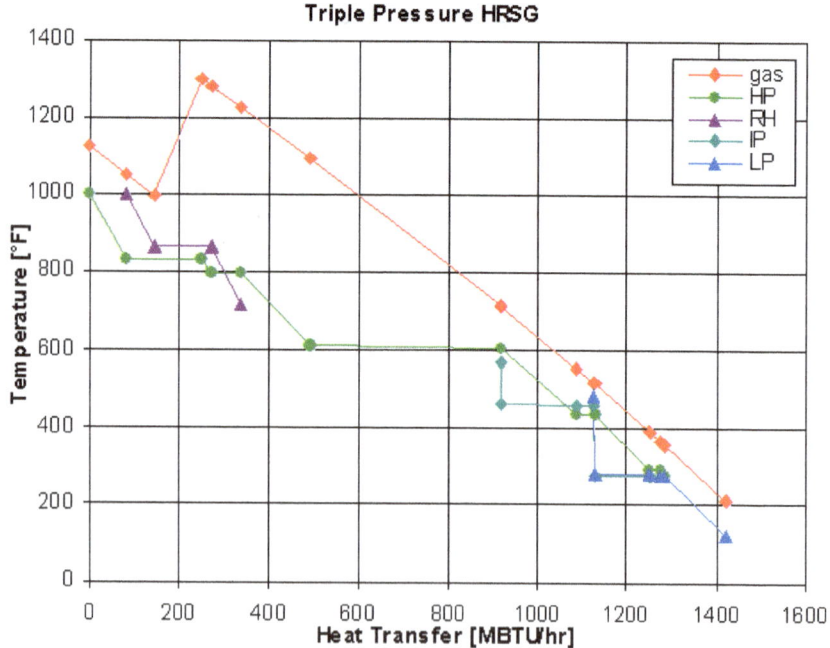

This spreadsheet also as an English and SI tab. The same data in SI units is shown in this next figure:

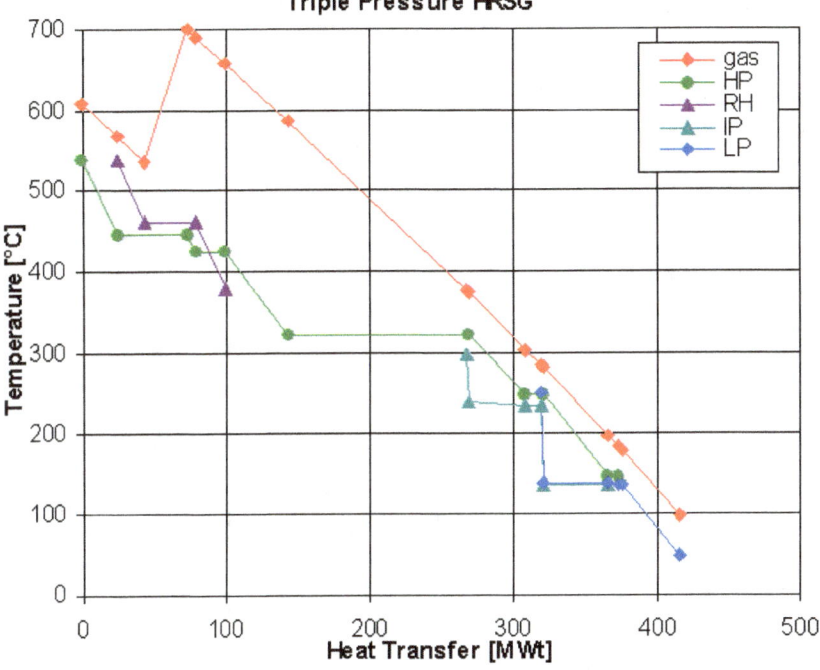

Chapter 23. Expansion Line

The Mollier diagram shows enthalpy vs. entropy and is an essential graphical representation of the expansion process that occurs in a steam turbine. A complete Mollier diagram can be found in Appendix D. In this chapter we will only consider part of the total figure, that is, the expansion line. The following figure (see expansion.xls in the examples\expansion folder) shows a typical expansion line:

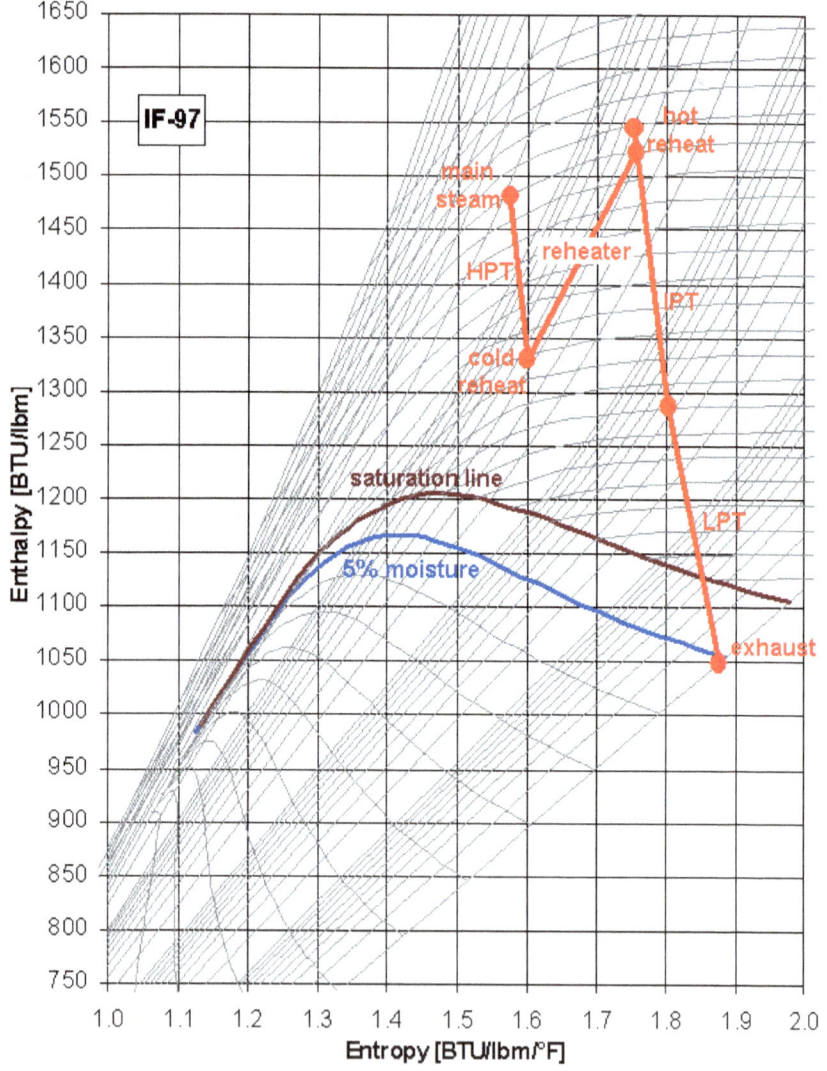

3

The reason a Rankine cycle uses reheat is to avoid dipping down too far into the saturation zone (below the brown saturation line). The power output is proportional to the total vertical drop of the red expansion line, so we want that to be as long as possible. Reheat moves the HRH point up along the isobar allowing for more drop before ending at the 5% moisture line in this case. If you want higher efficiency (power output divided by heat input), reheat is necessary. The practical HRSG must provide this.

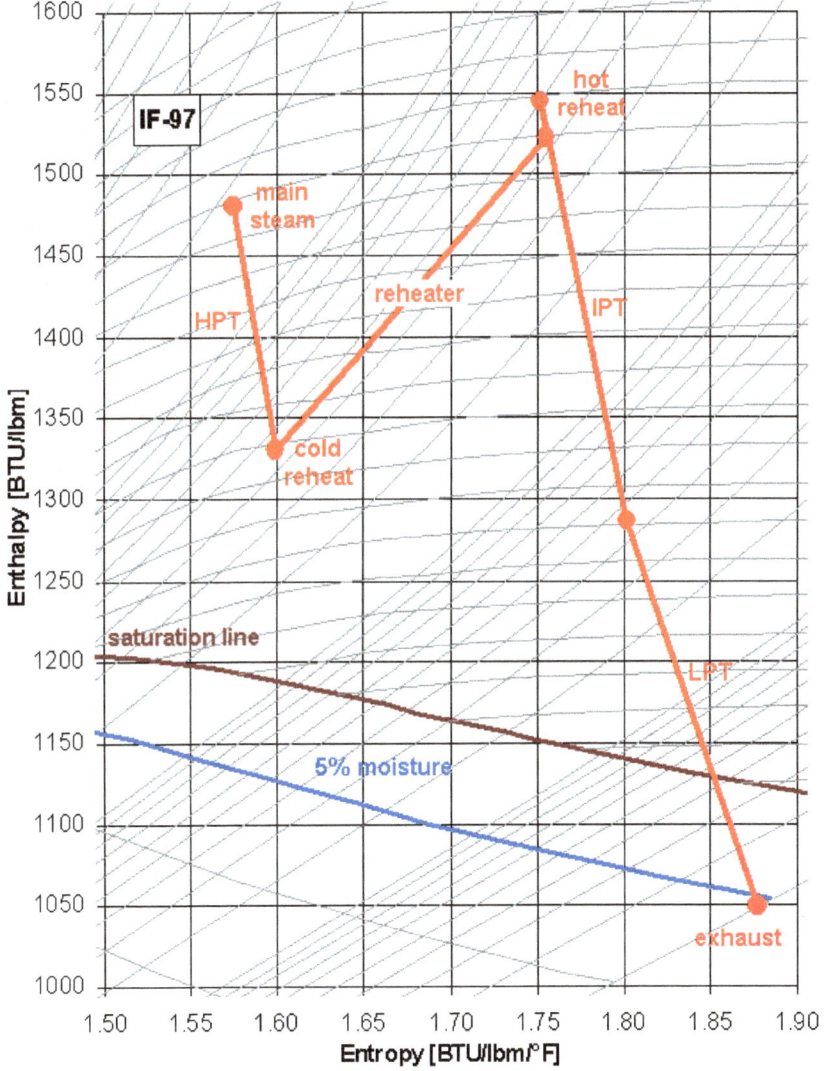

Notice that the HPT (high pressure turbine) drops down to the CRH (cold reheat) and may not come close to the saturation line. The IPT (intermediate pressure turbine) and LPT (low pressure turbine) expansion dips through the saturation line. The length of the HPT expansion (150.7 BTU/lbm) is significantly shorter (70% less) than the IPT+LPT expansion (495.3 BTU/lbm). It is often assumed that the high-pressure turbine does more work than the low-pressure turbine but the opposite is true. Steam turbines are volumetric expansion devices. The really big one (LPT) does most of the work. The little one (HPT) just sets things up for the LPT. The whole point of reheating the steam is to keep it out of the wet region (>5% moisture).

As illustrated in this figure, the CRH is still quite superheated (192°F). The exhaust is where you need to be concerned for moisture. This is why you will sometimes see the first heat exchange element (superheater) in a HRSG at the exit of the GT exhaust (and duct burner exit) is a reheater instead of a main steam one. If you must choose between higher HRH and higher MS (main steam) temperatures, the HRH is more critical in many systems, depending on the pressures and the steam turbine specifics. Again, it is often assumed that hotter main steam means more power output, but hotter reheat steam is more valuable in terms of available expansion line length and specific power output (enthalpy drop).

By the way... I have been asked more than once, "Why bother condensing the steam? Why not just run it through again and not waste all that heat?" Recall that the steam turbine is a volume expansion device? Well try compressing *steam* back to 1800 psia (or 12 MPa)! You really don't want to do that, as it will take more power than you got out of the turbine when expanding it. The differential work for an open system (e.g., a steam turbine or boiler feed pump), the differential work is:

$$dW = VdP = \frac{dP}{\rho} \qquad (23.1)$$

and *that's* why you want to expand vapors and compress liquids! Go to a power plant and look at the size of the components and the diameter of the shafts: low-pressure steam turbine vs. boiler feed pump (BFP). An LPT can be the size of a school bus and the corresponding BFP the size of a dishwasher.

Chapter 24. Element Arrangement

Optimal arrangement of elements means the most heat transfer with the least surface area and resulting material cost. This is accomplished by positioning the segments of the cold (steam) side process lines in the heat release diagram so that the distance between these and the hot (gas) side process line is fairly uniform over much of the range without creating pinch points. Of course, the width of the evaporator (flat) section is unavoidable, given the operating pressures (determined by the steam turbine swallowing capacity). We could view this like a tangram:

in the sense that we can move the sections around in order to obtain the desired shape, recognizing that there are a limited number of solutions and possible shapes.

The geometric shapes represented by the line segments in the heat release diagram are no more complicated than the ones above. We will use the three-pressure example (CCPP3.xls) to illustrate this concept. First, consider just the process lines:

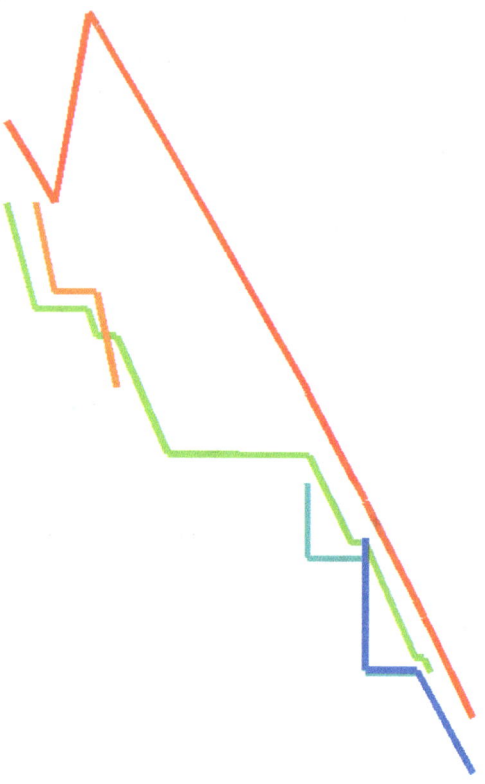

The horizontal shifts other than in the evaporator sections arise from having an intervening heat exchanger along a different path (HP, RH, IP, or LP). We can collapse these gaps in order to illustrate a point. Of course, the segments are no longer properly aligned and don't represent the actual pinch point(s).

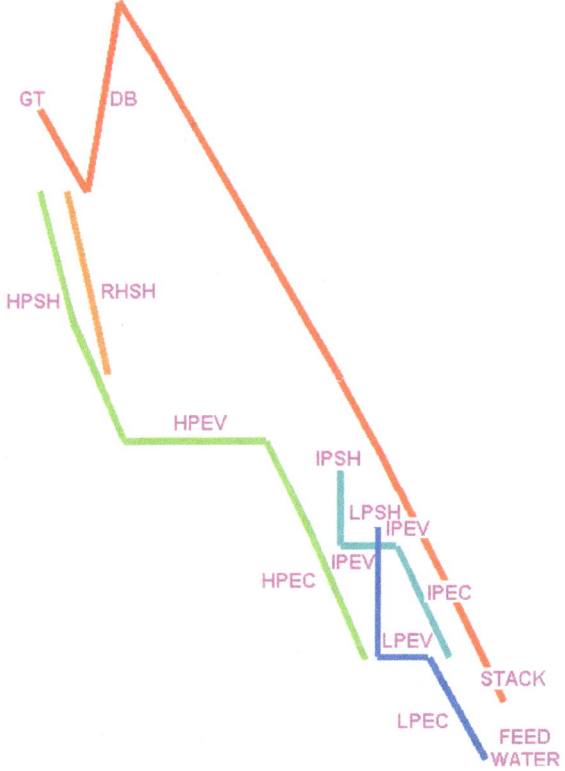

Except for the reheater, which has no evaporator section, the process lines all have the same shape: rise, flat, rise again at a different slope. The slope of each section is equal to minus the reciprocal of the mass flow rate times the specific heat. It doesn't matter how many economizer or superheater sections you have along a process line, the shape will be the same. The only difference will be the horizontal shifts (Q-axis discontinuities) representing the intervening components.

The more sections you split the HP economizer into (or the superheaters), the more fine-tuning available to minimize the pinch points. There is no obvious upper limit from a thermal point of view, but there certainly is from a mechanical and economic one. There's pressure drop with each manifold and from one section to another. Pressure drop generates entropy, reduces free energy (i.e., available for doing work), and is wasteful. Too many components increases expense and is not cost-effective.

Split Components

You will often see split components in a HRSG. There are several among the 421 examples in spreadsheet cases.xls. In such cases, the total gas flow is split into two streams—at least conceptually, for there may or may not be an actual baffle in the duct. This need not be an even 50/50 split. The following system has an absurd number of split components:

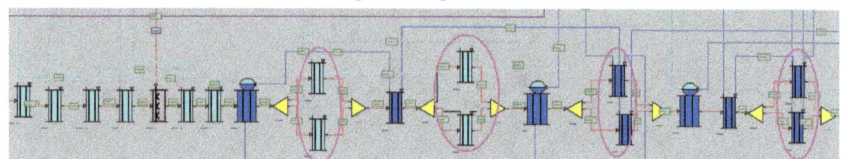

All three types of components (economizer, evaporator, and superheater) can appear in one of these splits. I have never seen a three-way split, although it is certainly possible, at least physically. Splits like this only make sense from a thermal perspective when the normal operating IP and LP steam flows are very much smaller than the HP steam flow. Even when this does occur, it's because the manufacturer wants to maintain a minimum number of rows of tubes in the direction of gas flow. This strategy can also make some sense from a mechanical or economic point of view if the manufacturer has only a limited number of designs for each type of element, fabricates them in advance, or obtained them from some other project. This latter situation is often called buying cheap stuff on the *gray market*.

You will almost never see the same exit gas temperature from two parallel components. The difference may be a few degrees up to tens. Any time you mix two different temperatures, entropy is generated—and cannot be reversed. When entropy is increased, the free energy (g=h-Ts) is reduced along with the available work. This does not make sense thermodynamically. Use split components if you must, but don't say, "Carnot made me do it!"[28]

First HP Economizer

How far down the gas side do you position the first HP economizer? That's a good question. Occasionally, you will see a design with an HP or IP economizer element downstream of the LP evaporator. CCPP3.xls contains one of these designs, which I put there for this very reason. The order (from GT to stack) is: HPSH1, RH1, DB, HPSH2, RH2, CRH, HPSH3, HPEV, IPSH, HPEC1, IPEV, LPSH, HPEC2, IPEC, LPEV, HPEC3, LPEC. Something you should notice is that the LP steam flow will be much lower (even zero) for duct-fired operation than without duct firing. This is all driven by the shifting pinch points caused by the bump up of the gas temperature line, as illustrated previously.

---

[28] Carnot refers to the theoretical maximum efficiency one can achieve with a heat engine operating between two temperatures. Named after Nicolas Léonard Sadi Carnot (1796–1832) French military scientist and physicist.

You must design the HRSG so that it will operate properly with and without duct firing. Sometimes there a trade-off is necessary to obtain an acceptable solution to both operating conditions. The first (from the feed water end) HP economizer component can shift the pinch point, avoiding excessive stack temperatures, which are wasted heat. You should also notice that the stack temperature will be lower for fired than unfired operation—again due to the shifting pinch point and trade-off of designs.

Optimization

It would be rather difficult to swap around the order and number of components in an Excel® spreadsheet such as CCPP3.xls. However, this is an easy task in the C programming language and we already have most of the code necessary. There might be something learned from swapping components around randomly, but not all combinations make sense. While I have seen at least one case where an HP economizer was upstream of an HP evaporator, this would not likely be an optimal solution. I have also seen designs where the HP evaporator is broken up into several sections and positioned between one or more superheaters. Once we develop a general code, we could investigate the merits of such an arrangement. You will find the code (CCPP3.c) in the folder examples\arrange. There are also two other similar programs that solve a combined cycle power plant, CCPP1.c and CCPP4.c in the folder examples\CCPP. The first uses SI units and the second uses English. The HRSG data statements in CCPP3.c can be modified and rearranged to handle other configuration, although there are limitations, which the code checks. The elements are defined in a data statement:

```
HRSG hrsg[]={
/*   name      st ty */
  {"HPSH1a",HP,SH},
  {"HPSH1b",HP,SH},
  {"RH1a"  ,RH,SH},
  {"RH1b"  ,RH,SH},
  {"DB"    ,NA,DB},
  {"HPSH2a",HP,SH},
  {"HPSH2b",HP,SH},
  {"RH2a"  ,RH,SH},
  {"RH2b"  ,RH,SH},
  {"CRH"   ,NA,CR},
  {"HPSH3a",HP,SH},
  {"HPSH3b",HP,SH},
  {"HPEV"  ,HP,EV},
  {"IPSHa" ,IP,SH},
  {"IPSHb" ,IP,SH},
  {"HPEC1a",HP,EC},
  {"HPEC1b",HP,EC},
  {"IPEV"  ,IP,EV},
  {"LPSHa" ,LP,SH},
  {"LPSHb" ,LP,SH},
```

```
{"HPEC2a",HP,EC},
{"HPEC2b",HP,EC},
{"IPECa"  ,IP,EC},
{"IPECb"  ,IP,EC},
{"LPEV"   ,LP,EV},
{"HPEC3a",HP,EC},
{"HPEC3b",HP,EC},
{"LPECa"  ,LP,EC},
{"LPECb"  ,LP,EC}};
```

The stream can be any one of: HP (high pressure), RH (reheat), IP (intermediate pressure), LP (low pressure), or NA (not applicable). The type can be any one of: SH (superheater), EV (evaporator), EC (economizer), DB (duct burner), or CR (cold reheat). You can comment out repeated elements, as illustrated above for HPEC1b, HPEC2b, and HPEC3b. You can also add more elements and even swap them around, as illustrated below for indices i and j:

```
void Swap(int i,int j)
  {
  HRSG h;
  if(i!=j)
    {
    h=hrsg[i];
    hrsg[i]=hrsg[j];
    hrsg[j]=h;
    }
  }
```

You can change the gas turbine exhaust flow and temperature (EGW, EGT), the superheat exit temperatures, duct burner temperature, and so forth to investigate various combinations. You could even swap the elements randomly and save each combination that improves the estimated net power output.

Typical program output is listed below:

```
3-pressure HRSG design
results file: arrange.csv
solving HRSG
GT flow=3600000
HPEV flow=780144
HP pressure=2580,1680,1500
IPEV flow=41781
IP pressure=520.0,480.0,460.0
RH pressure=460,452
CRH flow=750498
HRH flow=792279
LPEV flow=33062
LP pressure=148.00,48.00,46.00
HP temperatures=283,612,1000
6 HP economizers
6 HP superheaters
```

```
IP temperatures=279,463,565
2 IP economizers
2 IP superheaters
4 RH superheaters
LP temperatures=120,278,480
2 IP economizers
2 LP superheaters
name     mC     Pci   Pco   Tci   Tco    mH     Thi   Tho
HPSH1a  780144  1530  1500   935  1000  3600000  1125  1098
HPSH1b  780144  1560  1530   871   935  3600000  1098  1069
RH1a    792279   454   452   929  1000  3600000  1069  1043
RH1b    792279   456   454   858   929  3600000  1043  1016
DB           0     0     0     0     0  3614500  1016  1316
HPSH2a  780144  1590  1560   806   871  3614500  1316  1287
HPSH2b  780144  1620  1590   741   806  3614500  1287  1254
RH2a    792279   458   456   787   858  3614500  1254  1229
RH2b    792279   460   458   716   787  3614500  1229  1203
CRH     750498   460   460   725   716  3614500     0     0
HPSH3a  780144  1650  1620   676   741  3614500  1203  1163
HPSH3b  780144  1680  1650   612   676  3614500  1163  1101
HPEV    780144  1680  1680   607   612  3614500  1101   732
IPSHa    41781   470   460   514   565  3614500   732   730
IPSHb    41781   480   470   463   514  3614500   730   729
HPEC1a  780144  1830  1680   553   607  3614500   729   674
HPEC1b  780144  1980  1830   499   553  3614500   674   627
IPEV     41781   480   480   458   463  3614500   627   596
LPSHa    33062    47    46   379   480  3614500   596   595
LPSHb    33062    48    47   278   379  3614500   595   593
HPEC2a  780144  2130  1980   445   499  3614500   593   548
HPEC2b  780144  2280  2130   391   445  3614500   548   504
IPECa    41781   500   480   369   458  3614500   504   500
IPECb    41781   520   500   279   369  3614500   500   496
LPEV     33062    48    48   278   278  3614500   496   466
HPEC3a  780144  2430  2280   337   391  3614500   466   423
HPEC3b  780144  2580  2430   283   337  3614500   423   380
LPECa   854987    98    48   199   278  3614500   380   310
LPECb   854987   148    98   120   199  3614500   310   239
name    NTUc  NTUh     F    LMTD   UAc    UAh    UAf
HPSH1a  0.46  0.19  0.986  142.9  0.225  0.225  0.225
HPSH1b  0.36  0.16  0.991  179.9  0.186  0.186  0.186
RH1a    0.82  0.30  0.962   89.6  0.352  0.350  0.352
RH1b    0.54  0.20  0.983  134.8  0.229  0.228  0.229
HPSH2a  0.14  0.06  0.999  463.3  0.077  0.077  0.077
HPSH2b  0.13  0.07  0.999  497.0  0.079  0.079  0.079
RH2a    0.17  0.06  0.998  418.4  0.073  0.073  0.073
RH2b    0.15  0.06  0.999  463.8  0.067  0.067  0.067
HPSH3a  0.14  0.08  0.998  473.9  0.101  0.101  0.101
HPSH3b  0.13  0.13  0.997  488.0  0.148  0.148  0.148
HPEV                      1.000  267.3                1.554
IPSHa   0.27  0.01  1.000  190.6  0.008  0.011  0.008
```

```
IPSHb    0.21 0.01 1.000 240.5 0.007 0.011 0.007
HPEC1a   0.46 0.46 0.970 122.1 0.493 0.493 0.493
HPEC1b   0.44 0.39 0.974 124.7 0.418 0.418 0.418
IPEV               1.000 150.6             0.212
LPSHa    0.63 0.01 0.999 160.6 0.010 0.010 0.010
LPSHb    0.39 0.01 1.000 261.8 0.007 0.010 0.007
HPEC2a   0.57 0.48 0.961  98.2 0.499 0.499 0.499
HPEC2b   0.52 0.42 0.968 107.5 0.432 0.431 0.432
IPECa    1.11 0.05 0.991  81.3 0.050 0.050 0.050
IPECb    0.52 0.02 0.998 170.4 0.023 0.023 0.023
LPEV               1.000 202.2             0.151
HPEC3a   0.71 0.57 0.943  79.9 0.578 0.577 0.578
HPEC3b   0.62 0.49 0.956  90.7 0.492 0.491 0.492
LPECa    0.81 0.72 0.924 105.8 0.701 0.700 0.701
LPECb    0.74 0.66 0.935 114.7 0.631 0.630 0.631
total HP/RH conductance=6.0
total IP conductance=0.3
total LP conductance=1.5
estimated net power 124637 kW
```

The steam flows are adjusted to match the specified overall conductances and the net power output is estimated based on typical condenser conditions (expansion line end point enthalpy). This facilitates comparing various arrangements.

Cost Effectiveness

One of the first things to consider is estimated net power output divided by material costs (total UA), which is the cost-effectiveness. This answers the question, "How much surface area (material) is best?" To do this, just put the target overall conductances (total UAs) in a loop and let 'er rip! You can write the results to a CSV file and then pull it right into Excel®. One impact we expect to see with increasing UA is decreasing stack temperature, as more energy is being removed from the gas stream into the steam. The stack temperature should asymptotically approach that of the feed water. This next figure illustrates this:

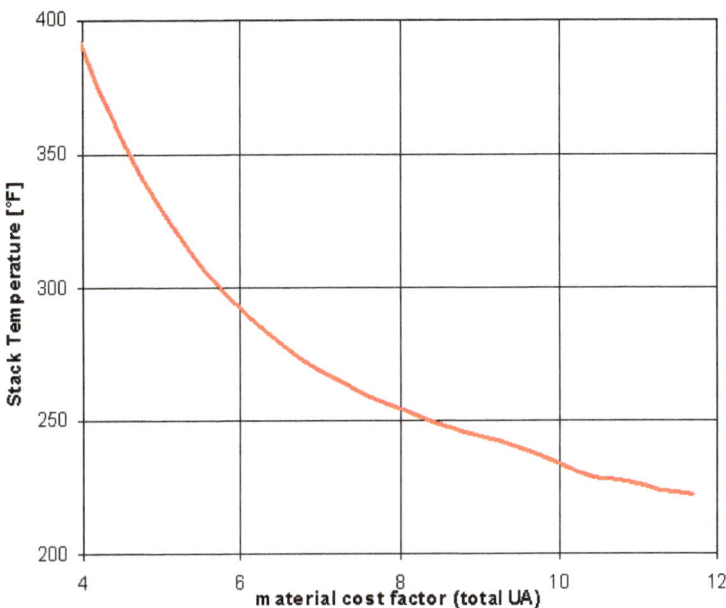

As more steam is produced with greater surface area, the estimated net power output should also increase asymptotically to the level that approaches recovery of the GT exhaust heat less turbine and generator efficiencies.

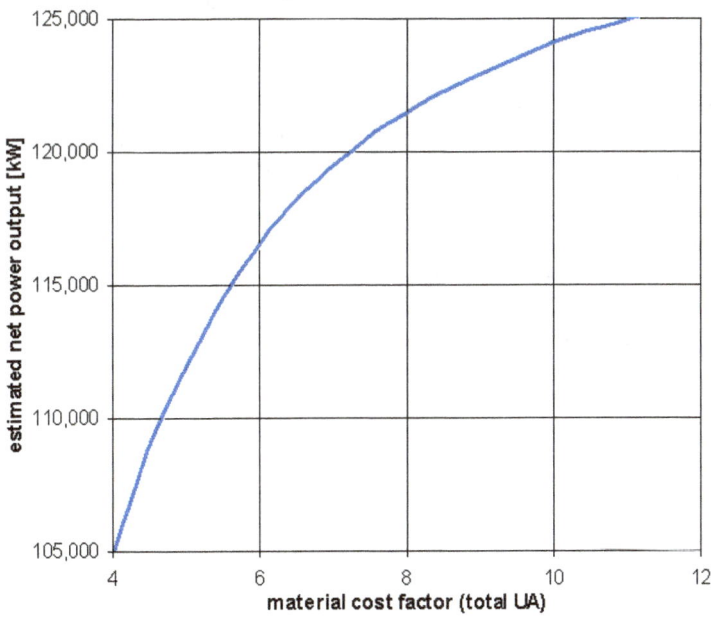

The HRSG is only part of the cost. There is also the steam turbine, generator, pumps, piping, condenser, cooling tower, control system, and much more. Using typical values for such a plant, we obtain the following curve of cost-effectiveness.

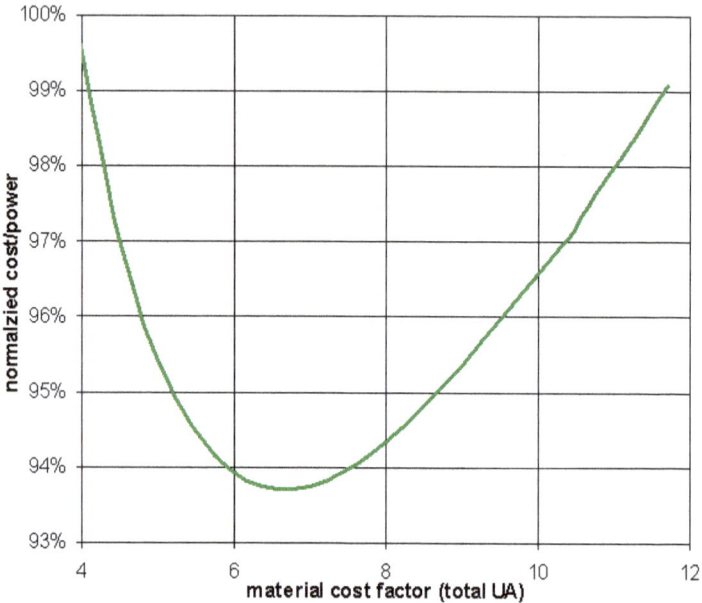

Paying twice as much for a HRSG with twice the UA to obtain 15% more steam flow is not cost-effective. Blowing 400°F/200°C gas into the sky is stupid, considering you could generate electricity, dry wood, heat a building, or bake a lot of pizza.[29] If your curve doesn't exhibit a most cost-effective point, there's something wrong, because this is the way the world works.

Sectionalized Components

One of the actual designs in cases.xls is shown on the next page. This particular design has several elements of the same type and stream immediately adjacent/sequential: LPEC1-4, LPEV1-3, HPEC1-2, IPEV1-3, HPEC4-5, HPEV1-4, and HPSH1-2. This is done for mechanical/physical considerations, not thermal. We can easily investigate using the program (arrange.c) by adding duplicate components.

---

[29] Yes, I've seen it and wondered how many birds would eventually die and what would happen if a helicopter flew over this hot blast on its way to the nearby hospital. It's humorous to point out that this design was cooked up by a professor at a prestigious university who has clearly never worked in industry but likely has many esteemed publications in the annuls of heat transfer.

GT
HPSH4
RH3
HPSH3
RH2
DB
HPSH2
HPSH1
RH1
HPEV4
HPEV3
HPEV2
HPEV1
HPEC6
SCR
IPSH
HPEC5
HPEC4
LPSH
IPEV3
IPEV2
IPEV1
HPEC3
IPEC
HPEC2
HPEC1
LPEV3
LPEV2
LPEV1
LPEC4
LPEC3
LPEC2
LPEC1
STACK

The following table shows the difference in performance resulting from splitting every one of the superheaters and economizers into two equal halves:

splitting superheaters and economizers				
result	units	1 each	2 each	diff.
main steam	lb/hr	757,897	763,261	0.71%
stack temp.	°F	256	240	-16
net power	kW	121,120	122,304	0.98%

Imagine the expense in piping and pressure drop required to get 0.7% more main steam, a 16°F drop in stack temperature, and 1% more net power? Manufacturers do this because they have elements already designed and ready to fabricate of various sizes.

<u>Rearranging Elements</u>

We will now consider rearranging the elements. This is easily accomplished with the code listed previously. The first example will be to swap the first four elements from HPSH1, HPSH2, RH1, RH2 (case 1) to RH1, RH2, HPSH1, HPSH2 (case 2). Then to HPSH1, RH1, HPSH2, RH2 (case 3) and RH1, HPSH1, RH2, HPSH2 (case 4). The results are:

case no.	Fms lb/hr	Tstk °F	power kW
1	763,261	239.8	122,304
2	762,822	239.9	122,283
3	764,885	239.5	122,385
4	765,129	239.4	122,397

The differences are so small as to be negligible. While there is little point swapping around the last main steam and reheat superheater sections, some improvement is possible by swapping around the IP/LP components and the HP economizer elements. We can adapt the code (arrange.c) to randomly swap two components, resolve, and compare to the previous results, keeping track of any improvements. After 2079 combinations, 1001 cases converged and 33 improvements were found. The initial configuration and results were listed 5 pages ago and the most improved case is listed below:

```
solving HRSG
GT flow=3600000
HPEV flow=795928
HP pressure=2580,1680,1500
IPEV flow=38581
IP pressure=520.0,470.0,460.0
RH pressure=460,452
CRH flow=765683
HRH flow=804264
LPEV flow=66254
LP pressure=98.00,48.00,45.00
```

```
HP temperatures=283,612,1000
6 HP economizers
6 HP superheaters
IP temperatures=279,461,565
2 IP economizers
2 IP superheaters
4 RH superheaters
LP temperatures=120,278,480
2 IP economizers
2 LP superheaters
name    mC     Pci   Pco    Tci   Tco     mH     Thi   Tho
RH2a    804264  454   452   929  1000  3600000  1125  1099
RH1b    804264  456   454   858   929  3600000  1099  1072
CRH     765683  460   460   725   717  3600000     0     0
RH1a    804264  458   456   788   858  3600000  1072  1045
DB           0    0     0     0     0  3614500  1045  1345
HPSH3a  795928 1530  1500   935  1000  3614500  1345  1319
LPSHa    66254   46    45   379   480  3614500  1319  1316
IPSHa    38581  470   460   513   565  3614500  1316  1315
HPSH1b  795928 1560  1530   871   935  3614500  1315  1287
HPSH2b  795928 1590  1560   806   871  3614500  1287  1257
RH2b    804264  460   458   717   788  3614500  1257  1231
HPSH2a  795928 1620  1590   741   806  3614500  1231  1197
HPSH1a  795928 1650  1620   676   741  3614500  1197  1156
HPSH3b  795928 1680  1650   612   676  3614500  1156  1093
HPEV    795928 1680  1680   607   612  3614500  1093   715
HPEC1b  795928 1830  1680   553   607  3614500   715   660
HPEC2b  795928 1980  1830   499   553  3614500   660   611
IPEV     38581  470   470   456   461  3614500   611   582
IPSHb    38581  480   470   461   513  3614500   582   581
HPEC2a  795928 2130  1980   445   499  3614500   581   534
IPECb    38581  500   480   368   456  3614500   534   531
HPEC1a  795928 2280  2130   391   445  3614500   531   486
LPSHb    66254   47    46   278   379  3614500   486   483
HPEC3a  795928 2430  2280   337   391  3614500   483   439
LPECa    66254   48    47   199   278  3614500   439   433
IPECa    38581  520   500   279   368  3614500   433   430
HPEC3b  795928 2580  2430   283   337  3614500   430   386
LPEV     66254   48    48   278   278  3614500   386   324
LPECb   900764   98    48   120   199  3614500   324   249
name    NTUc  NTUh    F    LMTD    UAc    UAh    UAf
RH2a    0.49  0.18  0.986  146.1  0.213  0.213  0.213
RH1b    0.37  0.14  0.992  190.7  0.162  0.162  0.162
RH1a    0.30  0.11  0.994  235.0  0.132  0.132  0.132
HPSH3a  0.18  0.07  0.998  364.1  0.089  0.089  0.089
LPSHa   0.11  0.01  1.000  886.8  0.004  0.012  0.004
IPSHa   0.07  0.01  1.000  776.2  0.002  0.012  0.002
HPSH1b  0.16  0.07  0.998  397.7  0.085  0.085  0.085
HPSH2b  0.15  0.07  0.998  433.5  0.084  0.084  0.084
RH2b    0.14  0.05  0.999  491.1  0.064  0.064  0.064
```

```
HPSH2a  0.15  0.08  0.998  440.2  0.092  0.092  0.092
HPSH1a  0.14  0.09  0.998  467.6  0.104  0.104  0.104
HPSH3b  0.14  0.13  0.997  480.6  0.153  0.153  0.153
HPEV                1.000  250.6                1.691
HPEC1b  0.52  0.54  0.960  107.8  0.575  0.575  0.575
HPEC2b  0.51  0.46  0.966  109.3  0.491  0.490  0.491
IPEV                1.000  137.9                0.215
IPSHb   0.57  0.02  0.998   92.4  0.017  0.017  0.017
HPEC2a  0.66  0.57  0.947   85.6  0.593  0.592  0.593
IPECb   0.76  0.03  0.996  115.8  0.032  0.032  0.032
HPEC1a  0.63  0.52  0.954   90.2  0.533  0.532  0.533
LPSHb   0.67  0.02  0.997  150.2  0.023  0.023  0.023
HPEC3a  0.58  0.48  0.960   96.2  0.481  0.481  0.481
LPECa   0.41  0.03  0.998  194.8  0.027  0.027  0.027
IPECa   0.87  0.03  0.995  102.3  0.035  0.035  0.035
HPEC3b  0.57  0.47  0.961   97.5  0.464  0.464  0.464
LPEV                1.000   71.9                0.853
LPECb   0.66  0.62  0.944  126.9  0.595  0.595  0.595
total HP/RH conductance=6.0
total IP conductance=0.3
total LP conductance=1.5
estimated net power 125955 kW
1001 converged, 1078 failed, 33 improved
```

Notice that the reheaters have been moved ahead of the HP superheaters. The IP and LP superheaters have also been moved up. Not surprisingly, the LP evaporator and economizer are still at the end. The differences are summarized in the following table:

Element Rearrangement			
Fms	Tdb	Tstk	kWnet
780,144	1316.3	239.0	124,637
786,893	1317.9	252.0	125,537
0.87%	1.6	13.0	0.72%

This rearrangement of elements yields a 0.87% increase in main steam flow, only a 1.6°F increase in duct burner exit temperature, a 13.0°F increase in stack temperature, and a 0.72% increase in estimated net power. Overall, it's not much. If you were hoping to double the main steam flow or pull the stack temperature down 100°F, that's not going to happen. Even with 33 steps of improvement over the initial configuration, only a very small increase in net power is achieved. This is at constant UA or heat exchange material cost. The heat release diagram for the final configuration can also be found in the spreadsheet arrange.xls and is shown on the next page.

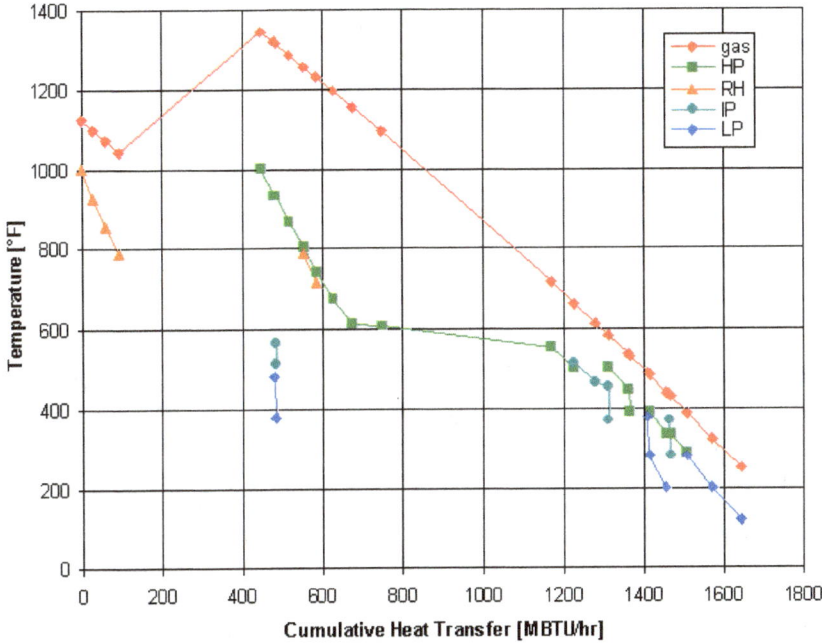

In spite of this being somewhat different from that of the initial configuration shown earlier in this chapter, the pinch point region between Q=1300 to 1500 MBTU/hr is very similar. As mentioned before, pinch points dominate the performance of a HRSG.

Chapter 25. Attemperation

Hotter is not necessarily better. As illustrated in the Mollier diagram and expansion line, you only need steam that is hot *enough*. Increasing the temperature beyond 1100°F/600°C can lead to material failures and doesn't provide a proportional increase in power output. More steam is always better, if it's hot enough and provided the equipment can handle it (i.e., adequate steam turbine swallowing capacity). I have tested more than one plant that had no reserve swallowing capacity, preventing them from making more money (selling more megawatt-hours) during times of peak demand, especially in cold months. This was an unwise design decision.

The worst case of under-performance I have ever seen is a system with far too much swallowing capacity—more than twice needed by the rest of the system. This resulted in low efficiency year round. This undesirable situation happened because a large international manufacturer (which will remain unnamed) winning the bid. This larger company had recently acquired a smaller company, wanted to integrate the two product lines, and utilize the existing stock. The company decided to simply pull one of the steam turbines off the shelf at this acquired company, not considering the consequences of this turbine being incorrectly sized for the application. This was a *business* decision, not an *engineering* one, so we know who got the short end of that stick! The turbine was actually quite well designed and fabricated, just the wrong size. As a result of this foolish decision, the company's reputation was damaged—in spite of having provided an excellent piece of equipment. If you need an alternator, a carburetor won't do the job, no matter how good it is.

Attemperation is accomplished by spraying feed water (liquid) into the superheated steam (vapor) to drop the temperature. This process naturally increases the mass flow rate. It is best to attemperate between the final two stages of superheat, which is one reason for having at least two HP and two RH superheaters in a HRSG. While it may be cheaper to construct and easier to maintain if attemperation is performed after the final superheat, as the steam leaves the HRSG, this isn't the best solution. You want this process to be sluggish and slowly responsive. It doesn't have to be a quick response. It may be more challenging to manage, but that's what control system logic is for. Adding feedwater before the final superheat assures complete mixing and also moderates temperature variations in the turbine. This adage is true: haste makes waste.

A note of caution... Should you get a bid package for a HRSG that indicates very large attemperation flows during normal operation, toss it in the trash. The components should be designed and fabricated to fit the application, which is driven by the gas turbine and steam turbine. You don't want to buy a HRSG that has been cobbled together with existing parts designed for some other system.

Chapter 26. Performance Testing

Let us begin by agreeing that it's not the HRSG manufacturer's fault if the GT isn't supplying sufficient flow and temperature. I've suffered through many arguments driven by a GT that provided too much flow and not enough temperature or vise versa. The HRSG design depends on both. Flow and temperature are not interchangeable. GT manufacturers love to talk about exhaust *energy*, meaning flow times temperature, but that's not how you design a HRSG. At the very least, there's an implied reference point for any and all enthalpies. Rate of energy supplied by the GT (actually thermal power) should at least be:

$$thermal\ power = \dot{m}\left(h_{EGT} - hR_{REF}\right) \quad (26.1)$$

but this still misses the point. Granted, gas turbines are volumetric flow devices, but HRSGs aren't—two completely different animals.

With this said, the steam flow produced by a HRSG is roughly proportional to the GT exhaust flow and the final superheat temperature is roughly proportional (though not in a multiplicative sense) to the exhaust temperature. But it's more complicated than this. Considering steam turbine requirements and attemperation, the final superheat temperature is somewhat fixed. In a performance test, you basically achieve it or you don't. The penalties per degree vary considerably, but these aren't particularly helpful and nobody comes away satisfied from a test failed. Net power output is approximately proportional to main steam flow so that penalties (or bonuses) based on flow rate are more meaningful than ones based on temperature. My advice is simple: Fix the GT and *then* retest the HRSG.

Code-Level Testing

Whether you are designing, fabricating, or purchasing a HRSG, you definitely want a defensible code-level performance test. The applicable code for testing HRSGs is ASME PTC-4.4, "Gas Turbine Heat Recovery Steam Generators," 2008. It is utterly pointless to perform a 4.4 test without also performing a test of the gas turbine to determine the exhaust flow. The applicable code for testing GT is ASME PTC-22, "Gas Turbines," 2014.

Some customers (end purchasers) are quite wary from being previously *burned* in an acquisition. This may motivate them to require all sorts of correction curves, adjustments, and test points. I can sympathize with this, much like buying a used car that looks nice, but falls apart as you're driving off the lot. Exorbitant penalties and endless requirements will not protect you from an experienced and dishonest car salesman. You can't possibly think of enough checks and tests. It is far wiser to contact previous purchasers of similar cars from the same dealer before laying out any cash. This also goes for buying a HRSG. Make sure the manufacturer knows what they're doing and has a proven track record.

There is a less common problem, which also must be considered and that is consistent information exchange. More than once I have seen a project that switched GTs (or STs) at some point and for various reasons (usually cost). This change was never communicated to the HRSG manufacturer (or the condenser or cooling tower manufacturer). The Engineering Procurement Contractor (EPC) dropped the proverbial ball and the HRSG, condenser, or cooling tower supplier faced penalties.

Accuracy is also important. I recall one case where the GT exhaust mole fractions of H_2O and CO_2 were reversed when transcribing information from one supplier to the other. I noticed this immediately, but that was after the plant had been built. Numerically, the H_2O and O_2 mole fractions are often closer than H_2O and CO_2, so that this mistake might be more likely overlooked. Water vapor has a much higher specific heat than any of the other constituents (see Appendix E). The end result of this discrepancy was insufficient energy entering the HRSG to produce the guaranteed steam, even if all were extracted such that the stack temperature were brought down to ambient. It was thermodynamically impossible to meet the guarantee. This was not the HRSG manufacturer's fault. The GT exhaust flow and temperature were considered in the guarantee, but not the composition. It wasn't the GT manufacturer's fault either. I know of two cases where the owner acquired GTs from the gray market and forced the change on the EPC. In one of those the HRSG manufacturer came out OK, but in the second one, they did not.

Single Pressure Combined Cycle Power Plant

You will find all of the file associated with a typical single pressure combined cycle power plant in the online archive in folder examples\CCPP named CCPP1.???. The two primary performance variables of interest are main steam flow and temperature. The other parameters are trivial (IP/LP steam flow and temperature. These are simply there to balance out the heat release diagram, provide non-condensable gas removal, and steady operation. They need to work properly, but the particular values don't really matter when considering performance. Reheat pressure drop is also important, but HRSGs rarely fail for this reason. If they ever do, you will have much bigger problems to worry with.

The main steam outlet temperature can vary and if it does, this will not be linear, as shown in the following figure:

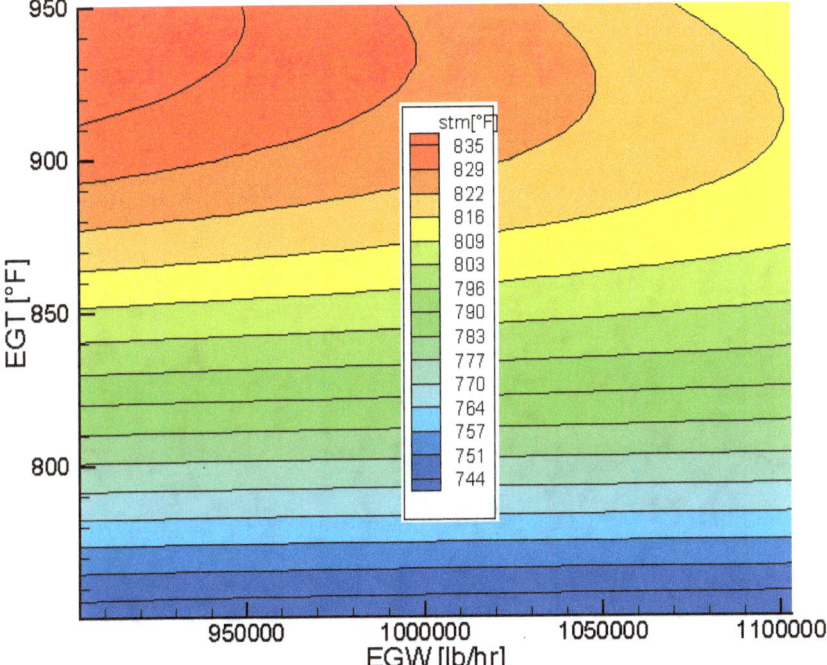

A simple linear correction curve will probably not do the job. This is the 21st century. Everything doesn't have to be a straight line. Excel® can handle a more complicated curve, as illustrated in the next chapter.

The main steam flow will be approximately bilinear (linear in two variables: z=a+bx+cy), as illustrated in this next figure:

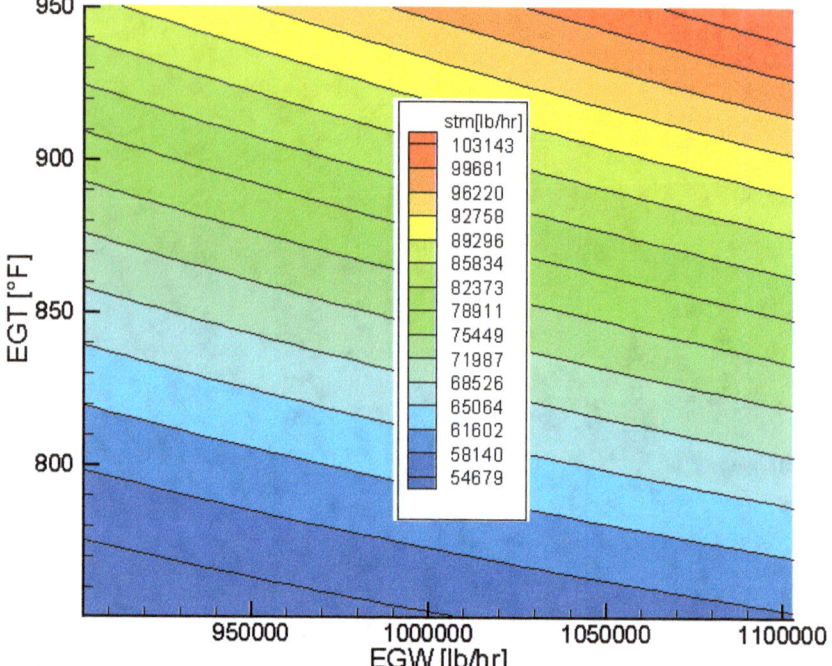

This is a simple relationship and easily corrected, that is, from the test point to the design point, as illustrated in the next chapter.

The net power output, which will be paying for the HRSG, is also approximately bilinear:

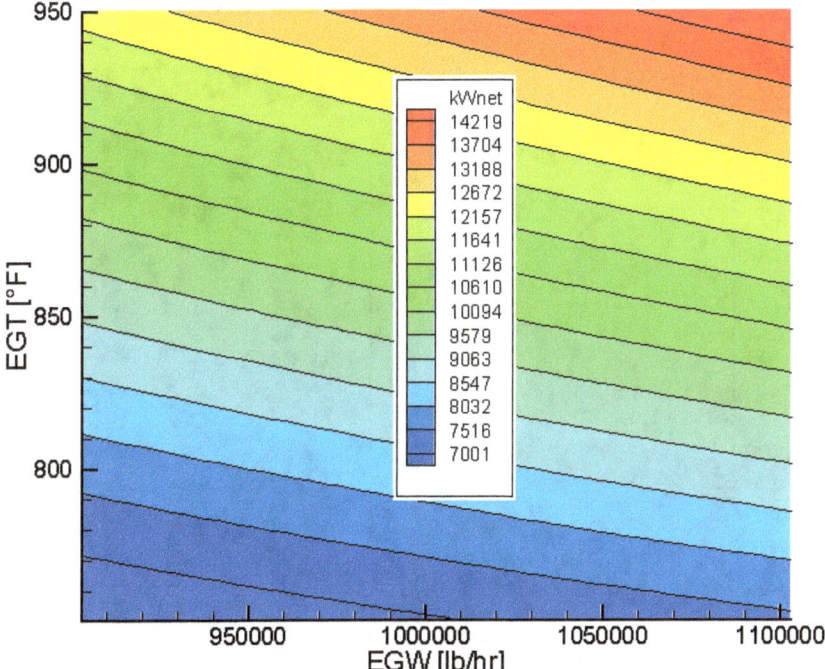

If you have GateCycle™ you can make adjustments to the model and see what happens. If you don't, you can work with the code (CCPP1.c) to try all sorts of things, including make it fit your design.

Three-Pressure Combined Cycle Power Plant

You will find all of the files associated with a typical three-pressure system (CCPP3.???) in the same folder. If you have GateCycle™ you can make runs or modify the code (CCPP3.c) to find the impact of various inputs and design parameters on output and performance measures.

Again, the variation in main steam temperature is nonlinear, though different from the previous system. This relationship will change with engine design., operation to control emissions, HRSG design, and ambient conditions (i.e., winter vs. summer).

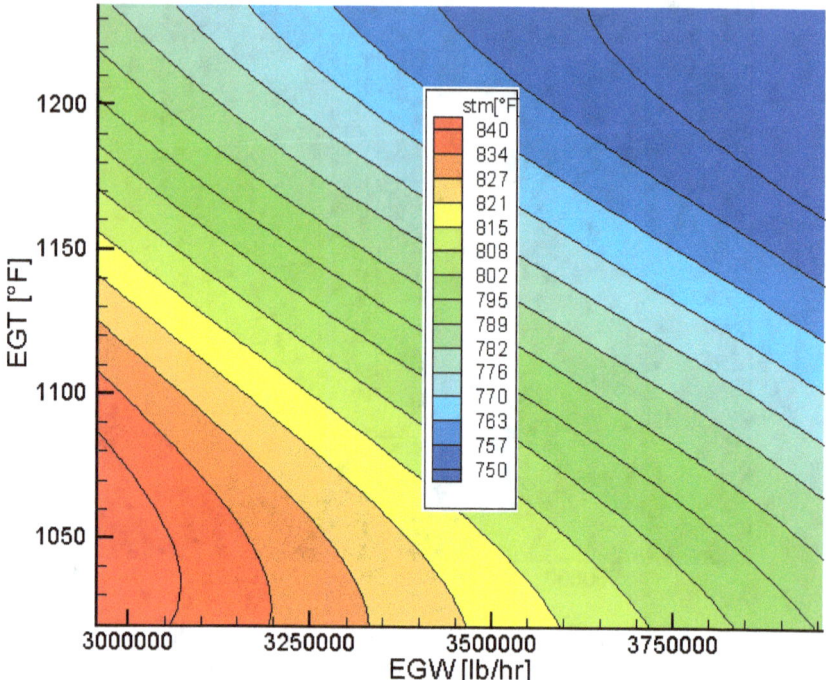

A linear correction will not suffice for this relationship either, but the steam flow and net power (of the steam tail) are again mostly bilinear.

The main steam flow in this case is almost exactly bilinear:

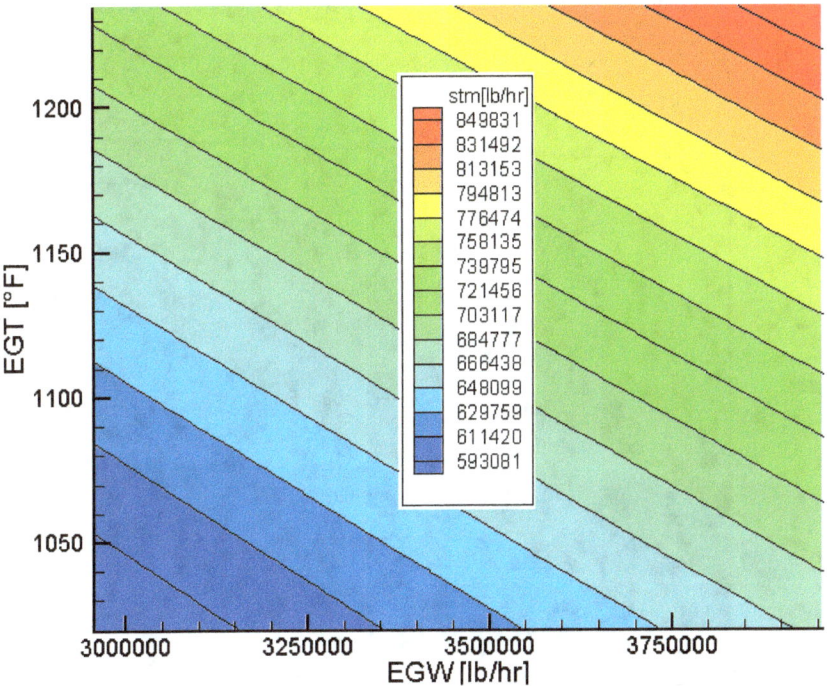

A bilinear correction or two linear ones would be adequate in this case.

The net power output of the steam tail is not quite bilinear. I would recommend using a second order correction ($z = a + b*x + c*y + d*x*y + e*x^2 + f*y^2$):

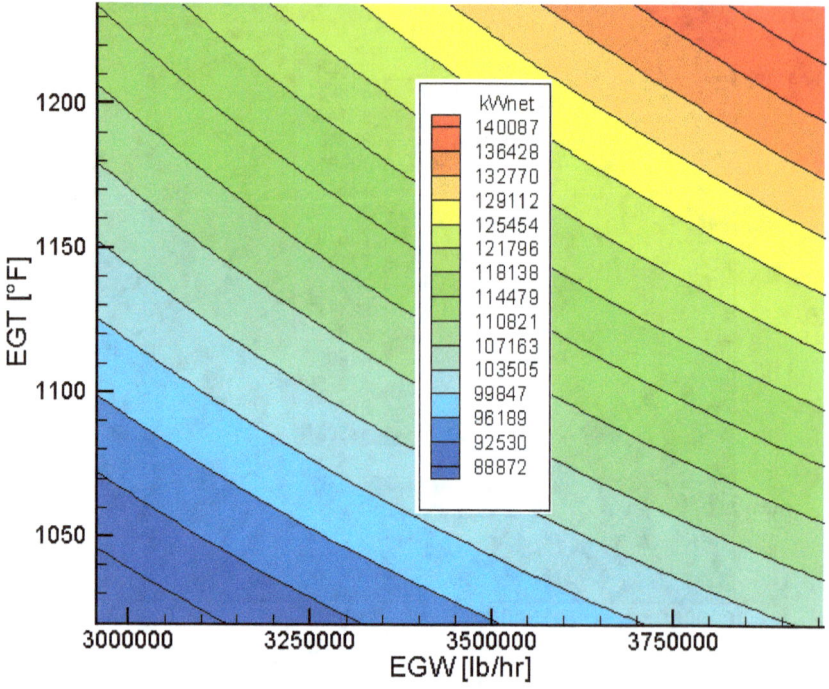

Chapter 27. GT and HRSG Heat Balances

As stated previously, you will want to perform a GT (PTC-22) and a HRSG (PTC-4.4) test. While the HRSG test does provide an estimate of the GT exhaust flow, there are far more assumptions and things taken for granted that are not necessarily the case. If the two don't agree, this provides awareness and a path forward to resolve the differences. A basic three-pressure system is illustrated below:

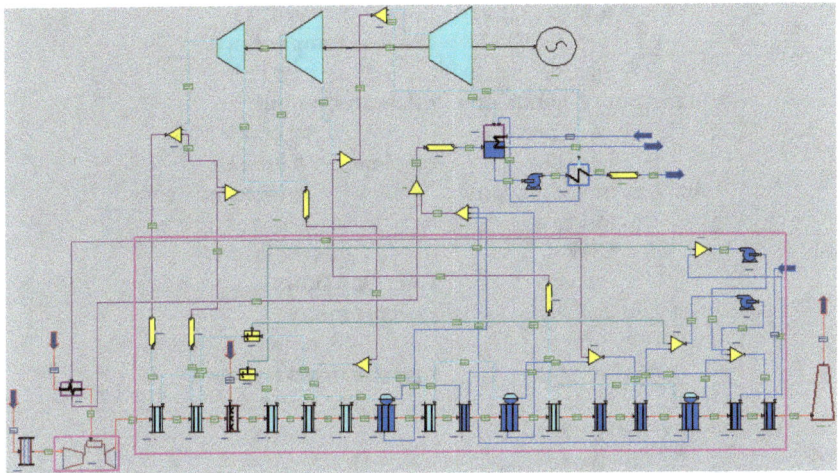

There are two boundaries (control volumes): 1) around the GT and 2) around the HRSG. We will first consider the GT. The most important thing to remember here is that the ASHRAE (moist air) properties are NOT consistent with the NASA (exhaust gas) OR the ASME/IAPWS (steam) properties. These have three different reference points and you can't just subtract the enthalpies without correcting for this. The same goes for the heats of formation (and consequently heats of combustion). These have a specific reference and you must not ignore this.

If your results (GT exhaust flow/HRSG heat loss) change in the slightest when you change the reference temperature, then your calculations are simply wrong! The reference temperature is an arbitrary value and the real world doesn't know or care what your reference temperature is.

You will find all of the necessary calculations in the online archive in folder examples\testing in the spreadsheet GT_heat_balance.xls. User inputs are blue and calculations are violet.

GT Heat Balance

	A	B	C	D
1				**GAS TURBINE HE**
2	**INPUTS**		**INPUTS Contd.**	
3	**reference temperature**		**injection flow & conditions**	
4	77	ref. temp.[°F]	10,000	flow [lbm/hr]
5	**ambient conditions**		150	pres.[psia]
6	14.540	amb. pres.[psia]	450	temp.[°F]
7	95	amb. temp.[°F]	**exhaust temperature**	
8	43%	amb. rel. hum.	1108	temp.[°F]
9	**bleed flow & conditions**		**generator output**	
10	5,000	flow [lbm/hr]	167,700	net power [kW]
11	600	temp.[°F]	**efficiencies & losses**	
12	**fuel gas mole fractions**		98.89%	generator efficiency
13	95.19%	Methane	100.00%	gearbox efficiency
14	2.48%	Ethane	0.82%	heat loss
15	0.44%	Propane	**CALCULATIONS**	
16	0.09%	IsoButane	**psychrometry**	
17	0.09%	Butane	0.015457	humidity [lb-wetr/lb-dry]
18	0.04%	IsoPentane	**fuel gas mass fractions**	
19	0.02%	Pentane	89.66%	Methane
20	0.06%	Hexane	4.38%	Ethane
21	0.51%	Nitrogen	1.13%	Propane
22	0.00%	Carbon Monoxide	0.32%	IsoButane
23	1.07%	Carbon Dioxide	0.32%	Butane
24	0.00%	Water	0.16%	IsoPentane
25	0.00%	Hydrogen Sulfide	0.10%	Pentane
26	0.00%	Hydrogen	0.31%	Hexane
27	0.00%	Helium	0.84%	Nitrogen
28	0.00%	Oxygen	0.00%	Carbon Monoxide
29	0.00%	Argon	2.77%	Carbon Dioxide
30	100.00%	total	0.00%	Water
31	17.031	mol. wt.	0.00%	Hydrogen Sulfide
32	**fuel gas flow & conditions**		0.00%	Hydrogen
33	76,149	flow [lbm/hr]	0.00%	Helium
34	440	pres.[psia]	0.00%	Oxygen
35	365	temp.[°F]	0.00%	Argon
36			100.00%	total

You will need to measure the barometric pressure, ambient temperature, and ambient humidity. Note that humidity sensors are not particularly accurate, which is why wet-bulb RTDs are most often used. A dew-point device is also quite accurate, but they don't work well dangling from a rope and swinging in the breeze.

You can enter zero for the compressor bleed if there is none. If there is water or steam injection, you must accurately measure the flow, pressure, and temperature. Measuring the exhaust temperature requires a rack of thermocouples or high temperature RTDs. As these are in a high velocity flow zone, special care is required affixing them and protecting the cables. Thermocouples are far more durable.

A precision power meter is required to determine the generator output and power factor. This is connected to special voltage and current transformers. The generator efficiency is obtained from curves, which are usually adequate for this purpose. The gearbox efficiency and heat loss cannot be measured in such a test, so you will need these values from the manufacturer, which they will have measured on a test stand.

Fuel gas flow is measured with a sharp-faced orifice plate or Coriolis meter. Do not accept anything else for fuel flow measurement, specifically, DO NOT accept an annubar or any other type of out-of-a-box flow meter, no matter how good someone tells you it is or how much it costs. While these devices may be accurate under all the right conditions, you don't know that to be the case! If you're not going to accurately measure the fuel flow, then don't bother performing a test because clearly you don't care (see Appendix I).

Fuel composition is measured with a calibrated gas chromatograph. You will need calibration gases and a unit that is much too sensitive to hang in a rack at a power plant between a fire hose and an axe. Online chromatographs (i.e., continuously operating) are OK for gross characterization and day-to-day monitoring of plant performance. These may even be quite accurate when they're first installed after calibration. They are not suitable for a performance test.

Fuel composition is entered by mole ratio, as this is customary. Mass fractions are needed for the calculations. The mass fractions are not inputs; rather, these are calculated from the mole fractions and molecular weights.

Exhaust gas mole fractions are calculated from the ambient plus fuel minus bleed plus water or steam injection. Both mass and mole fractions, as are mass and molar flow rates.

Steam enthalpies in the spreadsheet are calculated using ASME 1967 properties, which are supplied as macros. You could change these to something else, for instance, using the AllSteam Excel® Add-In. The gas enthalpies are calculated using NASA Glenn and also PTC-22, which should be consistent. These calculations are also implemented as macros. There are also 3800 entries on the NASA tab.

AT BALANCE CALCULATIONS

CALCULATIONS Contd.		CALCULATIONS Contd.	
enthalpy of dry air @Tin		**reactants [lbm/hr]**	
4.32	NASA enth.[BTU/lbm]	204,579	Carbon Dioxide
4.32	refd. enth.[BTU/lbm]	162,158	Water
enthalpy of dry air @Tbleed		640	Nitrogen
127.66	NASA enth.[BTU/lbm]	0	Sulfur Dioxide
127.66	refd. enth.[BTU/lbm]	0	Argon
enthalpy of dry air @Texh		0	Helium
258.93	NASA enth.[BTU/lbm]	291,227	stoichiometric O2
258.93	refd. enth.[BTU/lbm]	1,258,445	dry air [lbm/hr]
enthalpy of moisture @Tin		1,277,898	moist air [lbm/hr]
8.02	NASA enth.[BTU/lbm]	**products [lbm/hr]**	
8.02	refd. enth.[BTU/lbm]	205,183	Carbon Dioxide
enthalpy of moisture @Tbleed		162,158	Water
240.83	NASA enth.[BTU/lbm]	951,008	Nitrogen
240.83	refd. enth.[BTU/lbm]	0	Sulfur Dioxide
enthalpy of moisture @Texh		16,245	Argon
495.94	NASA enth.[BTU/lbm]	1	Helium
495.94	refd. enth.[BTU/lbm]	1,334,595	total [lbm/hr]
injection enthalpies		**enthalpy of products @Texh**	
vapor	state	290.72	NASA enth.[BTU/lbm]
1247.36	ASME enth.[BTU/lbm]	290.72	refd. enth.[BTU/lbm]
151.89	NASA enth.[BTU/lbm]	**air calculations**	
151.89	refd. enth.[BTU/lbm]	2,277,614	dry excess air (NASA) [lbm/hr]
fuel gas enthalpies & LHV		2,277,614	dry excess air (refd.) [lbm/hr]
164.00	NASA enth.[BTU/lbm]	0	error (refd.-NASA) [lbm/hr]
164.00	refd. enth.[BTU/lbm]	2,312,819	moist excess air [lbm/hr]
20,646	LHV [BTU/lbm]	3,595,717	moist total in air [lbm/hr]
0	LHV adj. [BTU/lbm]	3,676,866	moist total exh. air [lbm/hr]
20,646	adj. LHV [BTU/lbm]	**enthalpy of exhaust @Texh**	
power & heat loss		274.64	NASA enth.[BTU/lbm]
578.67	Shaft [MBTU/hr]	274.64	refd. enth.[BTU/lbm]
12.82	Heat Loss [MBTU/hr]		

This is a non-iterative calculation, that is, there is no button to push or residual to minimize. The entire system of algebraic equations are solved simultaneously to obtain the result, in this case, dry air flow. If you happen to have Maple™ or can read a Maple™ worksheet, the equations are in the same folder in file GT_heat_balance.mws. The final equation is Equation 27.1:

$$m_{excess\_dry} = \begin{pmatrix} m_{fuel}(LHV + h_{fuel, Tfuel} - h_{products, Texh}) - Q_{loss} - W_{shaft} \\ - \dfrac{(h_{air, Tbleed} - h_{air, Tin} + \omega(h_{H2O, Tbleed} - h_{H2O, Tin}))m_{bleed\_total}}{1 + \omega} \\ - m_{comb\_dry}(h_{products, Texh} - h_{air, Tin} + \omega(h_{H2O, Texh} - h_{H2O, Tin})) \\ - m_{inject}(h_{H2O, Texh} - h_{H2O, Tinj}) \end{pmatrix}$$
$$(h_{air, Texh} - h_{air, Tin} + \omega(h_{H2O, Texh} - h_{H2O, Tin}))$$

The calculations continue across the sheet. If you change any of the blue values, the results will quickly update. Excel® handles the sequential computations and formulas automatically.

EXHAUST (after injection)		Add-In Macro Test	
mass flow [lbm/hr]		3,676,892	exhaust flow [lbm/hr]
2,671,045	Nitrogen	73.2115%	Nitrogen
206,277	Carbon Dioxide	3.5988%	Carbon Dioxide
226,816	Water	9.6640%	Water
527,081	Oxygen	12.6479%	Oxygen
0	Sulfur Dioxide	0.0000%	Sulfur Dioxide
45,645	Argon	0.8773%	Argon
3	Helium	0.0005%	Helium
3,676,867	total [lbm/hr]	**mole fractions**	
mass fractions		73.2093%	Nitrogen
72.6446%	Nitrogen	3.5988%	Carbon Dioxide
5.6101%	Carbon Dioxide	9.6668%	Water
6.1687%	Water	12.6472%	Oxygen
14.3351%	Oxygen	0.0000%	Sulfur Dioxide
0.0000%	Sulfur Dioxide	0.8773%	Argon
1.2414%	Argon	0.0005%	Helium
0.0001%	Helium	100.0000%	total
100.0000%	total		
molar flow [moles/hr]			
95,349	Nitrogen		
4,687	Carbon Dioxide		
12,590	Water		
16,472	Oxygen		
0	Sulfur Dioxide		
1,143	Argon		
1	Helium		
130,241	total [mole/hr]		
28.231	mol. wt.		

The fuel properties are on the Fuels tab of the spreadsheet. The elements and required oxygen are shown, along with the lower heating value. Note that the higher heating value is a meaningless term. Higher heating is based on water vapor in the combustion products existing in the liquid state, which isn't going to happen unless you're operating on Antarctica in a blizzard. People who insist on using higher heating value clearly don't understand thermodynamics.

	A	B	C	D	E	F	G	H	I	J	K
1											
2		**Properties of Fuels (from NIST)**									
3											
4	Name	Formula	LHV	C	H	N	O	S	Ar	He	O2
5	Methane	CH_4	21,512	1	4	0	0	0	0	0	-2.0
6	Ethane	C_2H_6	20,429	2	6	0	0	0	0	0	-3.5
7	Propane	C_3H_8	19,922	3	8	0	0	0	0	0	-5.0
8	IsoButane	C_4H_{10}	19,590	4	10	0	0	0	0	0	-6.5
9	Butane	C_4H_{10}	19,658	4	10	0	0	0	0	0	-6.5
10	IsoPentane	C_5H_{12}	19,456	5	12	0	0	0	0	0	-8.0
11	Pentane	C_5H_{12}	19,497	5	12	0	0	0	0	0	-8.0
12	Hexane	C_6H_{14}	19,393	6	14	0	0	0	0	0	-9.5
13	Nitrogen	N_2	0	0	0	2	0	0	0	0	0.0
14	Carbon Monoxide	CO	4,342	1	0	0	1	0	0	0	-0.5
15	Carbon Dioxide	CO_2	0	1	0	0	2	0	0	0	0.0
16	Water	H_2O	0	0	2	0	1	0	0	0	0.0
17	Hydrogen Sulfide	H_2S	6,534	0	2	0	0	1	0	0	-1.5
18	Hydrogen	H_2	51,567	0	2	0	0	0	0	0	-0.5
19	Helium	He	0	0	0	0	0	0	0	1	0.0
20	Oxygen	O_2	0	0	0	0	2	0	0	0	1.0
21	Argon	Ar	0	0	0	0	0	0	1	0	0.0
22	per mole			1	2	2	1	1	1	1	1

Heating values and heats of formation are intimately linked. They also have a reference point (pressure and temperature). These are not independent. Do not change them unless you know what you're doing. The spreadsheet will perform the calculations for you. It is interesting to note that the heating values change when you change the reference condition on the first tab. If they don't, then you won't get the same answer and your calculations will be incorrect. NASA had to burn and/or ionize and split these substances in order to build the tables. You don't read heating value from a gas chromatograph.

M	N	O	P	Q	R	S	T	U	V	W	X	Y	Z
						based on NASA table						net	NIST
		reactants						products				NASA	NASA
by mole		by mass			by mole			by mass				LHV	diff.
O2	Fuel	O2	Fuel	Hf	H2O	CO2	SO2	H2O	CO2	SO2	Hf		
2.0	1.0	79.96%	20.04%	-400.7	2.0	1.0	0.0	45.02%	54.98%	0.00%	-4711.5	21,508	0.02%
3.5	1.0	78.83%	21.17%	-253.8	3.0	2.0	0.0	38.04%	61.96%	0.00%	-4577.2	20,427	0.01%
5.0	1.0	78.39%	21.61%	-220.5	4.0	3.0	0.0	35.31%	64.69%	0.00%	-4524.5	19,920	0.01%
6.5	1.0	78.16%	21.84%	-218.1	5.0	4.0	0.0	33.85%	66.15%	0.00%	-4496.4	19,588	0.01%
6.5	1.0	78.16%	21.84%	-203.2	5.0	4.0	0.0	33.85%	66.15%	0.00%	-4496.4	19,656	0.01%
8.0	1.0	78.01%	21.99%	-201.4	6.0	5.0	0.0	32.94%	67.06%	0.00%	-4478.9	19,454	0.01%
8.0	1.0	78.01%	21.99%	-192.3	6.0	5.0	0.0	32.94%	67.06%	0.00%	-4478.9	19,496	0.01%
9.5	1.0	77.91%	22.09%	-183.9	7.0	6.0	0.0	32.32%	67.68%	0.00%	-4466.9	19,392	0.01%
0.5	1.0	36.35%	63.65%	-1079.8	0.0	1.0	0.0	0.00%	100.00%	0.00%	-3844.1	4,343	-0.03%
1.5	1.0	58.48%	41.52%	-107.9	1.0	0.0	1.0	21.95%	0.00%	78.05%	-2821.3	6,535	-0.02%
0.5	1.0	88.81%	11.19%	0.0	1.0	0.0	0.0	100.00%	0.00%	0.00%	-5771.0	51,574	-0.01%

While it is generally assumed that the composition of air is uniform all over the Earth and at all times, this is not exactly true. It is a fair assumption though and probably better than some of the others, including: gear and heat loss as well as generator efficiency. Air composition is on a separate tab and linked to the calculations. Note that the total must be 100%. I have seen many discrepancies on the Web where various constituencies don't sum to unity. In this spreadsheet, the trace amounts of noble gases are lumped together. These don't react and the specific heats are flat, so it's not a significant consideration.

	A	B
1	\multicolumn{2}{l}{**Constituents of Air from the CRC Handbook**}
2	\multicolumn{2}{l}{**Chemistry & Physics, 1997 Edition.**}
3	air constituents by mole	
4	78.08400%	Nitrogen
5	20.94760%	Oxygen
6	0.93400%	Argon
7	0.03140%	Carbon Dioxide
8	0.00182%	Neon
9	0.00020%	Methane
10	0.00052%	Helium
11	0.00011%	Krypton
12	0.00005%	Hydrogen
13	0.00001%	Xenon
14	99.99971%	total
15	consolidated & normalized by mole	
16	78.08422%	Nitrogen
17	20.94766%	Oxygen
18	0.93594%	Argon, Neon, Krypton & Xenon
19	0.03160%	Carbon Dioxide & Methane
20	0.00057%	Helium & Hydrogen
21	100.00000%	total
22	dry air constituents by mole	
23	78.0842%	Nitrogen
24	20.9477%	Oxygen
25	0.9359%	Argon
26	0.0316%	Carbon Dioxide
27	0.0006%	Helium
28	100.0000%	total
29	28.965	mol. wt.
30	dry air constituents by mass	
31	75.5192%	Nitrogen
32	23.1418%	Oxygen
33	1.2908%	Argon
34	0.0480%	Carbon Dioxide
35	0.0001%	Helium
36	100.0000%	total

HRSG Heat Balance

The HRSG heat balance is in the same spreadsheet on a different tab. Again, the user inputs are blue and calculations violet. An energy balance is needed to calculate the heat loss from the HRSG. Don't assume this value, measure everything else and then calculate it. Even if you took infrared pictures all over the HRSG and stitched them together into a 3D shell, you still couldn't accurately calculate the heat loss, because there's no way to accurately estimate the convective and radiative heat transfer coefficients.

	A	B	C	D
1				HRSG HEAT BALA
2	INPUTS		INPUTS Contd.	
3	operating pressures		pump efficiencies	
4	245	LP FW pres.[psia]	75%	recirc. pump eff.
5	0	Recirc. dP [psi]	70%	IP pump eff.
6	81	LP Drum pres.[psia]	65%	HP pump eff.
7	78	LP Steam pres.[psia]	CALCULATIONS	
8	615	IP FW pres.[psia]	flows	
9	564	IP Drum pres.[psia]	781,530	total FW flow [lbm/hr]
10	557	CRH pres.[psia]	781,530	FW+Recirc. [lbm/hr]
11	527	HRH pres.[psia]	99,580	IP FW [lbm/hr]
12	2240	HP FW pres.[psia]	682,570	HRH [lbm/hr]
13	2098	HP Drum pres.[psia]	644,470	HP FW [lbm/hr]
14	1942	HP Steam pres.[psia]	0	total blowdown [lbm/hr]
15	operating temperatures		1,405,970	steam entering [lbm/hr]
16	182	Stack temp.[°F]	1,405,970	steam leaving [lbm/hr]
17	111	LP FW Inlet temp.[°F]	3,689,966	stack [lbm/hr]
18	633	LP Steam temp.[°F]	temperatures	
19	727	CRH temp.[°F]	313	LP Drum [°F]
20	1057	HRH temp.[°F]	314	IP FW [°F]
21	1056	Main Steam temp.[°F]	480	IP Drum [°F]
22	60	DB Fuel temp.[°F]	319	HP FW [°F]
23	operating flows		643	HP Drum [°F]
24	0	recirc. flow [lbm/hr]	steam enthalpies	
25	0	LP Blowdown [lbm/hr]	79.6	LP FW [BTU/lbm]
26	37,480	LP Steam [lbm/hr]	283.0	LP Blowdown [BTU/lbm]
27	0	IP Blowdown [lbm/hr]	1347.4	LP Steam [BTU/lbm]
28	41,450	IP FW to FGHX [lbm/hr]	285.2	IP FW [BTU/lbm]
29	58,130	IP Steam [lbm/hr]	464.0	IP Blowdown [BTU/lbm]
30	624,440	CRH flow [lbm/hr]	1369.7	CRH Steam [BTU/lbm]
31	0	HP Blowdown [lbm/hr]	1550.4	HRH Steam [BTU/lbm]
32	644,470	HP Steam [lbm/hr]	292.9	HP FW [BTU/lbm]
33	13,100	DB Fuel flow [lbm/hr]	683.6	HP Blowdown [BTU/lbm]
34			1510.8	HP Steam [BTU/lbm]

Feedwater flow, as measured by a calibrated nozzle of ASME grade, is far more accurate than anything you could possibly do with the steam. Always measure the liquid flows and calculate the vapor flows. See Appendix I for more details.

NCE CALCULATIONS

CALCULATIONS Contd. fuel gas enthalpies & LHV		CALCULATIONS Contd. stack mass fractions	
-8.65	NASA enth.[BTU/lbm]	72.3897%	Nitrogen
-8.65	refd. enth.[BTU/lbm]	6.5440%	Carbon Dioxide
20,646	LHV [BTU/lbm]	6.9028%	Water
0	LHV adj. [BTU/lbm]	12.9264%	Oxygen
20,646	adj. LHV [BTU/lbm]	0.0000%	Sulfur Dioxide
DB fuel [mole/hr]		1.2370%	Argon
800	Carbon Dioxide	0.0001%	Helium
1,548	Water	100.0000%	total
4	Nitrogen	**stack enthalpies**	
-1,566	Oxygen	26.72	NASA enth.[BTU/lbm]
0	Sulfur Dioxide	26.72	refd. enth.[BTU/lbm]
0	Argon	**heat transfer to steam**	
0	Helium	47.52	to LP [MBTU/hr]
stack [mole/hr]		214.27	to IP+RH [MBTU/hr]
95,353	Nitrogen	922.37	to HP [MBTU/hr]
5,487	Carbon Dioxide	**heat input from pumps**	
14,139	Water	0.00	recirc. [MBTU/hr]
14,906	Oxygen	0.23	IP [MBTU/hr]
0	Sulfur Dioxide	6.35	HP [MBTU/hr]
1,143	Argon	**ENERGY BALANCE**	
1	Helium	1009.81	GT exh. [MBTU/hr]
131,028	total [mole/hr]	-0.11	Fuel (sensible) [MBTU/hr]
28.162	mol. wt.	270.47	Fuel (LHV) [MBTU/hr]
stack mole fractions		6.57	pumps [MBTU/hr]
72.7729%	Nitrogen	1286.74	total in [MBTU/hr]
4.1875%	Carbon Dioxide	1184.16	steam [MBTU/hr]
10.7906%	Water	98.59	stack [MBTU/hr]
11.3764%	Oxygen	1282.74	total out [MBTU/hr]
0.0000%	Sulfur Dioxide	3.99	loss [MBTU/hr]
0.8720%	Argon	0.34%	loss
0.0005%	Helium		
100.0000%	total		

Chapter 28. Correcting Test Results

We have already seen the most important corrections in Chapter 26. Some contracts call for feed water temperature and even ambient corrections, but these are rarely meaningful. To investigate this we return to the single-pressure model (CCPP1). We can use GateCycle™ to vary the condenser cooling water temperature, which will result in a change in condenser backpressure and condensate temperature that will be felt at the LP economizer. Consider the following table from CCPP1.xls:

Tccw	Tfw	Fms	Tms	power
°F	°F	lb/hr	°F	kW
40	70.2	73,256	805.165	10,111
50	78.4	73,256	805.165	10,058
60	87.2	73,256	805.165	9,932
70	96.3	73,256	805.165	9,733
80	106.0	73,256	805.165	9,476
90	116.0	73,256	805.165	9,178
100	125.9	73,256	805.165	8,862
110	136.0	73,256	805.165	8,521
120	146.1	73,256	805.165	8,175

Even though the CCW temperature varies from 40°F to 120° and the feed water inlet temperature rises accordingly from 70.2°F to 146.1°F, there is no change in main steam flow (73,256 lb/hr) or main steam temperature (805.165°F). The ST generator output does vary, but this is not a result of changes in HRSG operation and so there should be no correction. That is, we cannot correct for feed water temperature based on this model. The same thing happens with the three-pressure model (see CCPP3.xls):

Tccw	Tfw	Fms	Tms	power
°F	°F	lb/hr	°F	kW
40	104.7	690,191	789.2	115,447
50	107.9	690,191	789.2	115,367
60	112.3	690,206	789.2	115,131
70	117.8	690,191	789.2	114,520
80	124.9	690,191	789.2	113,226
90	132.8	690,191	789.2	110,903
100	140.9	690,190	789.2	107,705
110	149.3	690,203	789.2	104,055
120	157.8	690,192	789.2	100,109

This problematic result (no impact from some parameter that should result in at least a small change) is one of the flaws in an otherwise excellent tool. The whole convergence and iterative calculation process within the program is disappointing. The GateCycle™ program was written so that it obtains final results very quickly but at the cost of accuracy. While there are adjustable convergence criteria, changing these is pointless. They're already set to the limits and it simply won't do any better. There is also the opposite problem of yielding differences when there should be none. For instance, if you run ambient temperatures of 50, 60, 70, 80, 90 you will get slightly different results than for 90, 80, 70, 60, and 50. Such should never occur with any software, unless it's a Monte Carlo simulation using random numbers.

The impact of feed water temperature on main steam flow is not zero, but it is quite small. I have been required on more than one occasion to provide a non-zero correction. While this may be challenging with GateCycle™, it is possible and may require composing special macros and other advanced techniques. The figure below is for an actual plant, which I tested and will convey the magnitude of the appropriate correction.

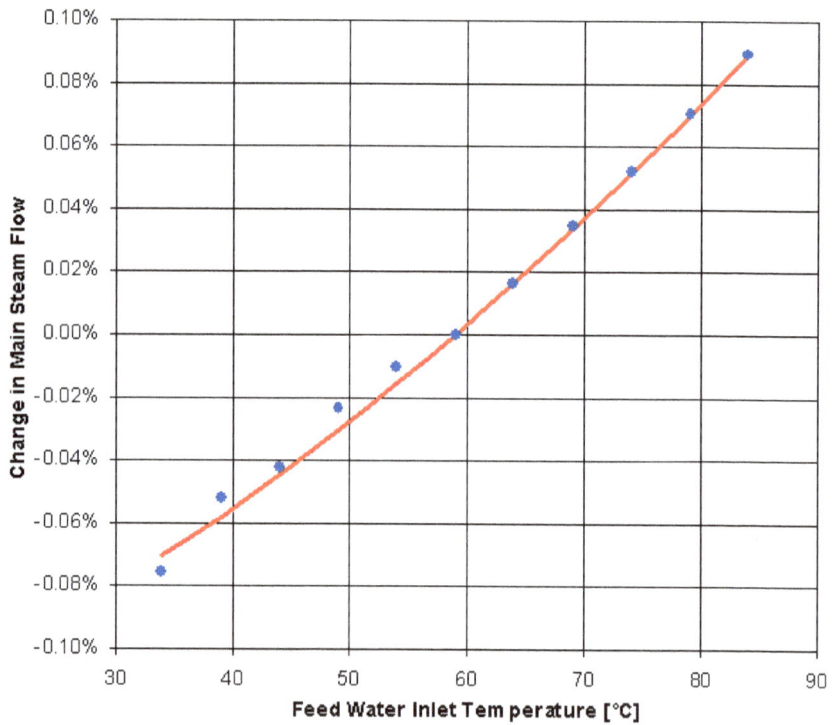

Single-Pressure HRSG Test

Typical summary test results are listed in the following table. The GT exhaust temperature and flow are determined by the PTC-22 test on a separate tab and brought in with linked formulas. The main steam flow and temperature are determined by the PTC-4.4 test and also brought in with links. The main steam flow is calculated from the measured feedwater flow and not a direct measurement of the steam, as this would be much less accurate. It is customary to average four one-hour test intervals.

SINGLE PRESSURE HRSG TEST							
INPUTS	units	Design	Test1	Test2	Test3	Test4	Avg.
GT exhaust flow	lb/hr	1,003,069	1,013,099	1,008,084	1,018,115	1,020,622	1,012,598
GT exhaust temp.	°F	850.2	848.2	852.6	847.4	848.9	849.5
main steam flow	lb/hr	73,275	74,008	73,641	74,374	74,557	73,971
main steam temp.	°F	805.1	834.4	809.0	777.2	834.2	812.0
CORRECTIONS	units	Design	Test1	Test2	Test3	Test4	Avg.
main steam flow	lb/hr	0	-191	-836	-324	-805	-431
main steam temp.	°F	0.0	1.4	-0.8	2.1	1.5	0.8
CORRECTED	units	Design	Test1	Test2	Test3	Test4	Avg.
main steam flow	lb/hr	73,275	73,817	72,805	74,050	73,752	73,540
main steam temp.	°F	805.1	835.8	808.2	779.3	835.7	812.8
TEST RESULT	units	Design	Test1	Test2	Test3	Test4	Avg.
main steam flow	-	N/A	PASS	FAIL	PASS	PASS	PASS
main steam temp.	-	N/A	PASS	PASS	FAIL	PASS	PASS

Note that the corrections are opposite performance calculations. Performance calculations answer the question: what is the expected flow or temperature under some particular conditions? Test corrections answer the question: what would the expected flow or temperature be at the guarantee conditions? If the as-tested GT exhaust flow is greater than expected, the measured steam flow is adjusted downward—likewise for the as-tested GT exhaust temperature. This process assures that one supplier is not penalized (or credited) for the product of another equipment supplier.

Overall Plant vs. Component Testing

As Dr. Keith Kirkpatrick[30] likes to say, "A 4 plus a 6 test does not equal a 46 test!" ASME PTC-4 and PTC-6 are the standard tests for conventional boilers and steam turbines, respectively. PTC-46 is a test for overall plant performance. Do not think that combining the first two will be an adequate replacement for the third. The same goes for a PTC-22 (gas turbine) and PTC-4.4 (HRSG) test. Each test is an evaluation of the equipment provided by a different supplier. The Engineering Procurement Contractor (EPC) should be responsible for everything, including the individual components (GT, HRSG, condenser, cooling tower, fire protection system, and everything else). Do not purchase a power plant thinking that all will be well and if it isn't, you'll just sue each and every supplier. You will have to operate it for years to come, long after the lawyers have retired to the beach.

In recent years, I have noticed a disturbing trend... Several gas turbine manufacturers have taken on the role of EPC, figuring that a lot of money has

---

[30] VP and Director of Testing for McHale Performance http://www.mchale.com/

been flowing into other pockets, which should have been theirs. I have yet to see this turn out well for the owner. What happens is common in human experience and not unique to power plant construction. One person who knows nothing of what another does, considers themselves oh so very much smarter and more hard-working, thinking they could do the other's job in their sleep. This is merely ignorance. It is shocking to see how much falls through the cracks when someone who has never done the work of an EPC takes on the roll.

This is a true story...
When we showed up to test one plant, everything was still in the crates. There wasn't so much as a parking lot, fence, guard shack, power line, sewer line, or porta potty! The inexperienced EPC thought the gas turbine supplier (a very reputable and experienced manufacturer) would set up their own equipment and not just ship it to the site and drop it off the back of a truck.

Three-Pressure HRSG Test

The following table shows typical test data and corrections for the three-pressure example (CCPP3.*):

THREE-PRESSURE HRSG TEST							
INPUTS	units	Design	Test1	Test2	Test3	Test4	Avg.
GT exhaust flow	lb/hr	3,456,937	3,455,844	3,459,590	3,454,478	3,456,881	3,456,746
GT exhaust temp.	°F	1127.0	1129.1	1127.6	1129.5	1127.6	1128.2
main steam flow	lb/hr	697,679	705,436	704,378	696,971	700,897	701,072
main steam temp.	°F	791.5	796.7	785.4	796.2	797.5	793.5
CORRECTIONS	units	Design	Test1	Test2	Test3	Test4	Avg.
main steam flow	lb/hr	0	-1,576	-798	-1,740	-479	-919
main steam temp.	°F	0.0	0.8	0.4	0.9	0.2	0.5
CORRECTED	units	Design	Test1	Test2	Test3	Test4	Avg.
main steam flow	lb/hr	697,679	703,860	703,580	695,231	700,418	700,153
main steam temp.	°F	791.5	797.5	785.8	797.1	797.7	793.9
TEST RESULT	units	Design	Test1	Test2	Test3	Test4	Avg.
main steam flow	-	N/A	PASS	PASS	FAIL	PASS	PASS
main steam temp.	-	N/A	PASS	FAIL	PASS	PASS	PASS

It is not common to fail one out of four test periods and pass the other three. This is only shown here for example. Typically, a HRSG either passes of fails miserably, without much in between. Most manufacturers build in a small margin, although some EPCs will presume this to be the case and may incorporate it or add a little of their own.

Chapter 29. Psychrometrics

Psychrometrics or psychrometry is the science of moist air, that is, the determination of physical and thermodynamic properties of air water vapor mixtures. The term comes from the Greek psuchron (ψυχρόν) meaning *cold* and metron (μέτρον) meaning *measurement*. This is where we begin the study of evaporative cooling.

The efficacy of evaporative cooling arises from the truly remarkable properties of the water molecule. This simple molecule appears in the solid, vapor, and liquid states, quite literally covering the surface of the Earth. Water has the highest latent heat of any naturally occurring substance and one of the highest specific heats. The only common molecule that comes close is ammonia, at a distant 60%. Next in line come the alcohols. Even so, ammonia and the alcohols don't exhibit all three states at nearly the same range of temperatures.

This remarkable behavior is due to the stress present in the unique structure of the water molecule. The hydrogen-oxygen bond is very strong, which draws them tightly together. The two hydrogen molecules repulse each other, which forces them apart. The end result is like a very tightly wound spring. This tug of war is illustrated in the following figure:

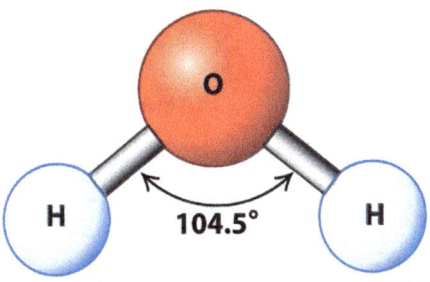

Mass Fraction of Water Vapor in Air by Temperature and Relative Humidity at 1 atm.											
°F	0%	10%	20%	30%	40%	50%	60%	70%	80%	90%	100%
0	0.00%	0.01%	0.02%	0.02%	0.03%	0.04%	0.05%	0.06%	0.06%	0.07%	0.08%
10	0.00%	0.01%	0.03%	0.04%	0.05%	0.07%	0.08%	0.09%	0.11%	0.12%	0.13%
20	0.00%	0.02%	0.04%	0.06%	0.09%	0.11%	0.13%	0.15%	0.17%	0.19%	0.21%
30	0.00%	0.03%	0.07%	0.10%	0.14%	0.17%	0.21%	0.24%	0.28%	0.31%	0.34%
40	0.00%	0.05%	0.10%	0.16%	0.21%	0.26%	0.31%	0.36%	0.41%	0.47%	0.52%
50	0.00%	0.08%	0.15%	0.23%	0.30%	0.38%	0.46%	0.53%	0.61%	0.68%	0.76%
60	0.00%	0.11%	0.22%	0.33%	0.44%	0.55%	0.66%	0.77%	0.88%	0.99%	1.10%
70	0.00%	0.15%	0.31%	0.46%	0.62%	0.78%	0.93%	1.09%	1.24%	1.40%	1.56%
80	0.00%	0.22%	0.43%	0.65%	0.87%	1.09%	1.30%	1.52%	1.74%	1.96%	2.19%
90	0.00%	0.30%	0.60%	0.90%	1.20%	1.50%	1.80%	2.11%	2.41%	2.72%	3.03%
100	0.00%	0.41%	0.81%	1.22%	1.63%	2.05%	2.46%	2.88%	3.30%	3.72%	4.14%
110	0.00%	0.54%	1.09%	1.64%	2.20%	2.76%	3.32%	3.89%	4.46%	5.04%	5.61%
120	0.00%	0.72%	1.45%	2.19%	2.94%	3.69%	4.44%	5.21%	5.98%	6.76%	7.54%

Because the latent heat of water is so large, evaporating even a little of it into the air caries with it a surprising amount of heat. The first table shows the mass fraction of water vapor in air for a range of temperature and relative humidity, which varies from 0 to 7.54%: The second table shows the energy fraction, that is, the fraction of the total energy (air plus water vapor) that is present in the water vapor alone:

°F	0%	10%	20%	30%	40%	50%	60%	70%	80%	90%	100%
\multicolumn{12}{	l	}{Energy Fraction of Water Vapor in Air by Temperature and Relative Humidity at 1 atm.}
0	1.1%	5.0%	9.6%	13.7%	17.5%	21.0%	24.1%	27.1%	29.8%	32.3%	34.6%
10	1.3%	5.5%	10.4%	14.9%	18.9%	22.6%	25.9%	29.0%	31.8%	34.5%	36.9%
20	0.9%	4.6%	8.7%	12.6%	16.1%	19.3%	22.3%	25.1%	27.7%	30.2%	32.4%
30	1.0%	4.9%	9.3%	13.3%	17.0%	20.4%	23.6%	26.5%	29.2%	31.7%	34.0%
40	1.3%	5.5%	10.4%	14.9%	18.9%	22.6%	26.0%	29.0%	31.9%	34.5%	37.0%
50	1.7%	6.4%	12.0%	17.1%	21.5%	25.6%	29.2%	32.5%	35.6%	38.3%	40.9%
60	2.3%	7.6%	14.2%	19.9%	24.9%	29.3%	33.3%	36.8%	40.0%	42.9%	45.6%
70	3.1%	9.1%	16.8%	23.3%	28.8%	33.7%	37.9%	41.7%	45.0%	48.0%	50.7%
80	4.3%	11.0%	19.9%	27.2%	33.3%	38.5%	43.0%	46.9%	50.3%	53.4%	56.1%
90	5.8%	13.2%	23.4%	31.5%	38.2%	43.7%	48.3%	52.3%	55.7%	58.7%	61.4%
100	7.8%	15.8%	27.4%	36.3%	43.3%	49.0%	53.7%	57.7%	61.1%	64.0%	66.6%
110	10.3%	18.7%	31.7%	41.3%	48.6%	54.4%	59.1%	63.0%	66.2%	69.0%	71.4%
120	13.2%	22.0%	36.4%	46.4%	53.9%	59.7%	64.3%	68.0%	71.1%	73.7%	75.9%

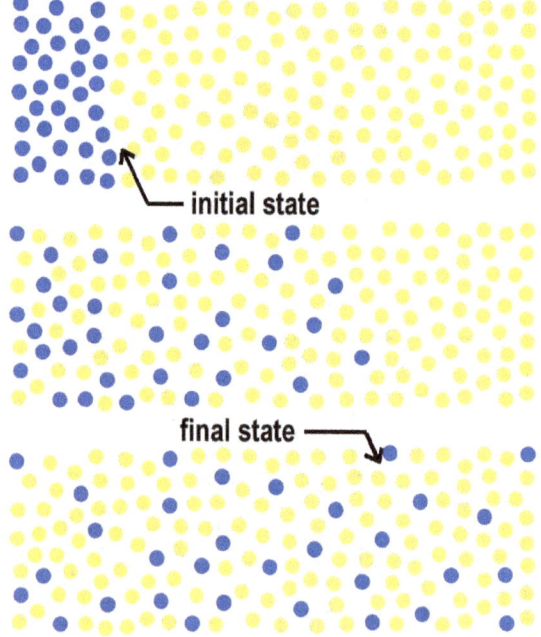

The states highlighted in blue have more energy in the water vapor than the air. Water vapor accounts for only 7.54% of the mass in the lower right corner, but 75.9% of the energy. Because moist air can contain this remarkable amount of energy, it can receive it through an evaporative process. In such a process, the water molecules must diffuse into the air. While the air-water diffusion coefficient isn't among the largest, it is large enough to readily facilitate natural and industrial processes. Diffusion is easily visualized by the spreading of dye in water. What happens on a molecular level is illustrated in the figure, where the blue dots represent water molecules and the yellow dots represent air molecules:

The properties of moist air are defined by the American Society of Heating, Refrigeration, and Air Conditioning Engineers (ASHRAE) in their Handbook of Fundamentals, which is published periodically. This is the de-facto worldwide standard. The reference point for these properties is as follows: 1) The enthalpy of dry air is zero at 0°F; and 2) The enthalpy of water vapor is zero at the triple point (i.e., 32.018°F). This difference in reference points for the dry air and water vapor is not a problem, as long as you are dealing with moist air at near atmospheric conditions and use this same formulation throughout.

This is not always the case, as we shall see in an application later on in the book. The problem comes when you mix moist air with some other gases or when the temperature is near or above 212°F/100°C. In such cases, a different formulation is necessary. The ASHRAE formulation is one of convenience. It's simply easier to handle calculations for applications that fit into this category. The ASHRAE formulation is not any less accurate for making this selection of reference conditions.

Dry-Bulb, Wet-Bulb, and Dew Point

The terms *dry* and *wet* are simply that and the term *bulb* arose in the era of mercury-in-glass thermometers. The dry-bulb is the ordinary air temperature that is cited in the evening weather report. The wet-bulb is found by attaching a wick (i.e., a little sock) on the end of a thermometer, keeping it wet, and gently blowing air across it. The *wet-bulb depression* is the difference between the dry-bulb and wet-bulb temperatures.

If the air is very dry, the water will quickly evaporate from the wick so that the temperature of the thermometer will drop. If the air is already very humid, the water will slowly evaporate, it at all, resulting in no drop in temperature. The recommended air velocity for wet-bulb measurement is 2.5 m/s (8.2 ft/sec). Such a device should be shielded from thermal radiation.

The purpose of wet-bulb measurement is to approximate the adiabatic (i.e., no heat transfer) saturation process. This means brining the air from it's ambient state to one of being saturated with water vapor without transferring any heat. In so much as the wet-bulb approximates the adiabatic saturation temperature, it is a measurement of the enthalpy. The temperature and enthalpy are independent properties; therefore, measurement of these two uniquely determines the state of the air, including the mass or mole fraction of the water vapor.

It is commonly observed that a hot dry day is not as uncomfortable as a hot humid one. For example, 110°F in Phoenix may be considered more comfortable than 90°F in New Orleans. This is because skin is porous and contains sweat glands, so that humans' perception of temperature is closer to the wet-bulb than the dry-bulb. It's the enthalpy, not the temperature, of air that determines how readily it will receive the heat your body must reject. A typical box-style psychrometer is shown in the next two figures:

Dew point is found by cooling the air to the point where condensation begins. For completely dry air, this will never happen, because there is no water vapor to condense. Dew point is simply a measure of the water vapor content and is not a measure of the enthalpy. Dew point is often included in the weather report, because it can be conveniently measured and the instruments don't require nearly the maintenance of a wet-bulber. Dew point is measured with a *chilled mirror hygrometer*, which is illustrated conceptually in this next figure:

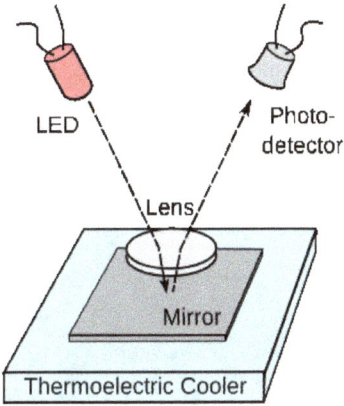

The actual device looks like the following:

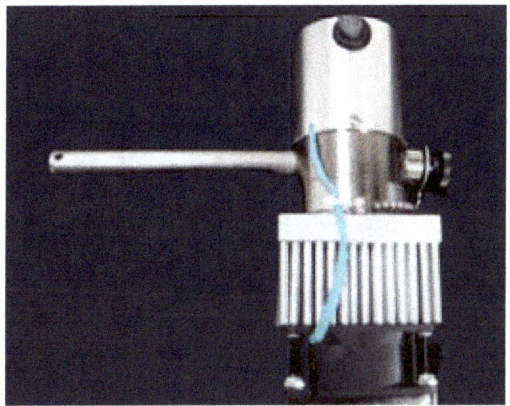

There is a little fan at the bottom (black barrel shape). On top of this is a finned aluminum heat sink. The mirror and photo detector are in the cylinder at the top. The thermocouple or RTD (resistance thermal detector) is also inside.

Under most conditions the dry-bulb will be higher than the wet-bulb, which will be higher than the dew point. The only time the three are the same is at saturation (i.e., 100% relative humidity). There are some odd cases, for instance, in a fog, the dew point is higher than the dry-bulb, as you would need to slightly heat the air to reach the point where condensation *first* begins.

Specific vs. Relative Humidity

The *specific humidity* is expressed as grains (7000 grains = 1 pound) per pound of dry air and is an antiquated term, though some instruments still report humidity in this form. The *humidity ratio* is simply the pounds (or kg) of water vapor per pound (or kg) of dry air. This is numerically equal to the specific humidity in grains divided by 7000. Humidity ratio is usually given the symbol "**W**" or "ω".

The most familiar term is *relative humidity*. This is the ratio of the actual water vapor content to the maximum (or saturation) value and is most often reported in percent. Completely dry air (devoid of any water vapor) has a relative humidity of 0%. When it's raining or foggy, the relative humidity is 100%. The amount of water vapor that is contained in air at saturation varies across orders of magnitude for the range of temperatures commonly present on the face of the Earth.

Chapter 30. Thermodynamic Properties of Moist Air

Precise determination of the thermodynamic properties of moist air in the West began with the work of Goff & Gratch, published in a series of papers between 1943 and 1949.[31,32] This work was continued by Hyland & Wexler between the years 1978 and 1983.[33,34,35] Nelson and Sauer further refined with their research between 1999 and 2001.[36] Most recently, Hermann, Kretzschmar, and Gatley have presented a more complete formulation coupled with the latest properties of steam.[37,38]

The basic formulation has been the same since the work of Goff & Gratch. The significant peculiarity of this approach is that the properties are all on a per pound of dry air basis, rather than on a per total (air plus water vapor basis). This facilitates some calculations, which was of greater concern in the 1940s than it is now. It doesn't matter, as long as you are consistent. For the most part these properties are used for atmospheric processes, which is not a problem. As mentioned in the previous chapter and in a subsequent example, it does matter at elevated temperatures (near or above 212°F/100°C).

Saturation Pressure and the Enhancement Factor

It is tempting to simply use the saturation pressure of steam along with Dalton's Law of Partial Pressures[39] to obtain values for the water vapor content in air at saturation, but this isn't accurate. The saturation pressure of steam is for

---

[31] Goff, J. A. and Gratch, S., "Thermodynamic Properties of Moist Air," *Heating, Piping & Air Conditioning*, pp. 334-348, 1945.

[32] Goff, J. A. and Gratch, S., "Low-Pressure Properties of Water from -160 to 212 F," ASHVE Trans., pp. 95-122, 1946.

[33] Hyland, R. W., Wexler, A., and Stewart, R., "Thermodynamic Properties of Dry Air, Moist Air and Water and SI Psychrometric Charts," ASHRAE RP-216 and RP-25, 1983.

[34] Hyland, R. W. and Wexler, A., "Formulations for the Thermodynamic Properties of the Saturated Phases of H2O from 173.15 K to 473.15 K," ASHRAE Trans., Vol. 89, pp. 500-519, 1983.

[35] Hyland, R. W. and Wexler, A., "Formulations for the Thermodynamic Properties of Dry Air from 173.15 K to 473.15 K, and of Saturated Moist Air from 173.15 K to 372.15 K, at Pressures to 5 MPa," ASHRAE Trans., Vol. 89, pp. 520-535, 1983.

[36] Nelson, H. F. and Sauer, H. J., "Formulation of High-Temperature Properties for Moist Air," *HVAC&R Research* Vol. 8, pp. 311-334, 2002.

[37] Herrmann, S., Kretzschmar, H.-J., and Gatley, D. P., "Thermodynamic Properties of Real Moist Air, Dry Air, Steam, Water, and Ice," *HVAC&R Research*, 2009.

[38] Herrmann, S., Kretzschmar, H.-J., and Gatley, D. P., "Thermodynamic Properties of Real Moist Air, Dry Air, Steam, Water, and Ice - Final Report," ASHRAE RP-1485, 2009.

[39] Dalton's Law of Partial Pressures states that, in a mixture of non-reacting gases, the total pressure is equal to the sum of the partial pressures exerted by each of the constituents and these individual contributions to the whole are each proportional to the mole fraction of that component.

H2O in the vapor state in equilibrium with H2O in the liquid state. This isn't the same as H2O in the vapor state in equilibrium with air in the gaseous state. An *enhancement factor*, *f*, is introduced to account for this difference. The enhancement factor is equal to the partial pressure of water vapor that should produce the observed content divided by the saturation pressure of steam at that same temperature. The values of *f* are close to unity and the symbol is appropriate, as this is simply a *fudge factor*.

Herrmann, Kretzschmar, and Gatley, present a very complicated equation in Section 3.4.2.1 of their report for *ln(f)* in terms of second and third virial coefficients[40], explaining that this arises from Henry's Law.[41] While this is interesting and may facilitate the calculation of *f* at elevated pressures without necessitating experiments, it is immaterial. All that is needed is to measure and tabulate the water content of air. An explanation as to why it is what it is, is not essential. This is the approach that Goff & Gratch and Hyland & Wexler took. The following table of f can be found in any edition of the *ASHRAE Handbook of Fundamentals*.

Enhancement Factor, f

T°F	Pressure [in.Hg]					
	10	15	20	25	30	32
0	1.0016	1.0025	1.0033	1.0040	1.0047	1.0051
20	1.0016	1.0024	1.0032	1.0039	1.0045	1.0048
40	1.0018	1.0025	1.0032	1.0038	1.0044	1.0047
60	1.0020	1.0026	1.0033	1.0039	1.0044	1.0047
80	1.0023	1.0029	1.0036	1.0041	1.0046	1.0049
100	1.0027	1.0033	1.0040	1.0045	1.0050	1.0053
120	1.0031	1.0037	1.0044	1.0050	1.0055	1.0057
140		1.0041	1.0048	1.0054	1.0059	1.0063

The variation of *f* with temperature at 1 atm. is shown in the following. For the purpose of calculations, the humidity ratio, **W**, is needed. This can be calculated from *f*, **Psat**, and the molecular weights by Equation 30.1:

$$W = \left(\frac{MW_{H2O}}{MW_{AIR}}\right)\left(\frac{fP_{SAT}}{P_{BARO} - fP_{SAT}}\right) \qquad (30.1)$$

The molecular weight of water is 18.01528 and of air is 28.9645. **Pbaro** is the barometric pressure. **Psat** is the saturation pressure of steam in the same units as the barometric pressure. The denominator in Equation 30.1 becomes zero when *fPsat=Pbaro*, which is why this formulation can't be used at elevated temperatures.

---

[40] The virial expansion of the equation of state was first proposed by Kamerlingh Onnes in 1901. It forms the basis for many developments in thermodynamics related to the properties of fluids. It is… $Z=PV/RT=1+B\rho+C\rho^2+\ldots$

[41] Henry's Law states that, at a constant temperature, the amount of a gas that will dissolve in a liquid is directly proportional to the partial pressure of that gas in equilibrium with that liquid. William Henry 1803.

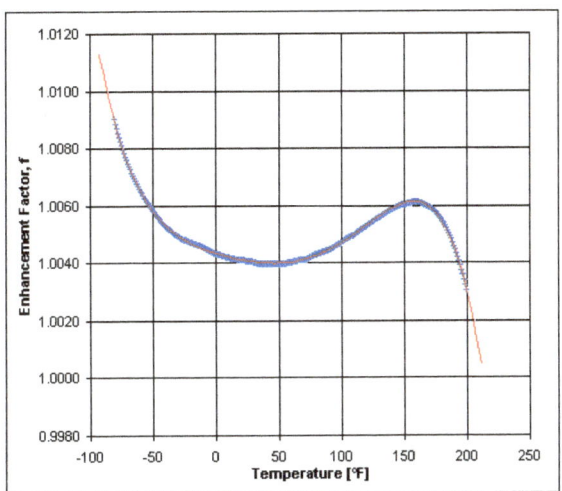

The variation of ***Psat*** and ***W*** with temperature at 1 atmosphere barometric pressure is shown in this next figure:

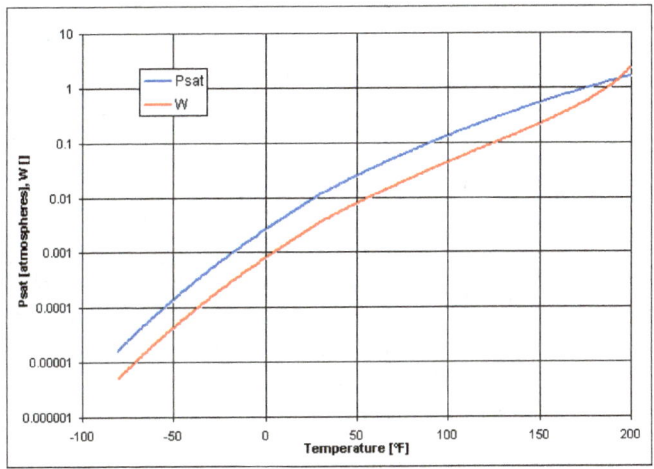

Enthalpy and Entropy

The enthalpy of moist air is also calculated on a per unit mass of dry air basis. Over the range of interest (-80°F to 212°F/-62°C to 100°C), the specific heat, ***Cp***, of air varies so little that 0.24 BTU/lbm/°F (1 kJ/kg/°C) is an adequate representation. The enthalpy of water vapor varies linearly over this range so that the following equation is adequate:

$$h_G = 1061 + 0.444T \quad (30.2)$$

Temperature is in degrees Fahrenheit and enthalpy is in BTU/lbm. Conversion to degrees Celsius and kJ/kg is trivial. It is very important to stress here that the appropriate enthalpy of water vapor is ***h_G*** and NOT ***h_{FG}*** (that is, the enthalpy of the saturated vapor, not the latent heat of vaporization or the difference between the vapor and liquid enthalpies). Although many will insist the latter is correct-this statement even appears in print-it is not true. Consider the case of steam at the critical point flowing into the room where you now sit. At the critical point $h_{FG}=0$. You should leave immediately. You'll be thoroughly cooked long before the temperature reaches 705°F, which it most assuredly will. Energy is entering the room, but $h_{FG}=0$.

The full equation for enthalpy is:

$$h = h_A + Wh_G = 0.24\,T + W(1061 + 0.444\,T) \tag{30.3}$$

The entropy is a little more complicated, because of the partial pressures:

$$s_A = 0.24\ln\left(\frac{T + 469.67}{469.67}\right) \tag{30.4}$$

$$s_G = 2.29688 - 0.003692687\,T + 0.0000055\,T^2 \tag{30.5}$$

$$s = s_A + Ws_G - R\ln\left(\frac{P}{14.696}\right) \tag{30.6}$$

A few values are listed in the following table:

T	Ws	Ps	f	Ha	Hg	Hs	Sa	Sg	Ss
-80	0.000004948	0.000236	1.009015	-19.221	1212.6	-19.215	-0.046	-11.7	-0.046
-60	0.00002121	0.001013	1.006593	-14.414	1037.2	-14.392	-0.034	0.7	-0.034
-40	0.00007929	0.003793	1.005314	-9.609	1046.8	-9.526	-0.022	2.0	-0.022
-20	0.0002632	0.012595	1.004719	-4.804	1052.4	-4.527	-0.011	2.1	-0.010
0	0.0007875	0.037671	1.004350	0.000	1060.3	0.835	0.000	2.2	0.002
20	0.0021531	0.102798	1.004101	4.804	1069.6	7.107	0.010	2.2	0.015
40	0.005216	0.24784	1.003959	9.609	1078.2	15.233	0.020	2.159	0.031
60	0.011087	0.52193	1.004025	14.415	1087.0	26.467	0.029	2.094	0.053
80	0.022340	1.03302	1.004285	19.222	1095.7	43.701	0.039	2.035	0.084
100	0.043219	1.93492	1.004706	24.031	1104.4	71.761	0.047	1.982	0.133
120	0.081560	3.45052	1.005246	28.842	1112.9	119.612	0.056	1.933	0.213
140	0.153538	5.88945	1.005830	33.656	1121.3	205.824	0.064	1.889	0.354
160	0.29945	9.6648	1.006120	38.474	1129.6	376.737	0.072	1.848	0.625
180	0.65911	15.3097	1.005510	43.295	1137.7	793.166	0.079	1.811	1.273
200	2.30454	23.4906	1.003033	48.121	1145.6	2688.205	0.087	1.776	4.180

In this table *a* denotes air, *g* denotes gas (or vapor), and *s* denotes saturation. Functions to calculate the properties of moist are may be found in Appendix K and are included in the on-line archive.

Chapter 31. Merkel's Equation

In 1925 Merkel proposed a theory relating the evaporation and sensible heat transfer occurring in a direct contact process such as cooling of water or humidification of air, to an air enthalpy difference.[42] Such a representation was suited (but not limited to) various types of cooling towers. The derivation was based on counterflow contact of water and air. In fact, there were six basic assumptions that were introduced at various points in the development to simplify the mathematics. This chapter is taken largely from a paper the author published with Al Feltzin in 1991, which I suggest for further reading.[43]

The model on which Merkel's theory was developed consists of a water droplet at temperature, T, surrounded by a thin air film (interface). He assumed that the air film is saturated and therefore is also at temperature T_A. Thus the film has humidity ratio, W_F, and enthalpy, h_F. Surrounding the air film is the bulk air mass at some lower temperature $T_A<T_F$, and humidity ratio, $W_A<W_F$, and an enthalpy $h_A<h_F$.

If a is the interfacial surface (ft²/ft³) and V is the contacting volume (ft³) then the interfacial surface area, $S=aV$ (ft) and the differential surface of the model droplet interfacial film is $dS=adV$ (surface element). The downward flux of water is given the symbol, L (lbs/hr/ft²) and the upward flux of air, G (lbs/hr/ft²). The two fluxes are opposing, that is, *counterflow*. The following figure illustrates this process and the variables:

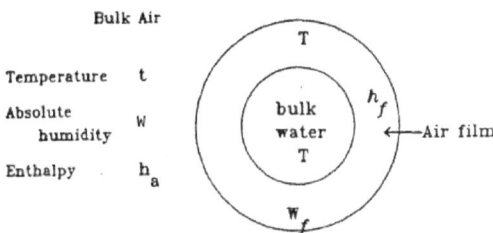

Heat is transferred from the water droplet to the bulk air through the interface by two means, sensible heat transfer (convection, due to a difference in temperature) and latent heat of evaporation (mass transfer by diffusion, due to a difference in concentration). Merkel assumed that the interface offers no resistance to heat transfer from the water droplet to the bulk air by either of these mechanisms. The sensible heat transfer rate by convection is given by:

$$dq_S = K_C(T_F - T_A)adV \qquad (31.1)$$

where K_C is the convective heat transfer coefficient, BTU/hr/ft²/F. The mass transfer rate is given by:

---

[42] Merkel, F. Verdunstungskuehlung, V.D.I. Forschungsarbeiten, No. 275, Berlin, 1925.
[43] Feltzin, A. E, and Benton, D. J., "A More Nearly Exact Representation of Cooling Tower Theory," CTI Technical Paper Number TP91-02.

$$dL = K_M (P_F - P_A) a dV \qquad (31.2)$$

where K_M is a diffusional mass transfer coefficient. P_F and P_A are the partial pressures of water vapor in the interfacial film at temperature, T_F, and bulk air at temperature, T_A, respectively. Next, Merkel assumed that the partial pressure of water vapor is proportional to humidity, that is:

$$P_F \propto W_F$$
$$P_A \propto W_A \qquad (31.3)$$

which can be substituted into Equation 31.2 to obtain:

$$dL = K_M (W_F - W_A) a dV \qquad (31.4)$$

The evaporative (latent) heat transfer rate due to diffusional mass transfer is given by:

$$dq_L = \lambda dL = \lambda K_M (W_F - W_A) a dV \qquad (31.5)$$

where λ is the latent heat of vaporization. The total transfer rate is then given by:

$$dq_{total} = [K_C (T_F - T_A) + \lambda K_M (W_F - W_A) a dV] \qquad (31.6)$$

At this point in the derivation, the concept of humid heat, C_S, (the heat capacity of an air-water vapor mixture) is usually introduced. By addition and subtraction of a term $C_S(T_W-T_A)$ to the right hand side of Equation 31.6 and algebraic manipulation (see Merkel Appendix J), we arrive at the following:

$$dq_{total} = K_M \left\{ (C_S T_F + \lambda W_F) - (C_S T_A + \lambda W_A) + C_S (T_F - T_A) \left[\frac{K_C - 1}{C_S K_M} \right] \right\} a dV \qquad (31.7)$$

Merkel also assumed that the Lewis Number of unity, that is:

$$L_E = \frac{K_C}{C_P K_M} = 1 \qquad (31.8)$$

This assumption causes the last term in Equation 31.7 to vanish. The terms $C_S T_F + \lambda W_F$ and $C_S T_A + \lambda W_A$ are close to, but not exactly, equal to Equation 30.3, reducing the heat transfer to:

$$dq = K(h_S - h_A) a dV \qquad (31.9)$$

The subscript on K has been dropped, as there is now only one. The subscript F (film) has been replaced with S (saturation), in keeping with the previous assumption. Conservation of energy requires that:

$$dq = d(LC_{PW} T_W) = d(Gh_A) \qquad (31.10)$$

where C_{PW} is the constant pressure specific heat of water in the liquid state. These two equations can be combined to form:

$$dq = C_{PW}(L dT_W - T_W dL) = G dh_A = K(h_F - h_A) a dV \qquad (31.11)$$

Dividing by $h_F - h_A$ and integrating yields:

$$KaV = C_{PW}\left\{\int_{T_{OUT}}^{T_{IN}} \frac{LdT}{(h_F - h_A)} + \int_{T_{OUT}}^{T_{IN}} \frac{TdL}{(h_F - h_A)}\right\} \qquad (31.12)$$

$$KaV = G\int_{h_{ain}}^{h_{aout}} \frac{dh_A}{(h_F - h_A)} \qquad (31.13)$$

Merkel went on to assume that the portion of the water evaporated was insignificant to the whole, or $dL=0$. After making this assumption, Equations 31.12 and 31.13 become:

$$\frac{KaV}{L} = C_{PW}\int_{T_{OUT}}^{T_{IN}} \frac{dT}{h_F - h_A} = \frac{G}{L}\int_{h_{ain}}^{h_{aout}} \frac{dh}{(h_F - h_A)} \qquad (31.14)$$

The third term in Equation 31.14 is of no further interest at this point, but will be utilized later. This leaves the classic form of Merkel's Equation:

$$\frac{KaV}{L} = C_{PW}\int_{T_{OUT}}^{T_{IN}} \frac{dT}{h_F - h_A} \qquad (31.15)$$

This is a nonlinear equation with no analytical solution and must be solved numerically. Merkel chose to use the 4-point Chebyshev method for its simplicity. In spite of these assumptions and simplifications, Merkel's Equation served the cooling tower industry well for decades.

Over the years, several approaches have been devised in an attempt to compensate for several of the above assumptions and approximations. Mickley[44] introduced temperature and humidity gradient, heat and mass transfer coefficients from water to interfacial film and from film to air. Baker & Mart[45] developed a hot water correction factor that reduced the scatter in test data. Temperature correction factors have been applied to demand curves in some cases and fill (characteristic) curves in others, this has resulted in a confusing situation at best. Effects of air temperature, barometric pressure, and salinity on fill characteristic KaV/L have been discussed by LeFevre.[46] Baker & Shryock discuss the method of integration at some length.[47]

---

[44] Mickley, H. S., *Chemical Engineering Progress*, Vol. 45, p. 739, 1949.
[45] Baker, R. and Mart, L T., *Refrigeration Engineering*, p. 965, 1952.
[46] LeFevre, M., CTI Paper 58, New Orleans, Louisiana, 1985.
[47] Baker & Shryock, *Journal of Heat Transfer*, Appendix B and Table lB, April, 1961.

Chapter 32. Counterflow Demand Curves

The Cooling Tower Institute[48] using Equation 31.15 and the 4-point Chebyshev, expanded on earlier KaV/L vs. L/G demand curves such as those utilized by Foster Wheeler Corporation[49] and J.F. Pritchard Company.[50] The CTI curves had the advantages of being computer generated and computer drawn, and made what had been very limited published data much more widely available and over a much broader selection of ranges and approaches. These curves served the industry for over 20 years. The following is typical:

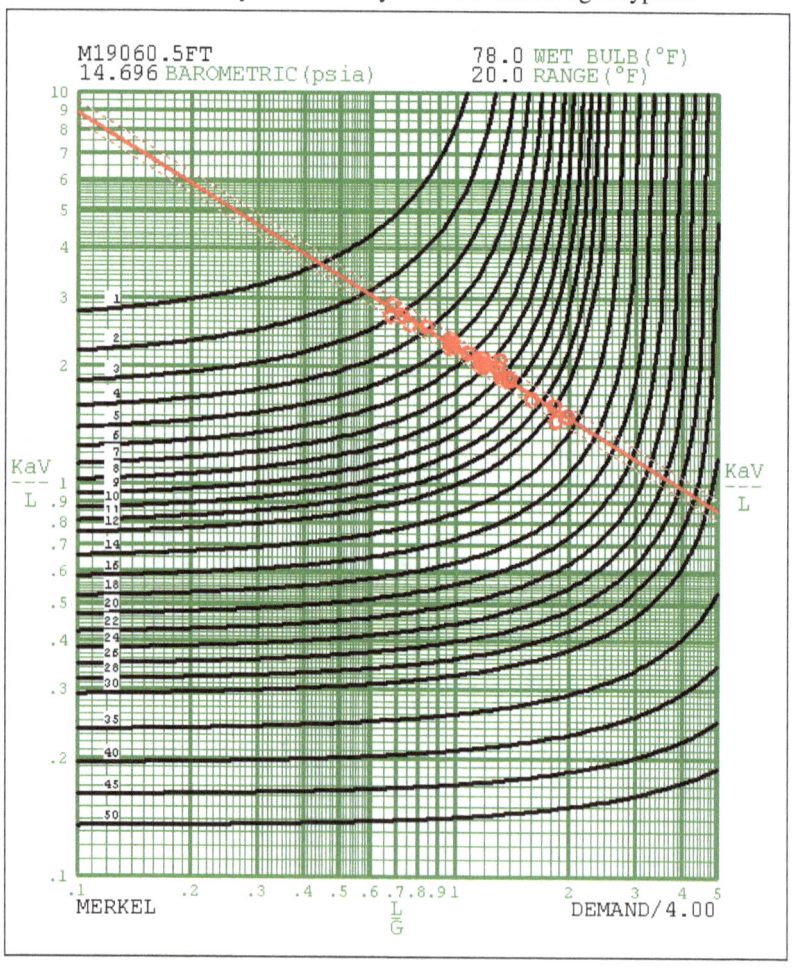

---

[48] *CTI Blue Book of Counterflow Demand Cooling Curves*, CTI, Houston, Texas, 1967.
[49] *Cooling Tower Performance: Bulletin CT432*, Foster Wheeler Corporation, 1943.
[50] *Counterflow Cooling Tower Performance*, F. Pritchard Company of California, 1957.

These *demand* curves were used to obtain graphical solutions of cooling tower performance, which was a tedious process. The loose-leaf notebook filled with such curves was an expensive item and hard to come by. The same curves can now easily be generated with an Excel® spreadsheet, as illustrated below. This spreadsheet is included in the on-line archive. A program to generate the professional grade curves is available free on-line from the author's web site.

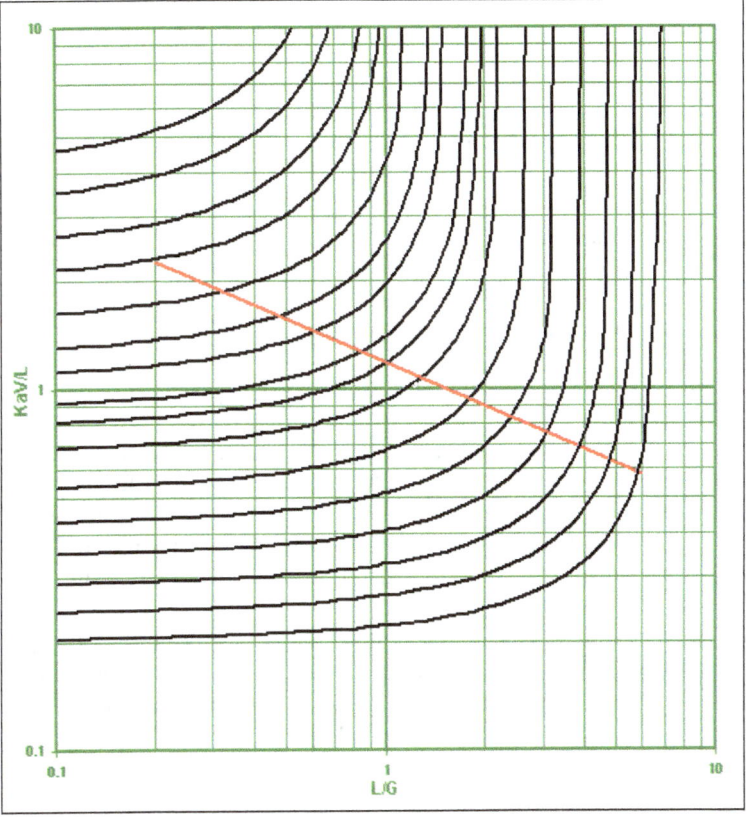

In these figures, the black curves are constant *approach* (cold water temperature minus ambient wet-bulb). The horizontal axis is the ratio of the water to air mass flows. The downward-sloping red line in the previous figure is the *supply* line, or the dimensionless mass transfer capacity provided by the packing material. Generating performance curves for a cooling tower using this graphical method was beyond tedious. This can also readily be done with an Excel® spreadsheet as illustrated in this next figure. This is on another tab in the same spreadsheet that comes with the archive.

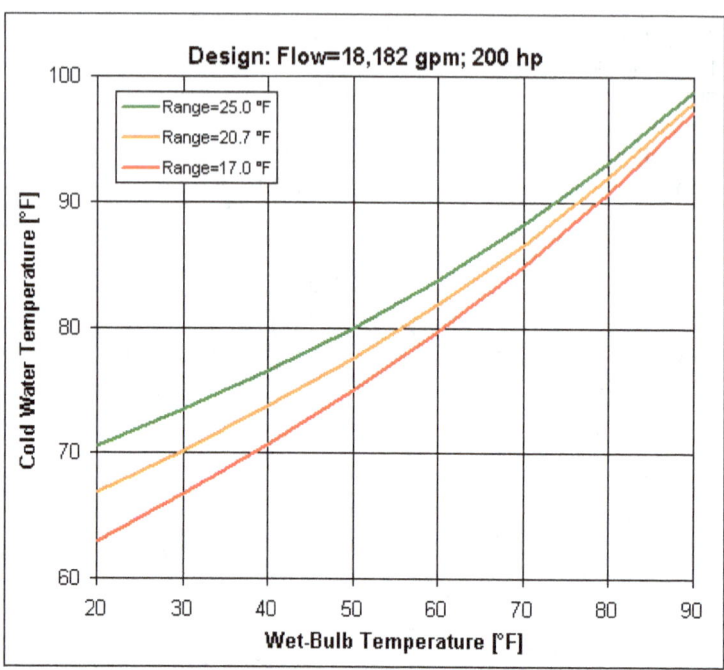

The *range* is the hot (entering) water temperature minus the cold (leaving) water temperature. Only the wet-bulb is needed here, because the airflow is provided by fans, making this a mechanical-draft cooling tower. The same formula and calculation can be used for natural-draft (buoyancy-induced airflow), as will be illustrated subsequently.

So far we have only considered the counterflow arrangement. There are operational and economic reasons for designing cooling towers with a crossflow arrangement. Solving the crossflow mass transfer equation requires a different approach, one that was quite a challenge in the 1960s with limited computational options.

Chapter 33. Crossflow Demand Curves

The counterflow problem can be expressed as a single integral equation. Although this is nonlinear, the solution is simple. This is not true for a crossflow problem. Not only must the integration be carried out over two dimensions, there are more computational problems that arise, none the least of which is a very strong tendency to over-shoot, which leads to instability and erroneous solutions. The crossflow calculations are also more complex to visualize. The following spreadsheet is arranged to help visualize the calculations:

Crossflow Cooling Tower Demand Calculations

user inputs								Tw				
50 Twb			Ta				120.0	120.0	120.0	120.0	120.0
120 Thw	50	102	113	117	119	119	92.6	107.7	114.5	117.5	118.9
2.00 L/G	50	78	97	108	113	117	81.7	95.2	104.3	110.6	114.7
1.38 KaY	50	70	86	97	105	111	74.7	87.1	96.2	103.2	108.5
calculations	50	65	79	89	98	104	69.8	81.1	89.9	96.9	102.6
83.2 Tcw	50	62	74	83	91	98	66.0	76.4	84.7	91.7	97.4
36.8 range									Hw		
33.2 approach			Ha				119.7	119.7	119.7	119.7	119.7
	20.3	75.2	99.7	110.7	115.7	117.9	59.6	87.2	103.7	112.2	116.3
	20.3	42.0	66.9	87.2	101.0	109.4	45.6	63.7	80.1	93.9	104.2
	20.3	34.3	50.5	66.8	81.8	94.2	38.4	52.1	65.3	77.7	88.9
	20.3	30.3	42.3	55.0	67.5	79.3	33.9	44.9	55.8	66.4	76.6
	20.3	27.8	37.2	47.5	57.9	68.2	30.8	39.9	49.1	58.3	67.3

Each box represents one computational cell in a 5x5 grid. The air flows in from the left and the water flows down from the top. There are several variables associated with each cell. These are arranged in groups and distinguished by colors.

The light blue cells under the heading "Ta" are the air temperature (wet-bulb). This group is 5x6, because for each cell there is an inlet and exit value, requiring one extra column. The light magenta cells under the heading "Tw" are the water temperature. This group is 6x5, because for each cell there is an inlet and exit value, requiring one extra row. The green and yellow cells contain the enthalpies for the air and water respectively. Hw is the enthalpy of saturated air at the water temperature, not the enthalpy of the water.

The difference between the air and water enthalpies in each cell is the potential driving the mass transfer in that cell. Each cell contains 1/25th of the packing material and surface area for the exchange. The product of the potential, the transfer coefficient, and surface area is equal to the heat transfer, Q. The exit water temperatures in the magenta section and exit air enthalpies in the green section are calculated from Q for each cell. The enthalpies in the yellow section corresponding to the exit water conditions are calculated from the water temperatures in the magenta section.

There are two tabs in the spreadsheet. One contains a 5x5 cell calculation and the other a 14x14. You can compare these two in order to get an idea of how many cells it takes to obtain a sufficiently accurate representation of this problem. The spreadsheet is included in the on-line archive and can be used to get a better understanding of the computer modeling process. The same free program that generated the counterflow demand curves will also generate crossflow curves, as illustrated in the next figure:

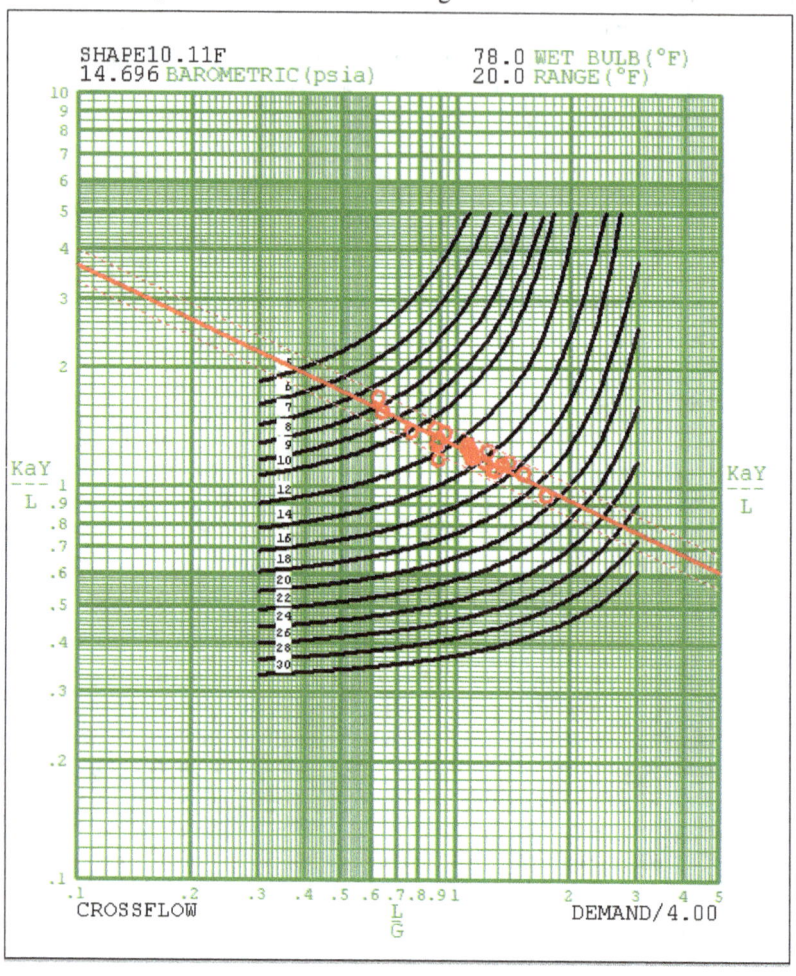

Chapter 34. A Closer Look at Crossflow

Modeling the heat and mass transfer process in an evaporative cooling tower is an exercise in applied mathematics wherein a system of differential equations is solved. There are many methods for solving such equations. The content of this chapter is a condensation of several publications that are available free on-line and which I suggest for further reading.[51,52,53,54] We begin with a schematic of the computational grid, in this case for a crossflow arrangement:

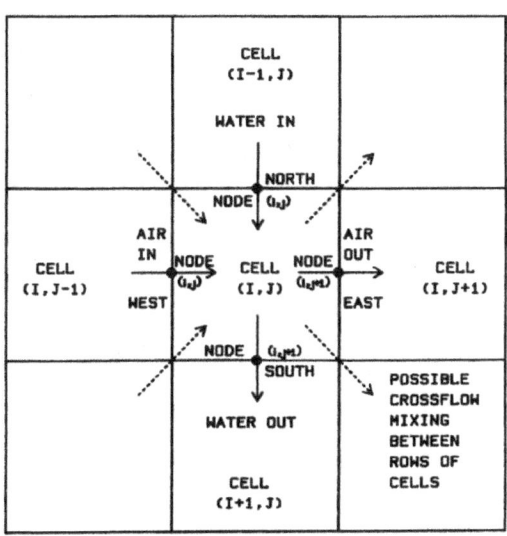

As in the multi-colored spreadsheet from the previous chapter, water enters the cells from the top and exits at the bottom. The air enters the cells from the left and exits to the right. At this point we will ignore redistribution of the airflow through the fill. We will also assume that both the air and water flux (mass flow per unit area) is uniform throughout the fill.

As the inlet conditions are required for the calculations and the exit conditions are the result, we simply step through the fill horizontally, shift down one row of cells, step through horizontally again, and continue this process to

---

[51] D. J. Benton, "A Numerical Simulation of Heat Transfer in Evaporative Cooling Towers," TVA Report No. WR28-1-900-110, September, 1983.

[52] D. J. Benton, "Development of the Finite-Integral Method," TVA Report No. WR2B-2-900-148, December, 1984.

[53] D. J. Benton, "Computer Simulation of Hybrid Fill in Crossflow Mechanical-Induced-Draft Cooling Towers," Proceedings of the ASME Winter Annual Meeting, New Orleans, Louisiana, December 9-14, 1984.

[54] D. J. Benton and W. R. Waldrop, "Computer Simulation of Transport Phenomena in Evaporative Cooling Towers," ASME Journal of Engineering for Gas Turbines and Power, Vol. 110(2), pp. 190-196, April, 1988.

the end. The simplest method for solving the differential equation is the 4th order Runge-Kutta method. Implementation is easy, requiring only a few lines of code:

```
void Cell(double X,double Y,double Ha,double*dHa,double
    Tw,double*dTw)
{
double Hw,Q;
Hw=fHtwb(Pbaro,Tw);
Q=KaY*(Hw-Ha);
dHa[0]=Q*LG;
dTw[0]=-Q;
}
```

The dependent variables, X and Y, are the position within the fill and aren't used in this case. There are two independent variables: 1) the enthalpy of the air and 2) the temperature of the water. These are passed in Ha and Tw, respectively. The differentials, dHa/dX and dTw/dY are returned in dHa and dTw, respectively. The heat transfer, Q, is out of the water and into the air. The ratio of the water to airflow rates, L/G, is applied to the air side.

As Merkel's approximations are assumed, the potential for driving the mass transfer is proportional to the enthalpy difference. The air enthalpies are initialized left face and water temperatures on the top face. The exiting water temperatures are averaged, yielding the cold water temperature, range, and approach. The following output is typical for a 9x9 grid:

```
grid: 9x9, KaY/L=1.38, L/G=2
************ Water Temps ************ avg. exit 89.0
120.0 120.0 120.0 120.0 120.0 120.0 120.0 120.0 120.0
109.2 111.6 113.4 114.9 116.0 116.8 117.5 118.0 118.5
101.3 104.9 107.8 110.1 111.9 113.4 114.6 115.6 116.4
 95.0  99.3 102.8 105.7 108.1 110.0 111.7 113.0 114.2
 89.9  94.6  98.5 101.8 104.5 106.8 108.7 110.4 111.8
 85.6  90.6  94.7  98.2 101.2 103.7 105.9 107.8 109.4
 82.0  87.1  91.4  95.0  98.1 100.8 103.2 105.2 107.0
 78.9  84.0  88.4  92.1  95.3  98.2 100.6 102.8 104.7
 76.2  81.3  85.7  89.4  92.7  95.6  98.2 100.5 102.5
 73.8  78.8  83.2  87.0  90.3  93.3  95.9  98.3 100.4
*************** Air Temps *************** avg. exit 105.4
 50.0  78.2  91.9 100.0 105.3 109.0 111.6 113.6 115.1 116.2
 50.0  72.5  85.2  93.5  99.4 103.7 107.0 109.5 111.7 113.2
 50.0  68.5  80.2  88.4  94.5  99.1 102.8 105.7 108.1 110.1
 50.0  65.6  76.3  84.2  90.3  95.1  99.0 102.2 104.9 107.2
 50.0  63.3  73.1  80.7  86.7  91.6  95.6  99.0 101.9 104.3
 50.0  61.6  70.5  77.7  83.5  88.4  92.5  96.0  99.0 101.6
 50.0  60.1  68.4  75.1  80.8  85.6  89.7  93.3  96.3  99.1
 50.0  58.9  66.5  72.9  78.4  83.1  87.2  90.7  93.9  96.7
 50.0  57.9  64.9  70.9  76.2  80.8  84.8  88.4  91.6  94.8
range=31.0, approach=39.0
```

It is important in developing such numerical solutions to differential equations to determine the appropriate grid size. The program has been written so that the grid is dynamically allocated. The number of horizontal and vertical cells was stepped from 5 to 25 by 5s and the exiting water temperature calculated to produce the following table:

Nx/Ny	5	10	15	20	25
5	89.64	89.22	89.12	89.07	89.04
10	89.21	88.97	88.92	88.89	88.88
15	89.11	88.92	88.87	88.85	88.84
20	89.06	88.89	88.85	88.84	88.83
25	89.04	88.88	88.84	88.83	88.82

The solution converges quickly with only a modest number of cells. The Runge-Kutta method for 2D may be found in Appendix L. The complete code is included in the on-line archive. This is the computational method used in the TEFERI code[55], which may be obtained through the Electric Power Research Institute (EPRI).

---

[55] Bourillot, C., "TEFERI, Numerical Model for Calculating the Performance of an Evaporative Cooling Tower," EPRI CS-3212-SR, August, 1983. Originally published by Electicite de France and translated by J. A. Bartz.

Chapter 35. Natural-Draft Cooling Towers

Predicting and presenting the performance of natural-draft cooling towers is considerably more complicated than mechanical-draft because there is an additional variable: ambient relative humidity. As a result of the Merkel approximation, the performance of mechanical-draft cooling towers can be accurately represented as depending on flow, range, and ambient wet-bulb. This is not true for natural-draft towers, as the heat and mass transfer impact the density, which drives the flow of air through the tower. The impact of ambient relative humidity may even be greater than that of range or flow in some circumstances.

The classical representation of natural-draft cooling tower performance is illustrated in the next two figures. The first shows the cold-water temperature vs. wet-bulb temperature for several values of ambient relative humidity. The second is the *range correction*, that is, the adjustment in cold-water temperature for range, also vs. ambient wet-bulb temperature.

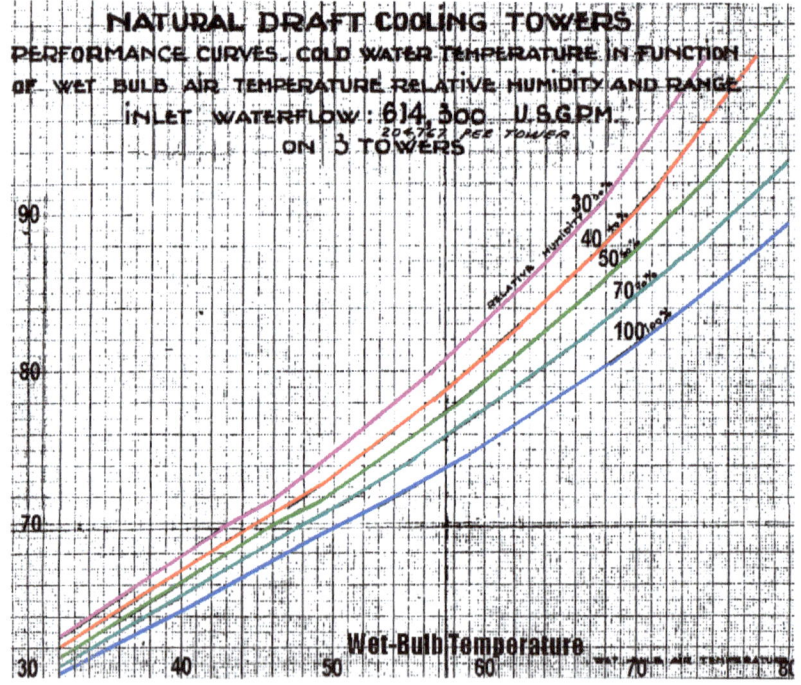

These curves are quite old (from the late 1960s), and are included here because such information is generally proprietary in nature. These curves, however, are for a U.S. government project (TVA's Paradise Steam Plant) and can be obtained through the Freedom of Information Act (FOIA).

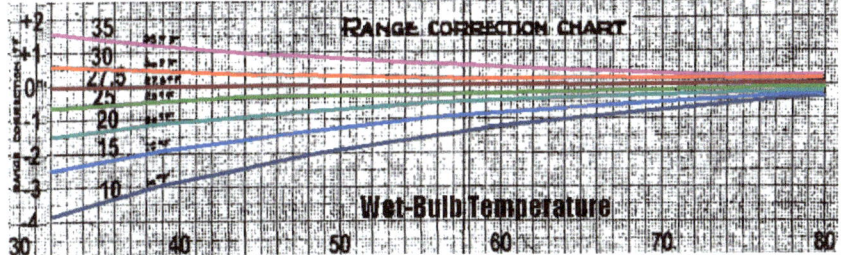

For this tower there are five sets of curves at five different water flows. These curves have been digitized and the range correction applied to each curve, resulting in over 15,000 points for cold-water temperature as a function of flow, range, ambient relative humidity, and ambient wet-bulb temperature. All of this data is included as an Excel® spreadsheet in the on-line archive. This relationship can be accurately approximated ($R^2=0.9969$) by a second-order multi-variable regression, as shown in this next figure:

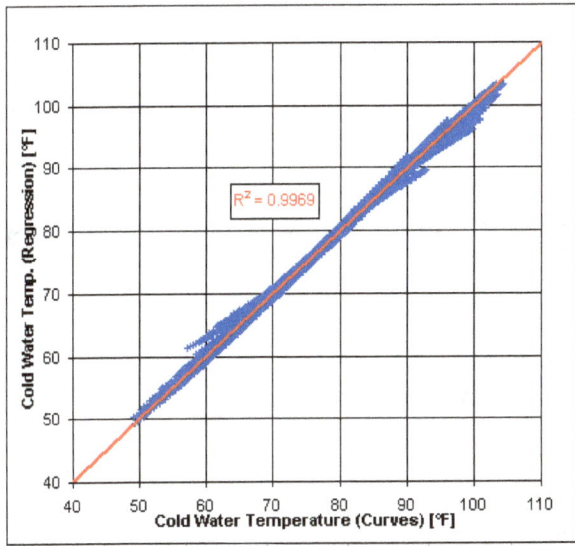

While such a regression is fully adequate for approximating the performance and correcting test data, it doesn't provide a meaningful visual presentation. A more creative presentation results from plotting curves on three sides of a cube, as illustrated in the next figure. Such a plot is usually drawn on a flat sheet of paper and then folded as indicated.

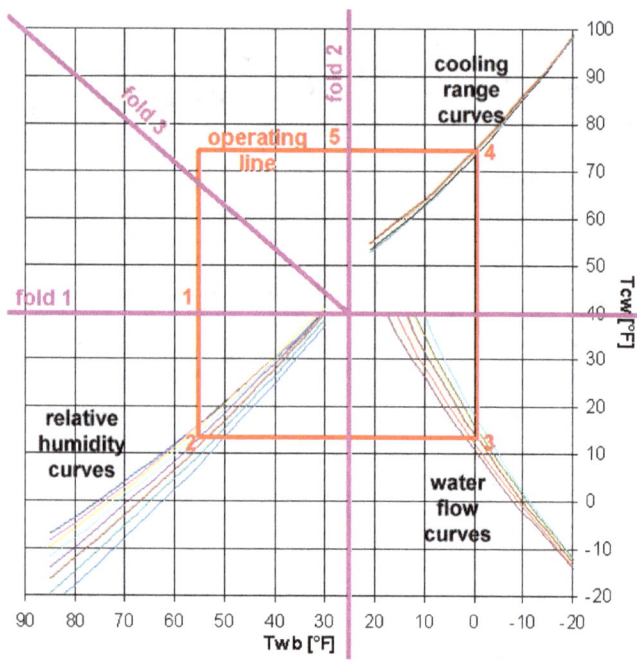

By folding along the lines as indicated above:

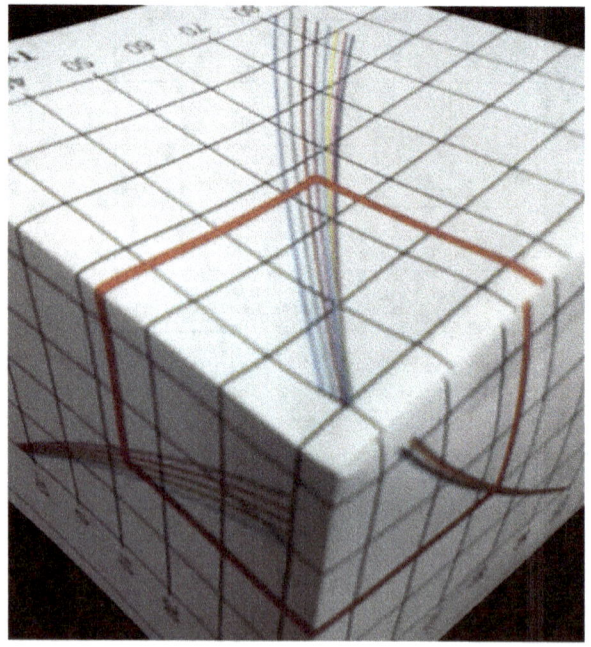

These curves representing the cooling tower performance are obtained by introducing two additional coordinates. The first will be given the symbol, *X*, and is the distance between points 1 and 2 on the previous graph. The second will be given the symbol, *Y*, and is the distance between points 2 and 3. These coordinates do not correspond to thermodynamic variables; they merely facilitate creation of the graph.

A typical operating point is found beginning at point 1 (the ambient wet-bulb) and drawing a vertical line down to point 2, where this intersects the ambient relative humidity curve. From point 2, a line is drawn horizontally to point 3, where it intersects the flow curve. From point 3, a line is drawn vertically to point 4, where it intersects the range curve. The cold-water temperature is then read from the scale at the right.

Each set of curves can be expressed in terms of a simple second order expansion. There are three sets of curves (relative humidity, flow, and range) and seven variables (wet-bulb, relative humidity, *X*, flow, *Y*, range, and cold-water temperature). Their inter-relation is given by the following equations:

$$X = a_1 + a_2 RH + a_3 Twb + a_4 RH^2 + a_5 RH \cdot Twb + a_6 Twb^2 \qquad (35.1)$$

$$Y = b_1 + b_2 flow + b_3 X + b_4 flow^2 + b_5 flow \cdot X + b_6 X^2 \qquad (35.2)$$

$$Tcw = c_1 + c_2 range + c_3 Y + c_4 range^2 + c_5 range \cdot Y + c_6 Y^2 \qquad (35.3)$$

There are eighteen constants in these three equations, but these are not all free. In order for the temperature scale of the three sets of curves to be the same, the coefficient b_3 must equal one. The coefficients a_1 and c_1 control the horizontal and vertical gaps, or the distance from each set of curves to the corner. A regression performed on these performance curves produces the following coefficients:

a1	70	b1	-1.0061	c1	70
a2	13.77244	b2	-11.3479	c2	3.433356
a3	-1.38779	b3	1	c3	-1.26823
a4	-14.8557	b4	1.756232	c4	-0.2345
a5	0.316842	b5	-0.24559	c5	0.131833
a6	0.002149	b6	-0.00385	c6	0.006758

This provides a convenient representation of an existing tower design, but doesn't explain how the tower is designed in the first place. This particular tower was designed by Marcel LeFevre using Merkel's method. This is evident from a plot of KaV/L vs. L/G. All 15,000+ performance points lie along a downward-sloping straight line on a log-log scale, as shown in this next figure:

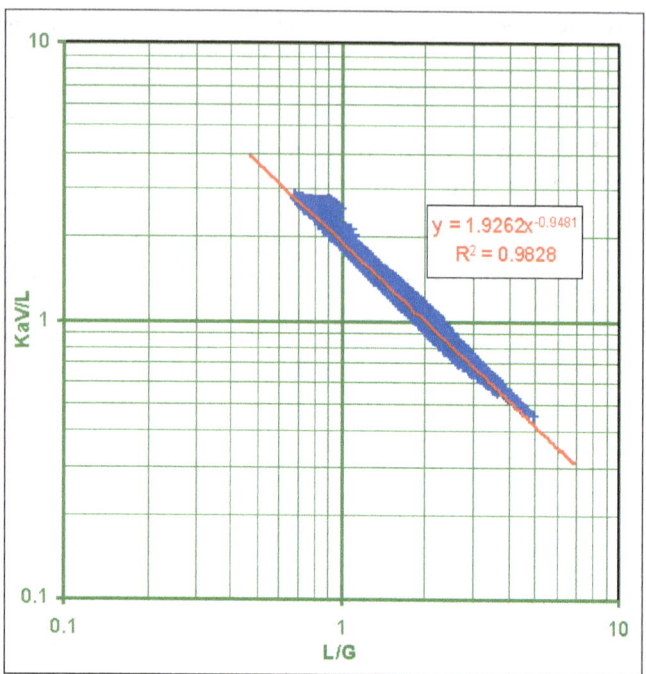

The slope and intercept are typical for the type of fill (packing) used (cement asbestos fiber board), which has since been replaced. The slope (-0.9481) is determined from small-scale test data and the intercept (1.9262) is calculated from test data and the depth of the fill (or packing), which in this design was 6 to 12 feet with an average of 11.1 feet.

This graph is still only part of the design, as the ratio of water to air mass flow rates, L/G, must be known in order to make this calculation. The exiting air from an evaporative cooling tower is saturated and at a temperature between the entering and exiting water temperatures. As a general rule, one-half may be assumed, although test data have shown a range of values between 35% and 65%. L/G can be calculated from an energy balance, by recognizing that:

$$G(h_{a,out} - h_{a,in}) = L \cdot C_{PW} \cdot \Delta T_W \tag{35.4}$$

Equation 35.4 can be rearranged to form:

$$\frac{L}{G} = \frac{(h_{a,out} - h_{a,in})}{C_{PW} \cdot \Delta T_W} \tag{35.5}$$

The flow of air through a natural-draft cooling tower is driven by the density difference, which acts on the air from the top of the drift eliminators (directly above the fill and spray nozzles) to the exit of the stack. This zone is shown in the following two figures. The first figure was taken without hot water on the tower and the second was taken with.

The assumed saturated condition is clear from this second figure, which is a dense fog.

While there is some dispute in the literature concerning the appropriate boundary condition at the top of the fill and at the exit of the tower, the author has personally measured the conditions at both locations and can assure the reader that the pressure is uniform radially at both elevations. More specifically, the pressure is constant and below atmospheric above the drift eliminators. It is also constant and equal to ambient at the exit of the tower, which is typically 400 to 500 feet above the ground.

> There is no sub-atmospheric pressure zone sitting on top of a natural-draft cooling tower sucking air out of it arising from the thermal plume. The pressure at the top is most assuredly atmospheric. If it weren't, there would be an inward radial flow, consistent with the Navier-Stokes equation.[56]

Recognizing that the flow of air through a natural-draft cooling tower is driven by the density difference, we can propose the following relationship:

$$\left(\frac{L}{G}\right)\Delta T_W \propto H\left(\frac{\Delta \rho}{\rho}\right) \qquad (35.6)$$

In Equation 35.6, **H** is the height over which the density difference acts (i.e., the total height minus the inlet) and **Δρ/ρ** is the relative difference in density. The validity of this proposed relationship is illustrated in the next figure:

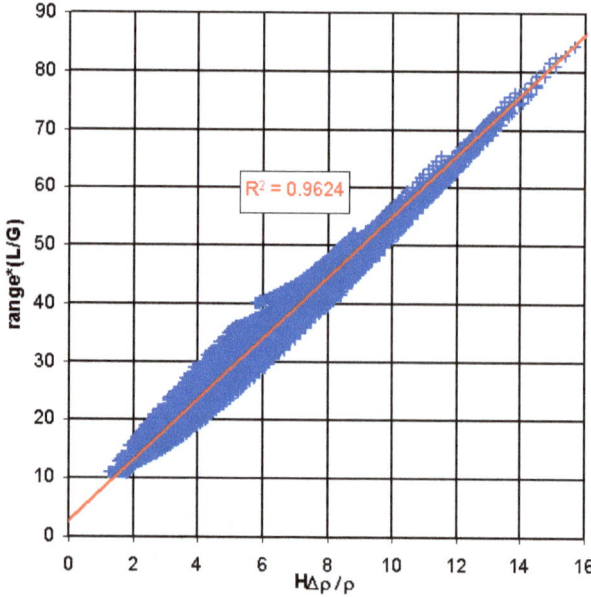

Given this linear relationship, plus the line on the previous graph of KaV/L vs. L/G, we see that only three constants (1.9262, 0.9481, 5.6337) are required to calculate all of the performance curves for this or any other natural-draft cooling tower. The slope of the **H** vs. **Δρ/ρ** line can be determined from a single

---

[56] Note that Bernoulli's equation only applies along a streamline, not across streamlines.

solution balancing the pressure drop through the fill and drift eliminators with the buoyancy to obtain a single operating point.

Dry vs. Wet Cooling

There were two natural-draft cooling towers at Schmehausen, Germany (pictured below). The smaller one on the left (made of concrete) is wet and the larger one on the right (made of steel) is a dry. Both were in operation when this picture was taken.

It cost ten times as much to construct the large dry cooling tower than the smaller wet cooling tower next to it. Besides being much smaller and cheaper, the wet cooling tower provided *far better performance*, hence the advantage of evaporative cooling. This dry cooling tower was demolished in 1991, because the materials were worth more as scrap than it took to build another wet tower.

Chapter 36. Heat Transfer from Falling Droplets

Evaporative heat transfer from falling droplets is an important industrial process. The *spray zone* in a cooling tower is only one such application. In this chapter we will consider this process from both a computational and empirical basis. This chapter is largely a condensation of a paper presented by the author and Bob Rehberg in 1986, which contains analysis and laboratory data.[57]

In a classic reference, Lowe & Christie[58] discuss a method developed by Nottage & Boelter[59] for calculating the mass transfer characteristic and pressure drop of droplets falling in a counterflow arrangement. The mass transfer characteristic, Ka/L'', of the droplets is defined as:

$$\frac{Ka}{L''} = \left(\frac{C_{PL}}{C_{PG}}\right)\left(\frac{\kappa_G}{\kappa_L}\right)\left(\frac{\Psi}{s_G}\right) \qquad (36.1)$$

and has units of 1/length. In this equation C_{PL} and C_{PG} are the liquid and vapor specific heats of water, respectively, κ_G and κ_L are the thermal conductivity of the liquid and vapor, respectively, and S_G is the velocity of air.

The *dynamical function*, Ψ, defined and tabulated by Nottage and Boelter, is a function of droplet diameter and has units of 1/time. The relative velocity of the falling droplet, s_A, is the difference between the free falling or terminal velocity, s_F, and the air velocity, s_G. The terminal velocity is a function of droplet diameter and is tabulated in the reference by Nottage and Boelter. The air velocity through the tower is defined as:

$$S_G = \frac{G''V_A}{3600} \qquad (36.2)$$

which has the units of length/time. The pressure drop per length l is given by:

$$dP = \left(\frac{\rho g L''}{3600 s_A}\right) dl \qquad (36.3)$$

and has the units of force/length². The pressure drop, as expressed in the number of velocity heads, dN, lost per foot is:

---

[57] Benton, D. J. and R. L. Rehberg, "Mass Transfer and Pressure Drop in Sprays Falling in a Freestream at Various Angles," International Association for Hydraulic Research, Fifth Cooling Tower Workshop, Palo Alto, California, September 29 - October 3, 1986.

[58] Lowe, H. J., and D. G. Christie, "Heat Transfer and Pressure Drop Data on Cooling Tower Packings and Model Studies of the Resistance of Natural Draught Towers to Airflow," International Division of Heat Transfer, Part V, pp. 333-950, American Society of Mechanical Engineers, New York, 1961.

[59] Nottage, H. B., and L. M. K. Boelter, "Dynamic and Thermal Behavior of Water Drops in Evaporative Cooling Processes," Transactlons of the American Society of Heating and Ventilating Engineers, Vol. 46, pp. 41-79, 1940.

$$\frac{dN}{dl} = \left(\frac{2}{s_G^2}\right)\frac{dP}{dl} \qquad (36.4)$$

and has the units of heads/length. Substitution of Equation 36.3 into 36.4 yields:

$$\left(\frac{1}{L''}\right)\frac{dN}{dl} = \frac{2g}{3600\,\rho\,s_A s_G^2} \qquad (36.5)$$

and has the units of length×time/mass. This development by Lowe & Christie has been used for years to model the heat transfer in the rain zone of cooling towers. A more general approach based on first principles and dimensionless correlations will be developed here that could also be used for other substances and orientations.

Theoretical Development

The mass transfer is determined from the Sherwood number, Sh:

$$Sh = \frac{K_A d}{\rho_A D_{AW} A_V} \qquad (36.6)$$

where κ_A is the thermal conductivity, d is the droplet diameter, ρ_A is the density of the air, D_{AW} is the air/water diffusion coefficient, and A_V is the surface area per unit volume. For flow over a sphere, the Sherwood number is given by:[60]

$$Sh = 2 + \left(0.4\,\mathrm{Re}^{\frac{1}{2}} + 0.6\,\mathrm{Re}^{\frac{2}{3}}\right) Sc^{0.4} \qquad (36.7)$$

where Re is the Reynolds number and Sc is the Schmidt number.

$$\mathrm{Re} = \frac{V_R \rho_A d}{\mu_A} \qquad (36.8)$$

where V_R is the velocity of the air relative to the droplet and μ_A is the dynamic viscosity of air.

$$Sc = \frac{\mu_A}{\rho_A D_{AW}} \qquad (36.9)$$

The surface area per unit volume, A_V, is given by:

$$A_V = \frac{A_S \phi_D}{V_I} \qquad (36.10)$$

where A_S is the surface area per droplet in length²/drop, φ_D is the flux of droplets in drops/length²/time, and V_I is the instantaneous droplet velocity. The relative velocity, V_R, in Equation 36.8 is given by:

---

[60] Krieth, F. and W. Z. Black, Basic Heat Transfer, Harper and Rowe, 1980.

$$V_R = \sqrt{(u_A - u_D)^2 + (v_A - v_D)^2} \qquad (36.11)$$

The heat and mass transfer for a single drop is computed from the Lagrangian Equations of Motion.[61] The horizontal and vertical position of a particle is given by:

$$x = \int_0^t u\,dt \qquad (36.12)$$

$$y = \int_0^t v\,dt \qquad (36.13)$$

The horizontal and vertical velocities are given by:

$$u = \int_0^t \frac{F_X}{m_D}\,dt \qquad (36.14)$$

$$v = \int_0^t \frac{F_Y}{m_D}\,dt \qquad (36.15)$$

The horizontal and vertical impulse are given by:[62]

$$J_X = \int_0^t F_X\,dt \qquad (36.16)$$

$$J_Y = \int_0^t F_Y\,dt \qquad (36.17)$$

and have the units of force×time. The horizontal force, F_X, is entirely due to drag and is given by:

$$F_X = \left(\frac{\pi d^2}{4}\right) C_D \frac{\rho_A |u_A - u_D|(u_A - u_D)}{2} \qquad (36.18)$$

The vertical force, F_Y, is due to drag and gravity and is given by:

$$F_Y = \left(\frac{\pi d^2}{4}\right) C_D \frac{\rho_A |v_A - v_D|(v_A - v_D)}{2} - (\rho_D - \rho_A) g \frac{\pi d^3}{6} \qquad (36.19)$$

---

[61] The Lagrangian approach follows the perspective of individual particles along their trajectories. This is in contrast to the continuum approach of Euler, most often used for fluids. This approach was introduced by the Italian-French mathematician and astronomer Joseph-Louis Lagrange in 1788.

[62] In classical mechanics, impulse is the integral of force over the time interval for which it acts. Impulse applied to an object produces a change in its linear momentum. Linear momentum is equal to mass times velocity. Force, impulse, and linear momentum are all vectors.

The difference in densities in the last term accounts for the buoyancy. The pressure drop across the zone is found by integrating Equations 36.12 through 36.17 along with Equation 36.4 to obtain:

$$NV' = \frac{\phi_D \sqrt{I_X^2 + I_Y^2}}{H\left(\dfrac{\rho_D V_D^2}{2}\right)} \qquad (36.20)$$

where **H** is the height of the zone. The force balance on a single drop is illustrated in the following figure:

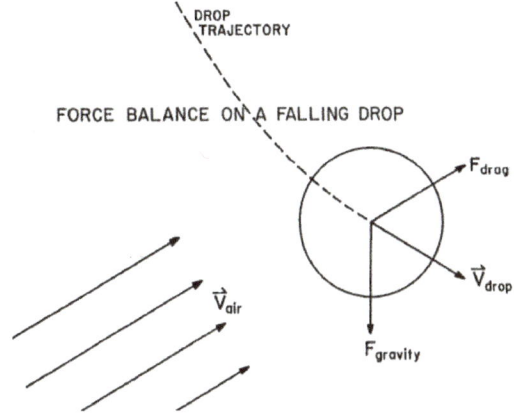

The program used to solve these equations and produce the graphs in the next section is listed in Appendix M.

Comparison to Experimental Results

There have been several experimental studies done on the mass transfer and pressure drop from single droplets falling through air. One study conducted by Missimer & Brackett at the TVA Engineering Laboratory in Norris, Tennessee, provides experimental data specifically on the transport phenomena occurring within the rain zone of a counter flow cooling tower. The test facility was designed to simulate the rain zone in the air flows toward the center of the tower in a crossflow configuration.[63]

The test section is 15 feet long, 6 feet high and 4 feet wide with a maximum water flux of 2085 lbm/hr/ft² and a maximum air flow rate of 15,000 ft³/min. The parameters measured during the experiments were the air and water flow rates, the hot water temperature entering the test section, the cold water temperature leaving the test section and the inlet wet bulb temperature. Given this data, the mass transfer characteristic, Ka/L", can be computed. Also, the

---

[63] Missimer, J. R., and C. A. Brackett, "Results of Model Tests of Heat Transfer in the Rain Zone of a Counterflow Natural-Draft Cooling Tower," TVA Report WR28-1-85-115, May 1985.

pressure drop across the test section was measured, thereby giving the number of velocity heads lost per foot of air travel, *NV'*.

The results of the initial 12 tests conducted at the TVA Rain Zone Facility are given in the following figure. In addition to the experimental data, the results of the present numerical analysis are also shown.

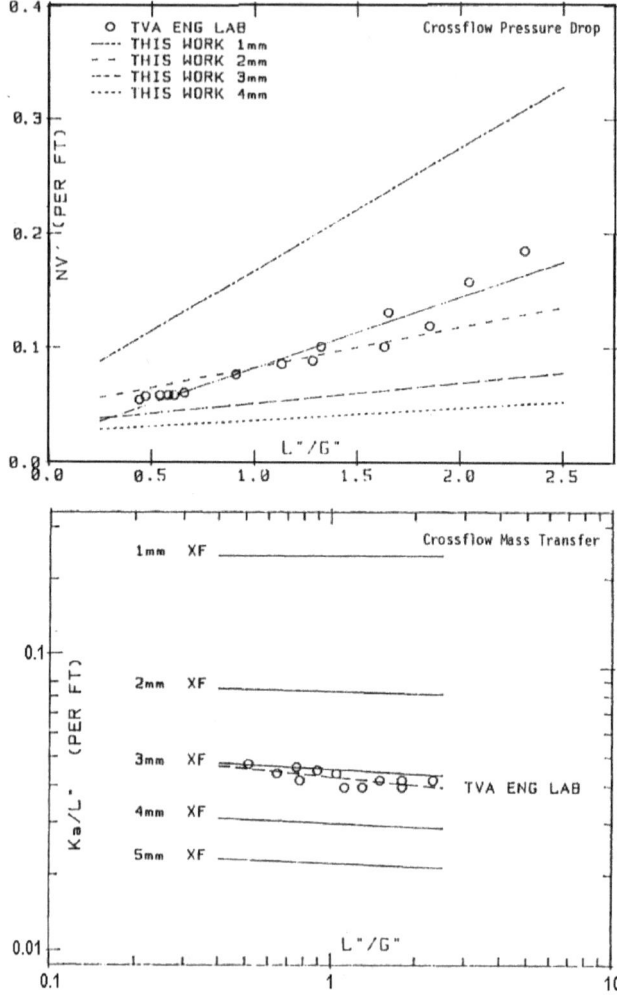

The theoretical curves show the dramatic effect of droplet size on the mass transfer. From this figure, it can be seen that a difference in droplet diameter of a few millimeters can produce an order of magnitude difference in the mass transfer characteristic. Accurate droplet sizes were not experimentally determined during the TVA rain zone testing. However, the theoretical curve for

3-mm droplets compare well with the experimental data, which is roughly the correct droplet diameter.

No experimental data for counterflow mass transfer were available. Two sets of curves are shown in the figure: the solid curves were computed based on the analysis discussed by Lowe and Christie, while the dashed curves were computed using the equations detailed above. As with the case of crossflow mass transfer, the dramatic effect of droplet size can be seen.

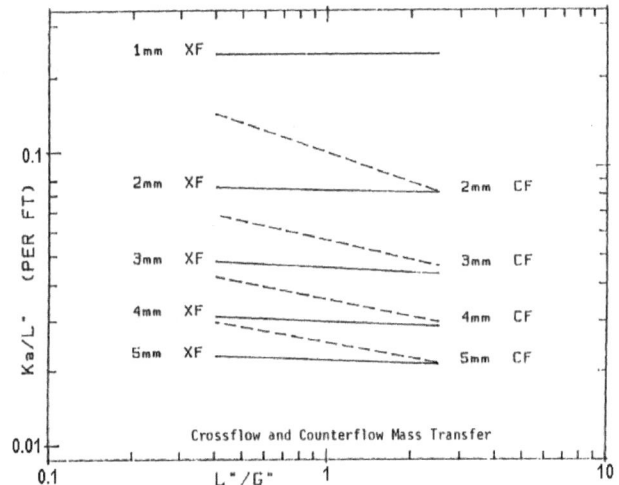

The figure also shows the decrease in mass transfer coefficient with increasing water loading (increasing L''/G''). An interesting comparison of crossflow and counterflow rain zone mass transfer is also shown in this figure. As expected, for a given droplet diameter the mass transfer coefficient is higher for the counterflow orientation.

Since the rain zone in a natural-draft counterflow cooling tower is a combination of both counterflow and crossflow, for a given droplet size, the mass transfer characteristic for the rain zone would presumably lie between the counterflow and crossflow curves. Therefore, the curves shown in this figure provide a theoretical upper and lower bound on the mass transfer actually occurring in the rain zone. Typical computed droplet trajectories are shown in this next figure:

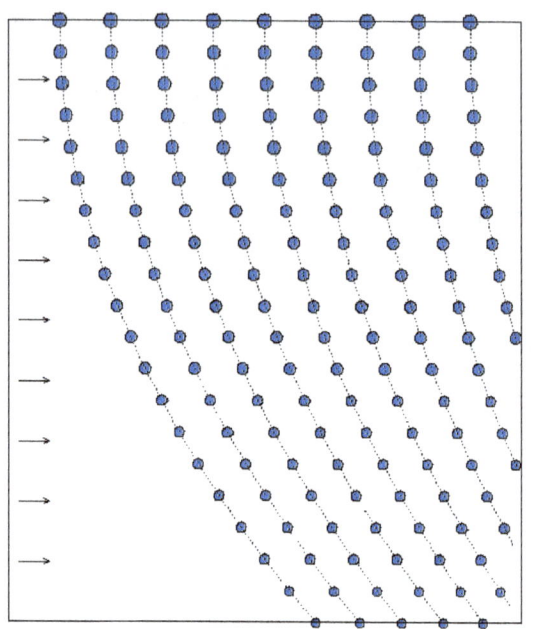

Chapter 37. Spray Cooling Systems

Large-scale open spray cooling systems were very popular in the early 1970s for their simplicity, reliability, and low cost. Such systems were even proposed and accepted as the ultimate heat sink for several nuclear plants. An Internet search will produce several U.S. Nuclear Regulatory Commission (NRC) documents containing descriptions and analyses for these systems from this era, but very few more recent publications.

By the late 1970s and early 1980s these systems came under attack by environmental activists, concerned that they were releasing harmful substances and/or radiation into the atmosphere, thereby endangering the public. While there was no basis for these concerns, spray systems are more visible, giving the false impression that something more significant is happening.

The spray cooling pond at the Rostov Nuclear Power Plant in Volgodonsk, Russia is shown in this next figure:

It can be challenging to quantify the cooling achieved by such systems. Fortunately, the supporting analysis has been published.[64,65] Frohwerk's patent contains an excellent description of spray systems.[66] The airflow induced by the spray process carries away the absorbed energy in the form of sensible and latent heat. The airflow and heat transfer are intimately linked. The conservation of energy can be expressed as follows:

---

[64] Berger, M. H. and Taylor, R. E., "An Atmospheric Spray Cooling Model," Proceedings of the 2nd AIAA/ASME Thermophysics and Heat Transfer Conference, Palo Alto, California, May 24-26, pp. 59-64, 1978.

[65] Chaturvedi, S. K. and R. W. Porter, "Thermal Performance of Spray-Canal Cooling Systems," Journal of Engineering for Power Vol. 102, No. 4, pp. 776-781, 1980. (available free on-line, search for 19770077431.pdf)

[66] United States Patent No. 3,622,074 issued to Paul A. Frohwerk of the Ceramic Cooling Tower Company on November 23, 1971. (available free on-line)

$$G\Delta h = LC_P \Delta T_W \qquad (37.1)$$

The Lagrangian droplet tracking model described in the last chapter can be run for a range of droplet diameter, *d*, and falling height, *h*, to produce the following graph of 1-effectiveness, or approach/(range+approach):

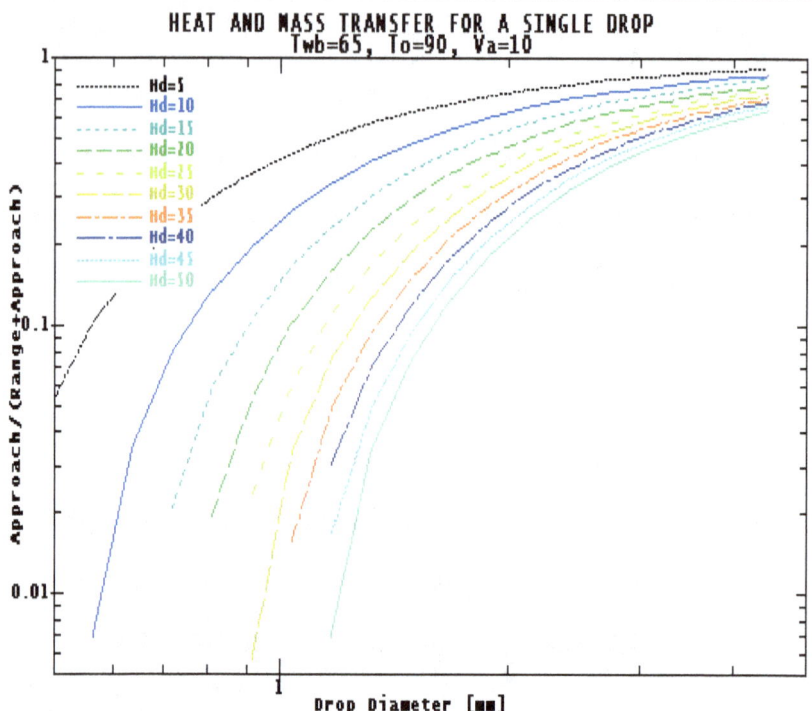

Based on regression, these curves can be approximated by the following formula for d≥0.5mm:

$$\varepsilon = 1 - \left\{ e^{[a+b\ln(h)+c\ln(d)]} - 1 \right\}^3 \qquad (37.2)$$

where a=-2.45591150714, b=0.631744225842, and c=-1.36867783072. As indicated at the top of the figure, this is for a single droplet. There are several manufacturers of high-quality industrial spray nozzles, including: Bete®, Lechler®, Phirex®, and Spray Systems®. The following is a Phirex® FogEx® nozzle used in fire suppression systems:

Spray nozzles produce a distribution of droplet sizes. The following distribution of droplet sizes is for a Bete® full-cone nozzle with 7 psig operating pressure.

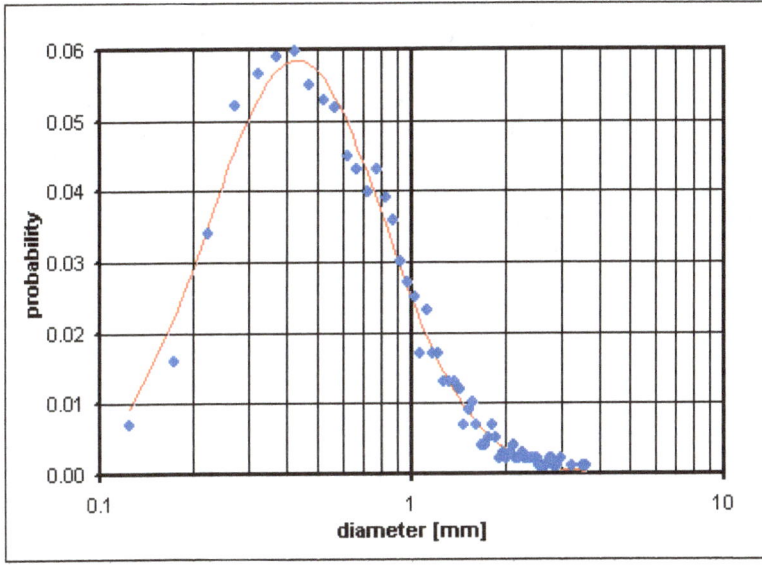

The blue dots are measured data and the red curve is a lognormal distribution having a mean of 0.43 mm and a standard deviation of 1.9 mm. The droplet size distribution can be applied to the preceding effectiveness curves[67] to arrive at a single curve of effectiveness vs. height for the droplets created by this nozzle at this operating pressure. This curve only applies to an average single droplet.

---

[67] The performance of the distribution of drop sizes is equal to the sum of the probability times the performance for each drop size, $\varepsilon = \sum Þ\varepsilon / \sum Þ$. The sum of the probabilities is equal to one.

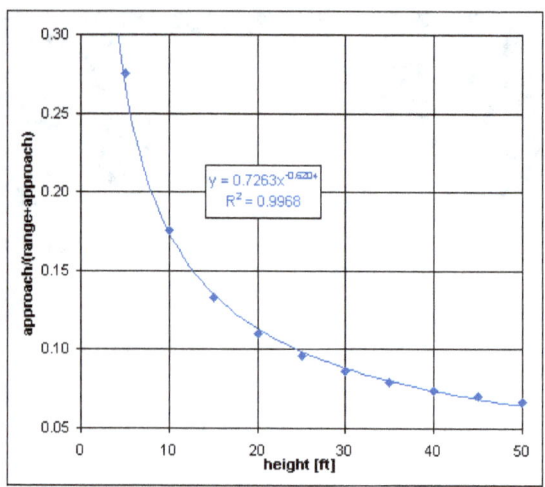

As water vapor (molecular weight ≈18) is lighter than air (molecular weight ≈29), evaporation from the spray will induce an upward draft of air, even in the absence of a crosswind. In the case of an open spray, as in the typical power plant application depicted here, the air flows radially inward and then upward. The streamlines formed by this flow may be approximated by the potential flow produced by two point sinks at $\pm h$, where h is some characteristic height. This forms a plane of symmetry at the ground, where there is no vertical velocity. This flow field is illustrated in the following figure:

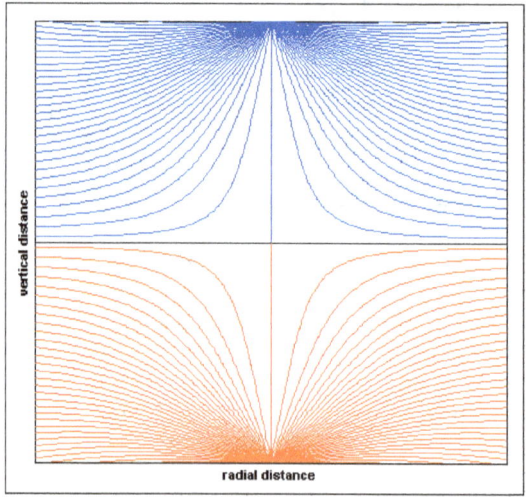

Floating spray modules have been deployed for decades in ponds and channels for supplemental cooling and aeration at steel mills, pulp mills, and wastewater plants. Utilization of these systems at power plants has been less common, at least in the U.S. Such systems can be quite cost-effective and there are still a few manufacturers, including ARWADH, whose 75 hp ThermoFlo® system is shown below (from their on-line brochure):

It is fortuitous that the same relationship for heat transfer and draft that governs the flow of air through a natural-draft cooling tower holds for a falling spray. This is particularly useful, as the problem of modeling this type of flow is much more complicated. This is primarily due to the fact that the flow in a cooling tower is confined by the structure; whereas, the flow induced by a spray is not. The fill also serves to guide the flow through the tower structure.

The following figure shows temperature data collected for the spray cooling pond at a paper mill:

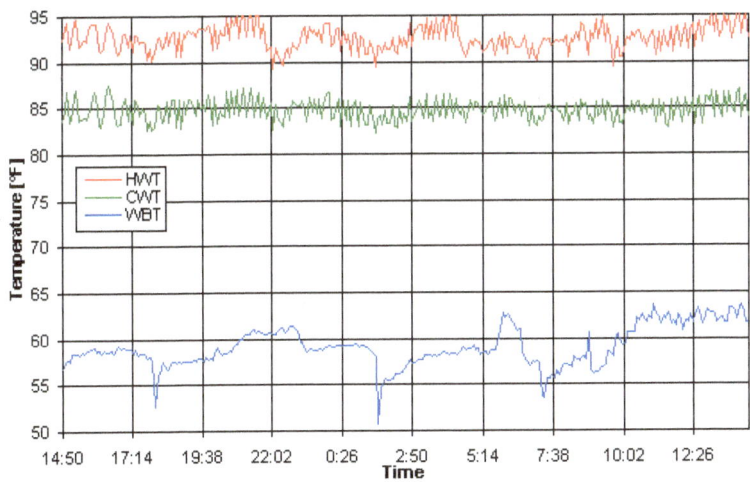

The water-to-air mass flow ratio, ***L/G***, can be calculated from Equation 37.1 and the mass transfer characteristic, ***KaV/L***, can be calculated from Equation 31.1. This next figure shows the results of these calculations:

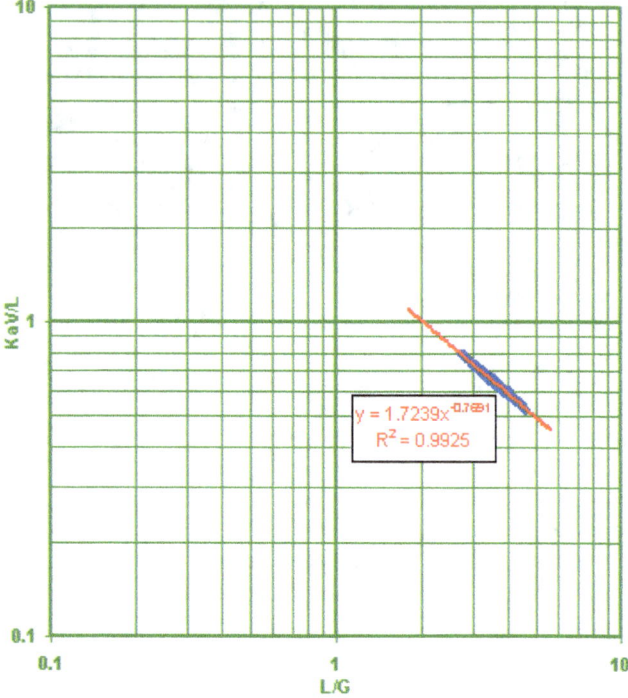

Data for this spray cooling system produces a remarkably tight cluster (R^2=0.9925) about the line ***KaV/L=1.7239/(L/G)$^{0.7691}$***, clearly illustrating the facility of Merkel's theory and this approach.

Chapter 38. Cooling Ponds

There was a great deal of interest in large cooling ponds during the early days of nuclear power plant design and construction in the United States. The U.S. Nuclear Regulatory Commission (USNRC) and the U.S. Environmental Protection Agency (USEPA) funded a series of studies extolling the efficacy of such ponds for waste heat removal.[68,69,70,71] Some of these articles may be found on-line. Several of these are included in the archive that accompanies this text.

Most of these studies refer back to the work of Langhaar.[72] While this is a very old reference, the properties of air and water have not changed. Langhaar's analysis was quite insightful and his calculations continue to be relevant and useful. This method has been updated and refined for presentation here. All of the calculations have been implemented in an Excel® spreadsheet that includes actual meteorological data and the performance of a certain nuclear plant that utilizes a lake for cooling.

All cooling pond calculations begin with the definition of an equilibrium temperature. This is the temperature a pond will eventually approach under steady conditions. The following figure is based on Langhaar's nomograph:

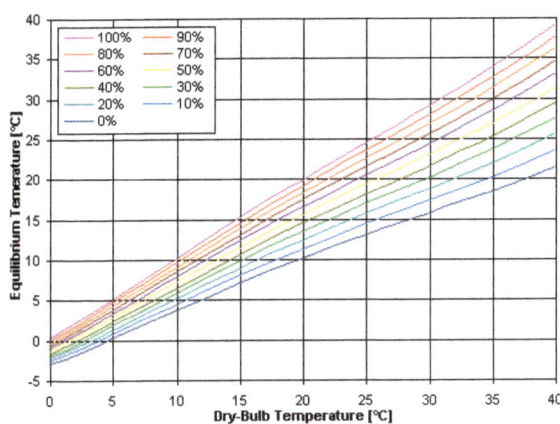

---

[68] Hogan, W. T., Liepins, A. A., and Reed, F. E., "An Engineering - Economic Study of Cooling Pond Performance," EPA Project 16130DFX05/70 Contract No. 14-12-521, May, 1970.

[69] Hadlock, R. K. and Abbey, O. B., "Thermal Performance Measurements on Ultimate Heat Sinks - Cooling Ponds," NUREG/CR-0008, February 1978, also published as a Batelle Pacific Northwest Laboratories Report PNL-2463.

[70] Codell, R. and Nuttle, W. K., "Analysis of Ultimate Heat Sink Cooling Ponds," NUREG-0693, November, 1980.

[71] Berger & Taylor include some helpful discussion as well (see Reference 34).

[72] Langhaar, J. W., "Cooling Pond Many Answer Your Water Cooling Problem," Chemical Engineering, August, 1953, pp. 194-199.

It is important to note that the equilibrium pond temperature is not simply the web-bulb; rather, it is based on empirical data. The following figure shows the difference between the equilibrium and web-bulb temperatures:

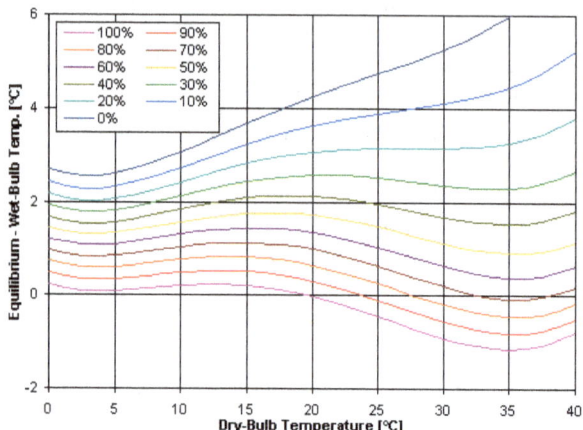

Wind and solar are strong influences on pond temperature, so that any model must contain corrections for these factors. The following figure shows the impact of wind speed on the equilibrium temperature with a Global Horizontal Irradiance (GHI) of 750 W/m². There are similar plots for 1000, 500, and 250 in the spreadsheet named Langhaar.xls. The functions to perform these calculations are also included as macros.

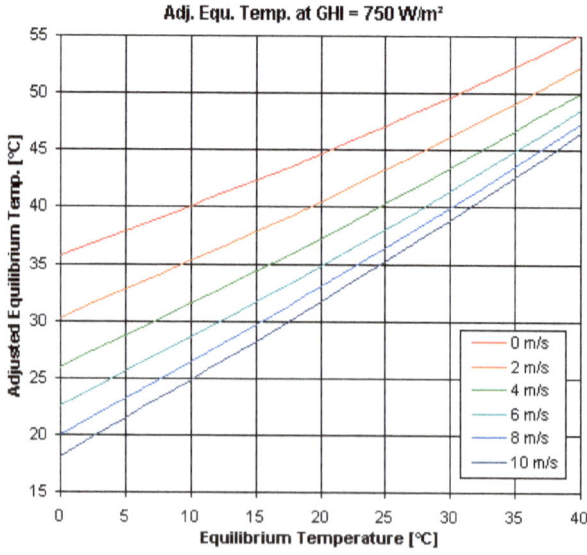

Langhaar's method includes two more parameters related to scale, given the symbols *P* and *Q*. The product *PQ=A/F* is the area divided by flow and is dimensional so that unit conversions are necessary for implementation. This has been implemented in the example spreadsheet, cooling_pond_simulation.xls, which models the performance of the nuclear plant and contains all of the necessary scaling factors.

The best way to illustrate Langhaar's model of pond performance is through an actual application where a large-scale pond is used to cool a power plant, in this case nuclear. The thermal performance of the plant is needed in order to capture the response of the heat rejection to the ambient conditions.

Performance functions for a particular nuclear plant is provided in Appendix N. This performance is typical of all nominal 1000 MWe Westinghouse pressured water reactors. Functions for the pond performance are provided in Appendix O. Functions that combine these two in order to model the response of the plant and pond are provided in Appendix P. All of this code is included in the on-line archive.

Pond area is determined from bathymetric surveys. Regression is used to fit area vs. elevation. Volume is determined by integrating the area with respect to elevation from the bottom up. The pond geometry is shown in this next figure:

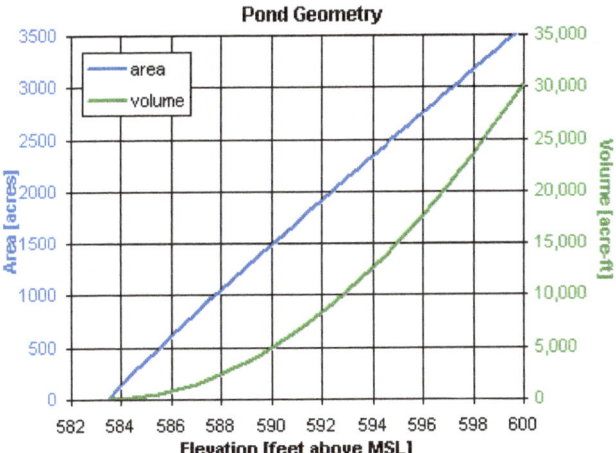

Hourly meteorological data was obtained from the National Climate Data Center. The dry-bulb and dew point temperatures are shown in the following figure:

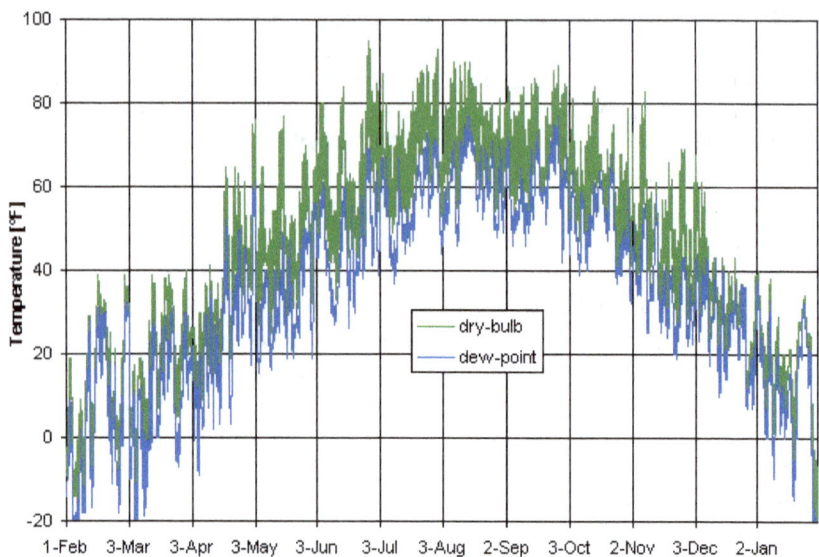

The Direct Normal Irradiance (DNI) and Global Horizontal Irradiance (GHI) is shown in this next figure:

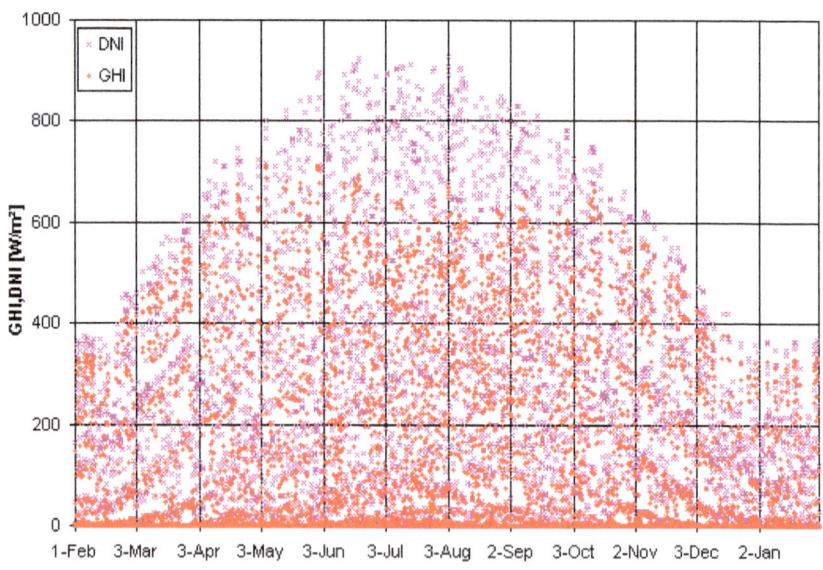

The calculated entering and exiting pond temperature is shown in this next figure:

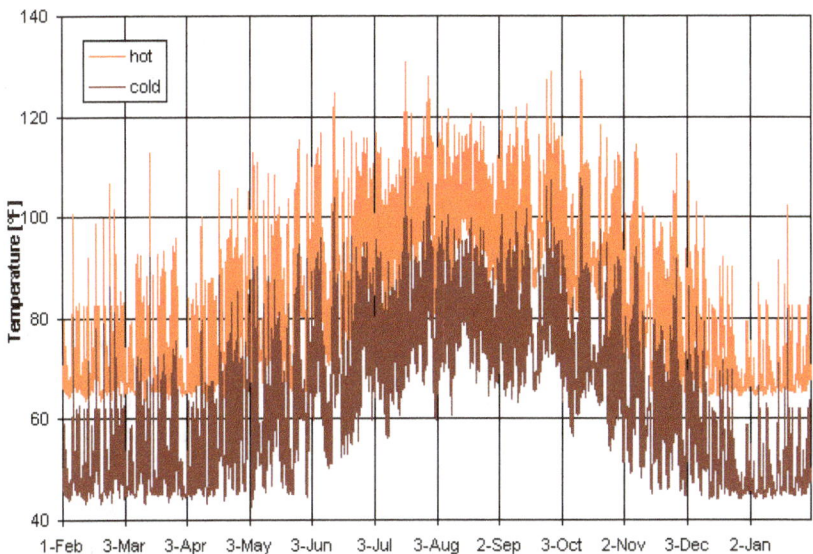

The generator output and zone three (highest) condenser pressure is shown in this next figure:

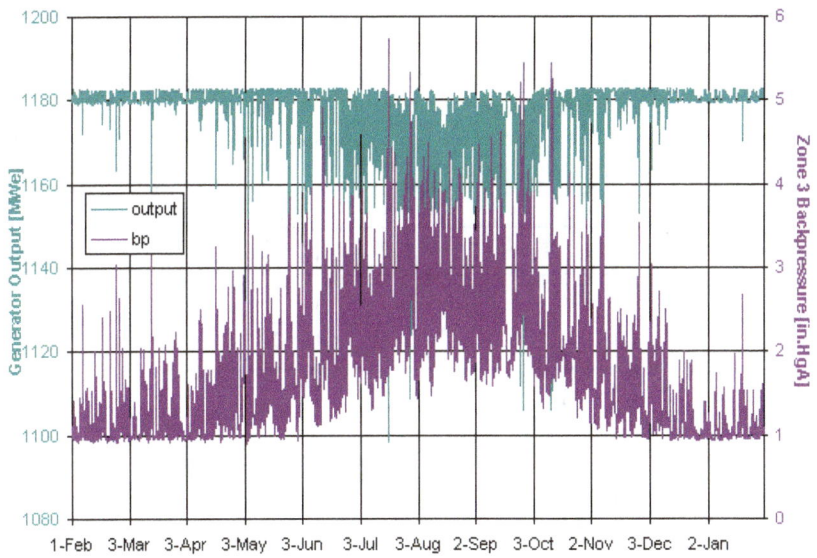

The simulation inputs and outputs are included in the archive in files nuke_input.csv and nuke_output.csv, respectively.

Chapter 39. The Lewis Number

In the derivation of Merkel's Equation, it was assumed that the coefficient of sensible heat transfer was proportional to that of mass transfer coefficient. This assumption was represented by Equation 31.8. The Lewis number is often expressed as:

$$L_E = \frac{Sc}{Pr} \qquad (39.1)$$

where **Sc** is the Schmidt number and **Pr** is the Prandtl number. The Schmidt number is equal to:

$$Sc = \frac{\mu}{\rho D} \qquad (39.2)$$

where μ is the dynamic viscosity, ρ is the density, and D is the diffusion coefficient or mass diffusivity. The Prandtl number is equal to:

$$Pr = \frac{\mu C_P}{\kappa} \qquad (39.3)$$

where C_P is the constant pressure specific heat and κ is the thermal conductivity. Equations 39.1 through 39.3 can be combined to form:

$$L_E = \frac{\kappa}{\rho C_P D} \qquad (39.4)$$

This expression of the Lewis number is a combination of molecular properties, thus, independent of any process peculiarities. This may be more accurately termed the *molecular* Lewis number, while Equation 31.8 may be more accurately termed the *turbulent* Lewis number. This ratio of properties for air is shown in the following figure:

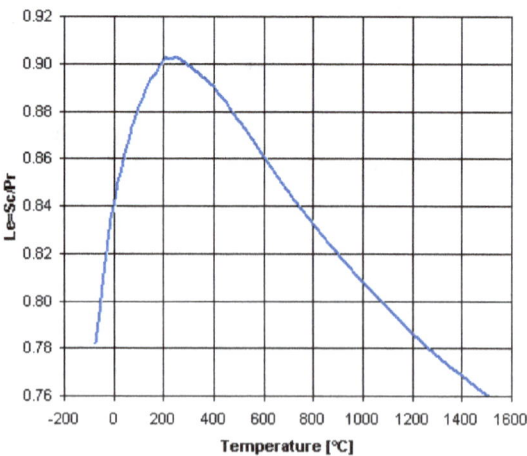

The molecular Lewis number for air never reaches unity, topping out at 0.902. At the conditions most often found in evaporative cooling, this ratio of properties is between 0.85 and 0.88. The turbulent Lewis number is of greater interest in evaporative cooling, but this can be difficult to determine and can't be directly measured, as the sensible and latent transfer processes can't be physically separated.

Little has been published on this subject, especially in the area of evaporative cooling, although there is at least one publication addressing this issue.[73] The method presented in this reference is a simple inverse heat or mass transfer problem with the two unknowns being the mass transfer coefficient and the sensible heat transfer coefficient. The ratio of these and the specific heat yields the experimental Lewis number.

It is easy enough to set up the problem and solve for the unknowns by minimizing the resulting errors. In this case the errors are the difference between the calculated and measured exiting water and air temperatures. As it turns out, if the exiting air is entirely saturated, the solution is degenerate, so the data set must be carefully selected to avoid this condition.

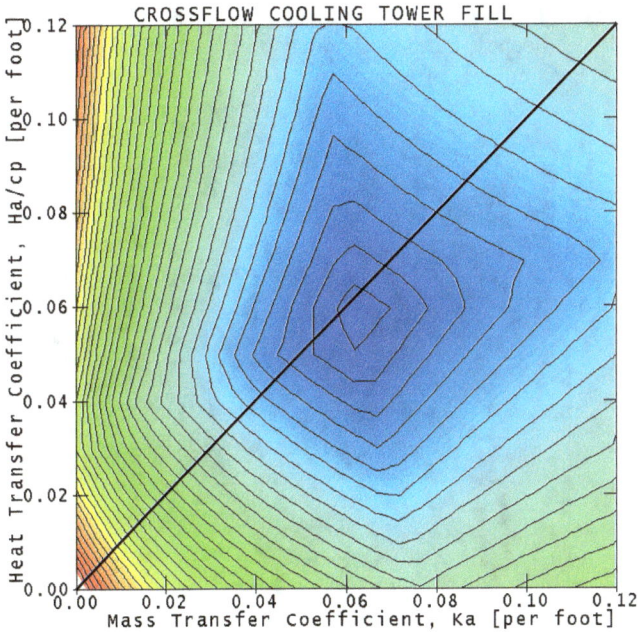

Perhaps even more interesting than the actual value of the experimental Lewis number is the sensitivity of the residuals to the unknown parameters. This

---

[73] Benton, D. J., "Determination of the Turbulent Lewis Number from Experimental Data for Wet Cooling Tower Fills," Cooling Tower Institute TP90-07, 1990.

peculiarity was only discovered by accident during the debugging of the software. The residual is given by:

$$r = \sqrt{(\Delta T_W)^2 + (\Delta T_A)^2} \tag{39.5}$$

Contours of this residual can be generated by varying the unknown coefficients over a range of values. Ideally, this would look something like the preceding figure. Small values of the residual are indicated by blue and large values by red. The actual magnitude of the residuals is immaterial. The smallest residual is located at the center of the roughly circular black contour at X=0.065, Y=0.055. This ratio indicates an experimental Lewis number of approximately 0.85, which is in line with the molecular value.

The diagonal black line corresponds to a Lewis number of unity. The center of this contour lies just below this line. This is a cross flow case and is reminiscent of a bull's-eye. In three dimensions this residual map would have the shape of a bowl.

As it turns out, this figure is not at all typical. In fact, it took some time to find a case within the data set that produced such a residual map. The following is more typical, but still ideal. It is a counter flow case. The contour having the smallest residual (deepest blue) is centered on the diagonal line, indicating an experimental Lewis number of 1.

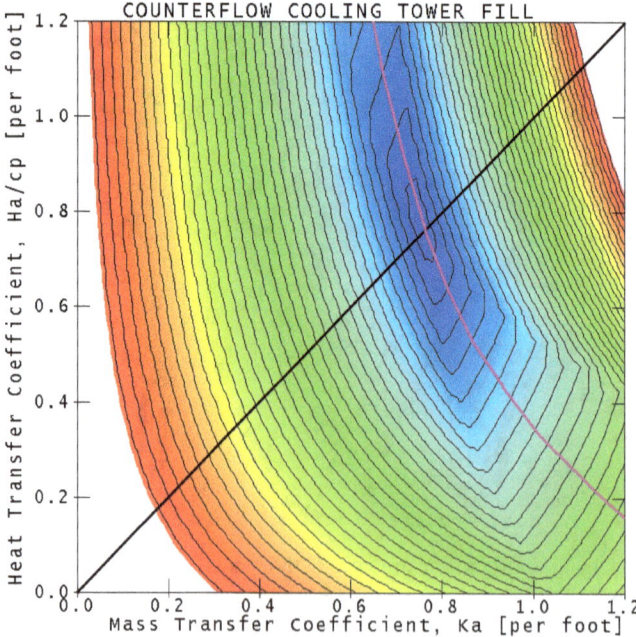

The optimum solution to this inverse combined heat and mass transfer problem corresponds to the smallest residual. This is a specific case of the more general problem of nonlinear minimization. The physical analogy to this mathematical process is one of finding the lowest elevation on a topographic surface. In this case, the longitude and latitude correspond to the two transfer coefficients, **Ka** and **Ha/Cp**, and the elevation corresponds to the residual.

This figure shows a very interesting character of this problem. Every value of X and Y inside a contour has the same or lower residual, making these equally valid solutions. The best solutions (i.e., the ones having the smallest residuals) lie along the magenta curve. In three dimensions this residual map would not have the shape of a bowl, rather it would be more like a gully, as shown in this next figure:

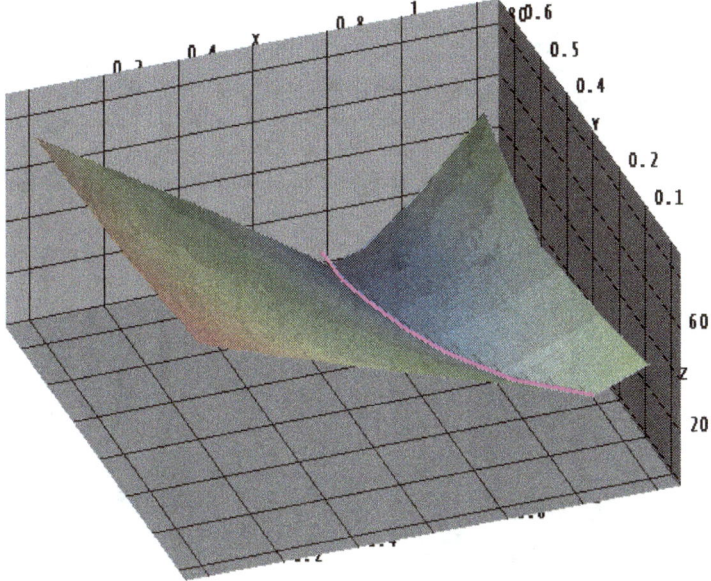

The sides of this residual surface are much steeper in one direction than the other. The locus of optimum solutions (i.e., the ones having the smallest residuals or lowest elevations) lie along the magenta arc. All solutions with the same shade of blue are equally valid in that they have the same residual.

It is tempting to assume the very best solution lies at the center of the inner most blue contour, but this overlooks the fact that these results are derived from experimental data, which is itself burdened with some level of uncertainty. No matter how careful measurements are made and how recently instruments are calibrated, these all have limited accuracy so that the data are not known exactly.

As seen in the two previous figures, the residual surface is much steeper across the magenta arc than it is a along the arc. The significance of this is that, while there may be considerable uncertainty as to the optimum distance along the arc, there is much less uncertainty as to the optimum distance perpendicular to the arc. For instance, given the accuracy of the data, we may be able to determine the distance along the arc to within ±5% and the distance perpendicular to the arc to within ±1% if the contours have an aspect ratio of 5:1.

The shape of the residual surface and contours of this particular data set (cooling tower packing of the *splash* type) vary considerably, as does the aspect ratio. Several of the cases had an aspect ratio of 50:1 or more, which is analogous to a very deep canyon. This is presents a significant problem when analyzing the data. This next figure is similar to many of the cases analyzed.

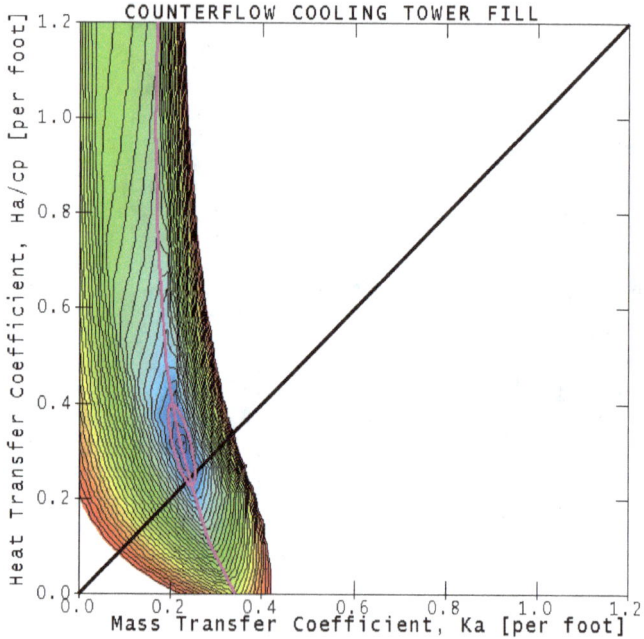

The smallest residual (i.e., best agreement with experimental data) corresponds to approximately **Ka=0.25** and **Ha/C$_P$=0.30** for an experimental Lewis number of **L$_E$=1.2**. Considering the uncertainty in the test data, the solution could be anywhere inside the magenta ellipse, that is **0.2≤Ka≤0.3 0.25≤Ha/C$_P$≤0.4**, such that the uncertainty of **Ha/Cp** is greater than **Ka**.

EPRI built a small-scale test facility for cooling tower fill at Houston Lighting and Power's Parish Station. Several counter flow and cross flow packings were tested and the results published in an EPRI report.[74] These data

---

[74] Bell, D. M., B. M. Johnson, E. V. Werry, P. B. Miller, D. E. Wheeler, K. R. Wilbur, D. J. Benton, and J. A. Bartz, "Cooling Tower Performance Prediction and Improvement," EPRI RP2113 1988.

were used to determine the experimental Lewis number as described in TP90-07. The following figures show the cross flow and counter flow data, respectively:

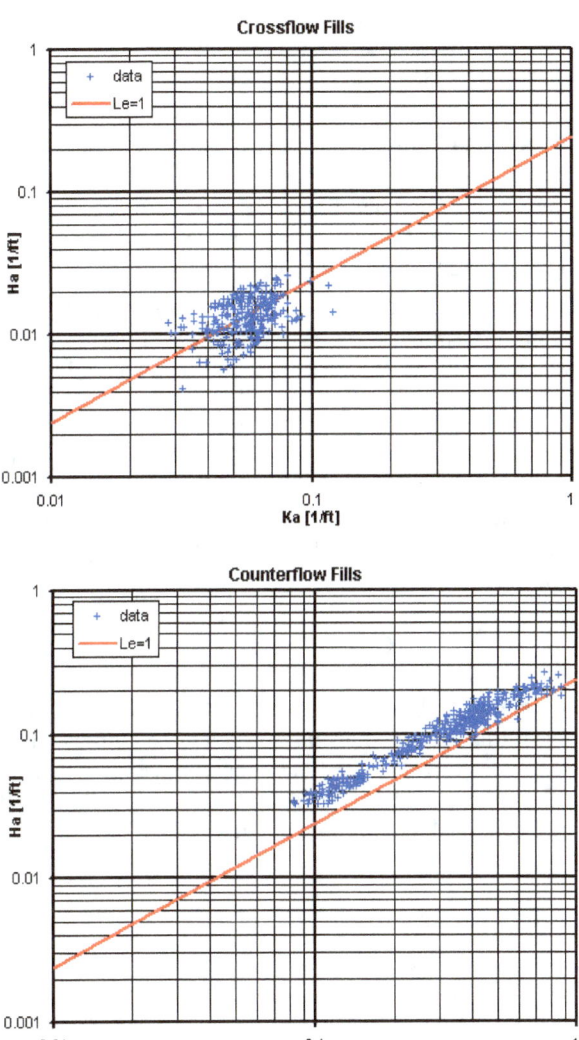

There is considerable scatter in the cross flow graph. The average experimental Lewis number for this data is 1.03±0.59 and for the counter flow is 1.40±0.37. It is therefore reasonable to use a value of unity for cross flow analysis, but not for counter flow. Both the cause and significance of this difference is still unknown.

Chapter 40. Air into Water

It is truly remarkable that the same approach presented in Chapter 36 for water droplets evaporating into air can be used to analyze air bubbles absorbing into water. This final chapter has been included to give perspective on the evaporative cooling process within the greater context of mass transfer. This discussion and data are based on a report written for the U. S. Army Corps of Engineers (USACE).[75]

The goal of the study presented in this report was to develop and utilize a numerical model of the nitrogen supersaturation process occurring in the plunge pool beneath dam at the Jennings Randolph Lake Project. A literature review was conducted and the model developed by the USACE Waterways Experiment Station (WES) was found to be the most promising.[76]

This model was based on Roesner and Norton's work with dissolved gas levels downstream of a spillway.[77] was developed by Roesner and Norton, who began with a simple mass transfer model that can be expressed as follows:

$$C_d = C_s - (C_s - C_u) e^{-Kt} \qquad (40.1)$$

where Cd is the downstream concentration, Cs is the saturation concentration, Cu is the upstream concentration, K is the mass transfer coefficient, and t is the residence time in the stilling basin. This equation forms the basis of the WES model. Hibbs and Gulliver utilized this equation in computing the effective saturation concentration, Ce:[78]

$$C_e = C_s \left(1 + \frac{d_e \gamma}{P_a} \right) \qquad (40.2)$$

where Cs is the saturation concentration (taken to be 100%), de is the effective bubble depth, γ is the specific weight of water, and Pa is the atmospheric pressure. Geldert et al. provide three field data sets: Ice Harbor, The Dalles, and Little Goose. The measured effective saturation concentrations and the values computed using Equation 40.2 are shown in this next figure:

---

[75] Benton, D. J., "Benton, "Modeling of Nitrogen Supersaturation at Jennings Randolph," Advanced Technology Systems Report for the USACE, September, 1999. This report was part of the larger Section 1135(b) Study for Jennings Randolph Lake conducted under the Water Resources Development Act (WRDA) of 1986.

[76] Geldert, D. A., J. S. Gulliver, and S. C. Wilhelms (1998), "Modeling Dissolved Gas Supersaturation Below Spillway Plunge Pools," *Journal of Hydraulic Engineering*, May 1998, pp. 513-521.

[77] Roesner, L. A., and W. R. Norton, "A Nitrogen Gas (N₂) Model for the Lower Columbia River," Report No. 1-350, *Water Resources*, 1971.

[78] Hibbs, D. E. and J. S. Gulliver, "Prediction of an Effective Saturation Concentration at Spillway Plunge Pools," *Journal of Hydraulic Engineering*, Vol. 1, No. 3, pp. 940 949, 1997.

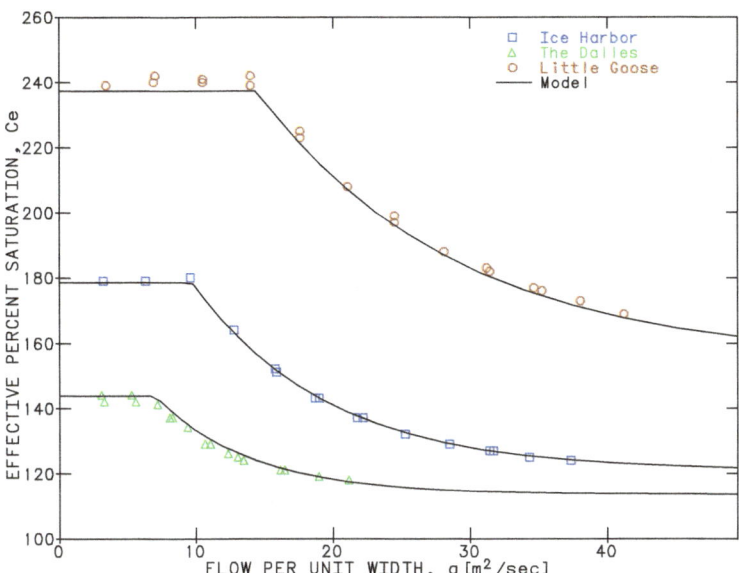

The effective depth is computed from the bubble half-life depth (i.e., the length traveled over the half-life), **hb**, by Equation 40.3:

$$d_e = h_2 + (h_1 - h_2) e^{\left(1 - \frac{\beta h_b}{L_s}\right)} \quad for \quad \frac{\beta h_b}{L_s} > 1 \qquad (40.3)$$

$$d_e = h_1 \quad for \quad \frac{\beta h_b}{L_s} \leq 1$$

where β is an empirical constant equal to 2.2, h_1 is the effective bubble depth in the stilling basin (presumed to be 2/3 of the stilling basin depth, **hs**), h_2 is the effective bubble depth in the river (presumed to be 1/2 of the river depth, **hr**), and **Ls** is the length of the stilling basin. The bubble half-life depth is computed from the discharge per unit width, **q**, and the bubble rise velocity, **vr** (presumed be constant at 0.25 meters/second), by Equation 40.4.

$$h_b = \frac{q}{v_r} \ln(2) \qquad (40.4)$$

Geldert et al. reasoned that the mass transfer included a bubble component into the water and a surface component out of the water. The rate of change of the concentration, **C**, is then given by Equation 40.5:

$$\frac{dC}{dt} = K_L a_b (C_e - C) + K_L a_s (C_s - C) \qquad (40.5)$$

where K_L is the mass transfer coefficient, a_b is the bubble interfacial area per unit volume and a_s is the surface interfacial area per unit volume. The solution of this differential equation is given by Equation 40.6.

$$C_d = C_e - (C_e - C_u)\left\{ e^{-(K_L a_b t_b + K_L a_s t_s)} + \vartheta \right\}$$
$$\vartheta = \frac{K_L a_s t_s}{K_L a_b t_b + K_L a_s t_s}\left(\frac{C_e - C_s}{C_e - C_u}\right)\left[1 - e^{-(K_L a_b t_b + K_L a_s t_s)}\right] \quad (40.6)$$

where t_b is the residence time for the bubbles and t_s is the exposure time for the surface transfer. Geldert et al. presumed that the combination $K_L a_s t_s$ would be a dimensionless constant on the order of unity for any particular application. The void fraction, φ, is computed using Equation 40.7:

$$\phi = \frac{v_j \lambda}{v_j \lambda + q} \quad (40.7)$$

where λ is an empirical constant on the order of 0.2 meters and v_j is the effective velocity of the plunging jet of water. Geldert et al. did not provide a means of obtaining v_j, simply stating that this was "computed by a standard water surface profile technique." Geldert et al. used the void fraction and an empirical correlation to obtain the dimensionless bubble transfer group, $K_L a_b t_b$, given by Equation 40.8:

$$K_L a_b t_b = \alpha\phi \frac{(1-\phi)^{1/2}}{(1-\phi^{5/3})^{1/4}} W_e^{3/5} R_q^{2/3} S_c^{-1/2} R_r^{-1} \quad (40.8)$$

where α is an empirical constant on the order of unity, We is the Weber number (Equation 40.9), Rq is the Reynolds number for the flow (Equation 40.10), Sc is the Schmidt number for air/water (Equation 40.11), and Rr is the Reynolds number for the rising bubbles (Equation 40.12).

$$W_e = \frac{\rho q^2}{\sigma d_j} \quad (40.9)$$

where ρ is the density of water, σ is the surface tension, and d_j is the effective depth of the plunging jet ($d_j = q/v_j$).

$$R_q = \frac{q}{\upsilon} \quad (40.10)$$

where υ is the kinematic viscosity of water.

$$S_c = \frac{\nu}{D} \quad (40.11)$$

where D is the air/water diffusion coefficient.

$$R_r = \frac{2 d_e v_r}{\upsilon} \quad (40.12)$$

These equations form the original WES model. Modifications are required in order to complete the model and to obtain good agreement with field data. The original WES model also lacks an explicit calculation for the plunging jet velocity, v_j. In order to fill this gap in the model, a computer program was developed to "back out" the jet velocity implied by the data points for the three sites given in the WES report. These values were then compared to all of the dimensionless quantities that can be formed from the site parameters. The best correlation obtained ($R^2=0.92$) is given in Equation 40.13.

$$\frac{v_j}{\sqrt{g\,h_t}} = 0.15 \left(\frac{\dfrac{q}{h_t}}{\sqrt{g\,h_t}} \right)^{0.23} \qquad (40.13)$$

where g is the gravitational acceleration and ht is the total (effective) head. The agreement between the data and this equation is illustrated in the following figure.

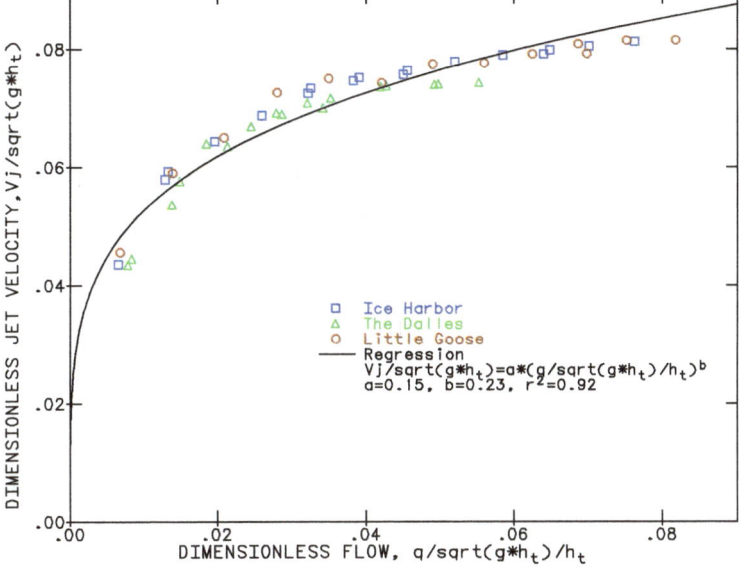

The bubble transfer group, $K_L a_b t_b$, can then be computed from Equation 40.8 and this correlation for the jet velocity (Equation 40.13). The results are illustrated in this next figure:

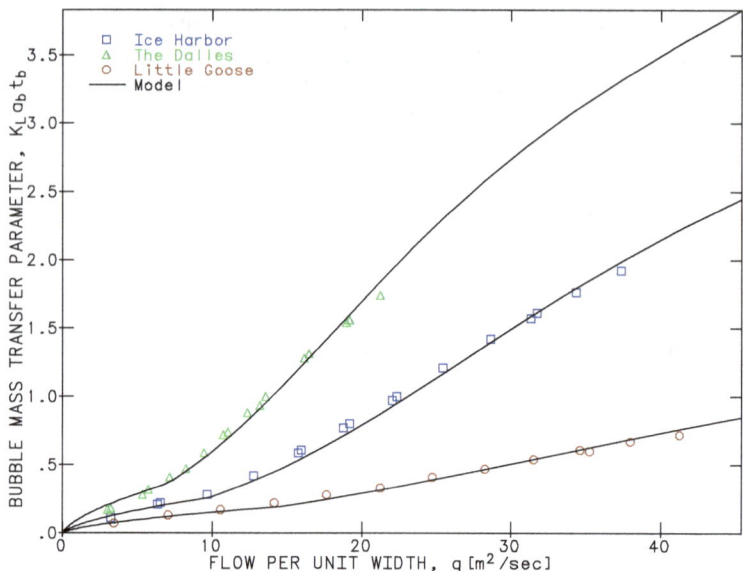

This same model can be applied to the Jennings Randolph site. The data and model results for all four sites are illustrated in the following figure:

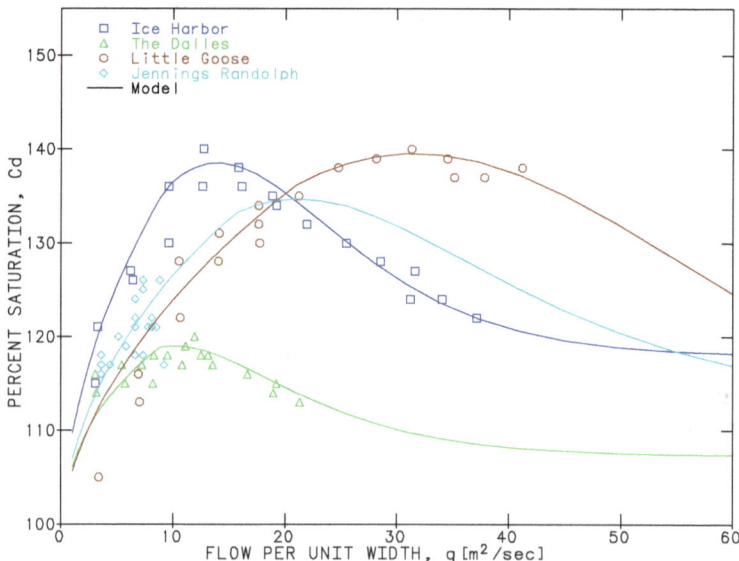

As stated previously, this is a zero-dimensional empirical model. The Modified WES Model is significantly different than a one-, two-, or three-dimensional finite difference or finite element model in which the domain is subdivided into computational cells. Any direct applications of this model are limited to the variables that appear in the various equations, for instance, a single value must represent the depth of the stilling basin, *hs*. If the depth of the stilling basin changes significantly over its length, this model will only accommodate a single number for the average or effective depth. The source code implementing this model is provided in Appendix Q.

Appendix A. Crossflow Program

The following program implements the unmixed crossflow calculation discussed in Chapter 3 with and without iterative implicit temperature differences plus the closed-form implicit algorithm. Some sections containing trivial code have been omitted. The complete program is included in the on-line archive.

```
double CpH=4.0;
double CpC=1.0;
double mH=100.;
double mC=500.;
double Thi=50.;
double Tho=20.;
double Tci=15.;

double Ts(double Tn,double Tw,double UA,int nx,int ny)
  {
  double alpha,beta,delta,gamma;
  beta=UA/mH/CpH;
  alpha=mC*CpC/mH/CpH;
  delta=2*alpha*nx*ny;
  gamma=(ny*ny+nx*nx*alpha*alpha+delta)*beta*beta
    +delta*delta;
  return(Tn+((Tn-Tw)*(ny+nx*alpha)*beta*beta-beta*(Tn-
    Tw)*sqrt(gamma))/delta/ny);
  }

double solve(int nx,int ny,int m)
  {
  int i,iter,x,y;
  double*dT,dT1,dT2,*dQ,*Tc,tco,*Th,tho,U1,U2,UA;

/* allocate arrays */

  dT=alloc(nx*ny,sizeof(double));
  dQ=alloc(nx*ny,sizeof(double));
  Th=alloc(nx*(ny+1),sizeof(double));
  Tc=alloc((nx+1)*ny,sizeof(double));

/* initialize arrays */

  for(x=0;x<nx;x++)
    Th[x]=Thi;
  for(y=0;y<ny;y++)
    Tc[(nx+1)*y]=Tci;

/* initialize UA */

  tco=Tci+mH*CpH*(Thi-Tho)/mC/CpC;
  U1=mH*CpH*(Thi-Tho)/LMTD(Thi-tco,Tho-Tci);
```

```
      U2=8.*U1;
      for(iter=0;iter<32;iter++)
        {
        UA=(U1+U2)/2.;

/* step through grid */

        for(y=0;y<ny;y++)
          {
          for(x=0;x<nx;x++)
            {
            if(m>=0)
              {
              dT[nx*y+x]=fmax(0.,Th[nx*y+x]-Tc[(nx+1)*y+x]);
              dQ[nx*y+x]=(UA/nx/ny)*dT[nx*y+x];
              Th[nx*(y+1)+x]=Th[nx*y+x]-dQ[nx*y+x]/
    CpH/(mH/nx);
              }
            else
              {
              Th[nx*(y+1)+x]=Ts(Th[nx*y+x],Tc[(nx+1)*y+x],
    UA,nx,ny);
              dQ[nx*y+x]=(Th[nx*y+x]-Th[nx*(y+1)+x])
    *CpH*(mH/nx);
              dT[nx*y+x]=dQ[nx*y+x]/(UA/nx/ny);
              }
            Tc[(nx+1)*y+x+1]=Tc[(nx+1)*y+x]+dQ[nx*y+x]
    /CpC/(mC/ny);
            }
          }

/* implicit temperature correction */

        for(i=0;i<m;i++)
          {
          for(y=0;y<ny;y++)
            {
            for(x=0;x<nx;x++)
              {
              dT1=Th[nx*y+x]-Tc[(nx+1)*y+x];
              dT2=Th[nx*(y+1)+x]-Tc[(nx+1)*y+x+1];
              dT[nx*y+x]=fmax(0.,(2.*dT1+dT2)/3.);
              dQ[nx*y+x]=(UA/nx/ny)*dT[nx*y+x];
              Th[nx*(y+1)+x]=Th[nx*y+x]-
    dQ[nx*y+x]/CpH/(mH/nx);
              Tc[(nx+1)*y+x+1]=Tc[(nx+1)*y+x]+dQ[nx*y+x]/
    CpC/(mC/ny);
              }
            }
          }
```

```
/* solve for exit temperatures */

    for(tho=x=0;x<nx;x++)
      tho+=Th[nx*ny+x];
    tho/=nx;
    for(tco=y=0;y<ny;y++)
      tco+=Tc[(nx+1)*y+nx];
    tco/=ny;

/* use bisection search algorithm, adjusting UA to match
   Tho */

    if(tho>Tho)
      U1=UA;
    else
      U2=UA;
    }

/* release memory */

  free(dT);
  free(dQ);
  free(Th);
  free(Tc);

  return(UA);
  }

void test4(int m)
  {
  int n;
  for(n=5;n<=100;n++)
    printf("%i %lG\n",n,solve(n,n,m));
  }

int main(int argc,char**argv,char**envp)
  {
  test4(-1);
  return(0);
  }
```

The following is a sample of the output:

```
nx ny    UA
10 10 1724.26
10 20 1941.27
10 30 2024.87
10 40 2069.15
20 10 1887.29
20 20 2150.29
20 30 2253.31
20 40 2308.28
30 10 1948.66
30 20 2230.29
30 30 2341.32
30 40 2400.72
40 10 1980.85
40 20 2272.56
40 30 2387.94
40 40 2449.75
```

Appendix B. Moisture Separator/Reheater Program

The analysis of two-phase flow inside tubes presented in Chapter 14 is only a part of the code necessary to model a moisture separator/reheater. The entire code consists of several sections, some of which will be listed here. The entire code, along with a pre-compiled Windows® executable, is included in the on-line archive.

The program, called MISER, first checks the inputs, then establishes the tube connections, then initializes the variables, then solves the equations, iterating until convergence is reached. The two main sections of interest are the two-phase flow calculations within a computational element of tube and the combined tube-side/shell-side calculation.

Tube-Side Module

```
/ *************************************************************
    Determine the two-phase flow frictional pressure
      gradient
    using Chisholm's method for separated flow
    (this method is reportedly valid for steam above 435
      psia but it
    is about the best two-phase pressure drop correlation
      available
    so it will be used anyway...)

    Determine the two-phase heat transfer coefficient for
      horizontal
    pipe flow using  the Taitel-Dukler method

    For more deatils see "Prediction of Horizontal
      Tubeside
    Condensation of Pure Components Using Flow Regime
      Criteria,"
    G. Breber, J. W. Palen, and J. Taborek, 18th ASME
      National
    Heat Transfer Conference, San Diego, 1979.


    a........ tube cross sectional area [sq.ft.]
    c........ constant (see Collier equation 2.72, p.50)
    c2....... constant (see Collier equation 2.72, p.50)
    cbar..... constant (see Collier equation 2.75, p.51)
    cpf...... constant pressure specific heat of saturated
      liquid
    cpg...... constant pressure specific heat of saturated
      vapor
    d........ tube diameter [ft]
    dpdx..... two-phase frictional pressure gradient
    dpdxf.... frictional pressure gradient based on liquid
      alone
```

dpdxg.... frictional pressure gradient based on vapor alone
dt....... the temperature difference across the condensing vapor
ff....... friction factor - liquid alone
fg....... friction factor - vapor alone
fm....... friction factor - mixture
flow..... flow [#/hr]
fp....... two-phase flow pattern (ie. annuar, wavy, slug etc.)
fr....... Froude number
g........ gravitational acceleration (32.174 [ft/sec**2])
gm....... mass flux [#/sq.ft./hr]
gf....... mass flux of liquid flowing alone [#/sq.ft./hr]
gg....... mass flux of vapor flowing alone [#/sq.ft./hr]
gstar.... critical mass flux [#/sq.ft./hr] (see Collier p.50)
gsubc.... Newton's constant (32.174 [lbm-ft/lbf/sec**2])
hf....... enthaply of saturated liquid
hg....... enthaply of saturated vapor
h2p...... two-phase heat transfer coefficient [BTU/hr/f/sq.ft.]
tf....... thermal condictivity of the saturated liquid [BTU/hr/f/ft]
tg....... thermal condictivity of the saturated vapor [BTU/hr/f/ft]
gamma.... constant (see Collier equation 7.72, p.50)
b........ Blasius exponent (see Collier p.50)
dvf...... viscosity of saturated liquid [#/ft/hr]
dvg...... viscosity of saturated vapor [#/ft/hr]
dvm...... viscosity of mixture [#/ft/hr]
phif..... two phase multiplier (see Collier equation 2.50, p.34)
phif2.... square of the two phase multiplier (phif**2)
prf...... Prandtl number of the saturated liquid
psi...... two phase multiplier (see Collier equation 2.73, p.51)
ref...... Reynolds number for liquid flow only
reg...... Reynolds number for vapor flow only
rem...... Reynolds number of mixture
slip..... slip ratio (vapor velocity/liquid velocity)
x........ quality
xx....... Lockart-Martinelli factor (see Collier equation 2.67, p. 37)
ug....... linear momentum flux [#/ft/hr**2]

```
vf...... specific volume of saturated liquid
    [cu.ft./#]
vg...... specific volume of saturated vapor
    [cu.ft./#]
alpha.... void fraction (cross-sectional area of vapor
    flow/
         cross-sectional area of total flow)
*************************************************/

void phase2(double *p,double *x,double *flow,double
    *a,double *d,double *rem,double *fm,double
    *dpdx,double *dt,double *h2p,double *slip,double
    *alpha,double *ug,double *fp,int *ier)
{
int izone;
double
    b,c,c2,cbar,cpf,cpg,dpdxf,dpdxg,dvf,dvg,dvm,f,ff,fg,f
    r,ftp,gamma,gf,gg,gm,gstar,hcg,hcs,hf,hfg,hg,hl,hv,ph
    if,phif2,prf,prg,psi,ref,reg,rhof,rhog,t,tf,tg,tsat,v
    ,vf,vfg,vg,xx;
static BOOL rough=FALSE;
/* fetch thermodynamic and transport properties */
    tsat=fTsat(*p);
    vf=VfofT(tsat);
    vg=VgofT(tsat);
    hf=HfofT(tsat);
    hg=HgofT(tsat);
    dvf=MUfofT(tsat);
    dvg=MUgofT(tsat);
    tf=TKfofT(tsat);
    tg=TKgofT(tsat);
    cpf=CpfofT(tsat);
    cpg=CpgofT(tsat);
    vfg=vg-vf;
    rhof=1./vf;
    rhog=1./vg;
    hfg=hg-hf;
    prf=cpf*dvf/tf;
    prg=cpg*dvg/tg;
/* determine frictional pressure gradient for subcooled
    liquid */
    if((*x)<=0.0001)
      {
      gf=(*flow)/(*a);
      ref=gf*(*d)/dvf;
      *rem=ref;
      ff=fblas(ref,roughness);
      *fm=ff;
      *dpdx=-4.*ff*sq(gf/3600.)*vf/2./gsubc/(*d)/144.;
      *slip=0.;
```

```
            *alpha=0.;
            *ug=sq(gf)*vf;
            *fp=flopat(5);
   /* set two-phase heat transfer coefficient equal to the
      forced convection
      heat transfer coefficient (see Breber, Palen, and
      Taborek p.7) */
            *h2p=1.86*(tf/(*d))*pow(ref*prf,0.333333);
            if(ref>3000.)

            *h2p=0.024*(tf/(*d))*pow(ref,0.8)*pow(prf,0.333333);
   /* calculate the two-phase pressure drop for a
      liquid/vapor mixture (see Collier p.50) */
            }
         else if((*x)>0.9999)
            {
   /* determine frictional pressure gradient for
      superheated vapor */
            gg=*flow/(*a);
            reg=gg*(*d)/dvg;
            *rem=reg;
            fg=fblas(reg,roughness);
            *fm=fg;
            *dpdx=-4.*fg*sq(gg/3600.)*vg/2./gsubc/(*d)/144.;
            *slip=1.;
            *alpha=1.;
            *ug=gg*vg;
            *fp=flopat(6);
   /* set two-phase heat transfer coefficient equal to the
      forced convection
      heat transfer coefficient (see Breber, Palen, and
      Taborek p.7) */
            *h2p=1.86*(tg/(*d))*pow(reg*prg,0.333333);
            if(reg>3000.)

            *h2p=0.024*(tg/(*d))*pow(reg,0.8)*pow(prg,0.333333);
            }
         else
            {
            gamma=0.75;
            if(rough)
               gamma=1.;
            gstar=1.47E6;
            if(rough)
               gstar=1.1E6;
            b=0.;
            if(rough)
               b=0.25;
   /* determine the effective viscosity (see Collier p.30)
      */
```

```
      dvm=1./((*x)/dvg+(1.-(*x))/dvf);
/* determine frictional pressure gradient for liquid
   alone */
      gf=*flow*(1.-(*x))/(*a);
      ref=gf*(*d)/dvf;
      ff=fblas(ref,roughness);
      dpdxf=-4.*ff*sq(gf/3600.)*vf/2./gsubc/(*d)/144.;
/* determine frictional pressure gradient for vapor
   alone */
      gg=(*flow)*(*x)/(*a);
      reg=gg*(*d)/dvg;
      fg=fblas(reg,roughness);
      dpdxg=-4.*fg*sq(gg/3600.)*vg/2./gsubc/(*d)/144.;
/* determine Lockhart-Martinelli parameter (see Collier
   p.37) */
      xx=sqrt(dpdxf/dpdxg);
/* (see Collier p.50) */
      gm=(*flow)/(*a);
      c2=fmin(2.*gamma,gstar/gm);
/* (see Collier p.51) */
      cbar=sqrt(vg/vf)+sqrt(vf/vg);
/* (see Collier p.50) */
      c=(gamma+(c2-gamma)*sqrt(vfg/vg))*cbar;
/* sub-critical flow (see Collier pages 50 & 51) */
      if(gm>gstar)
        {
/* super-critical flow (see Collier p.52) */
        t=pow((*x)/(1.-(*x)),(2.-
        b)/2.)*pow(dvf/dvg,b/2.)*sqrt(vf/vg);
        psi=(1.+c/t+1./sq(t))/(1.+cbar/t+1./sq(t));
        phif2=(1.+cbar/xx+1./sq(xx))*psi;
        }
      else
        phif2=1.+c/xx+1./sq(xx);
      *dpdx=dpdxf*phif2;
/* determine the effective friction factor and Reynolds
   number */
      gm=(*flow)/(*a);
      *rem=gm*(*d)/dvm;
      v=vf+(*x)*vfg;
      *fm=-(*dpdx)*144. *(*d)*gsubc*2./v/sq(gm/3600.)/4.;
/* calculate the two-phase multiplier */
      phif=sqrt(phif2);
/* calculate slip ratio using Zivi's method (ASME-JHT
   May, 1964 pages 247-252) */
      *slip=pow(vg/vf,0.333333);
/* calculate void fraction using zivi's method (ASME-JHT
   May, 1964, pages 247-252) */
      *alpha=1./(1.+((1.-(*x))/(*x))*sq(*slip));
/* determine the linear momentum flux */
```

```
          *ug=sq(gm)*(sq(*x)*vg/(*alpha)+sq(1.-(*x))*vf/(1.-
          (*alpha)));
   /* Determine the two-phase flow regime and calculate the
          two-phase heat transfer coefficient. First determine
          the froude number (see Breber, Palen, and Taborek
          p.3) */
          fr=sqrt(sq(gg/3600.)/(*d)/g/rhog/(rhof-rhog));
   /* determine the two-phase flow regime (see Breber,
          Palen, and Taborek p.7) */
          izone=1;
          if(xx<=1.25&&fr<=1.)
            izone=2;
          if(xx>1.25&&fr<=1.)
            izone=3;
          if(xx>1.25&&fr>1.)
            izone=4;
          *fp=flopat(izone);
   /* calculate the liquid heat transfer coefficient
          (see Breber, Palen, and Taborek p.7) */
          hl=1.86*(tf/(*d))*pow(ref*prf,0.333333);
          if(ref>3000.)
            hl=0.024*(tf/(*d))*pow(ref,0.8)*pow(prf,0.333333);
   /* calculate the heat transfer coefficient for
          condensing vapor (see Collier pages 328 & 331) */
          f=0.31*pow(reg,0.12);
          hv=f*pow(rhof*(rhof-
          rhog)*(g*3600.*3600.)*hfg*cube(tf)/(*d)/dvf/(*dt),0.2
          5);
   /* determine the two-phase heat transfer coefficient
          (see Breber, Palen, and Taborek p.7) */
          ftp=pow(phif2,0.45);
          hcs=1./(1./(hl*ftp)+1./hv);
          fg=0.79;
          hcg=fg*hv;
   /* heat transfer coefficient within zones i-iv
          (see Breber, Palen, and Taborek p.7) */
          if(izone==1)
            *h2p=hcs;
          if(izone==2)
            *h2p=hcg;
          if(izone==3)
            *h2p=hcs;
          if(izone==4)
            *h2p=hcs;
   /* heat transfer coefficient between zone boundaries
          (see Breber, Palen, and Taborek p.7) */
          if(fr>.5&&fr<1.5)
            *h2p=hcs*(fr-0.5)+hcg*(1.5-fr);
          if(xx>.5&&xx<1.5)
            *h2p=hcs*(1.5-xx)+hcg*(xx-0.5);
```

    }
}

Shell-Side Module

```c
void bundle(int is1,int is,int js,int it,int jt1,int
    jt,int ib)
{
int ibis;
double
    a0,ai,ao,at,bdt,cpg,dlt,dpdx,dpf,dpm,dt,dvg,gm,gmax,h
    ai,hao,haw,hc,hdk,hs1,ht1,pr,ps1,pt1,qmax,qmin,slip,t
    g,tsat,tubes,ua,ug1,us1,vs1,wall;
/* solve the integral equations for a cell in the tube
    bundles */
/* initialize new values using old values */
pt1=pt[it-1][jt1-1];
ht1=ht[it-1][jt1-1];
ug1=ug[it-1][jt1-1];
ps1=ps[is1-1][js-1];
vs1=vs[is1-1][js-1];
hs1=hs[is1-1][js-1];
us1=us[is1-1][js-1];
/* determine the heat exchange and flow areas */
tubes=(double)(ntr[it-1]-ntp[it-1]);
dlt=tl[ib-1]/(double)(NC-2);
if(jt==1||jt==NC)
    dlt=di[ib-1];
ai=M_PI*di[ib-1]*dlt*tubes;
a0=M_PI*d0[ib-1]*dlt*tubes;
ao=a0*af[ib-1];
at=M_PI*tubes*sq(di[ib-1])/4.;
/* determine the velocity */
ut[it-1][jt-1]=ft[it-1]*vt[it-1][jt1-1]/at/3600.;
if(rev[it-1])
    ut[it-1][jt-1]=-ut[it-1][jt-1];
/* determine the two-phase flow frictional pressure
    drop and heat transfer coefficient */
dt=fmax(1.,tt[it-1][jt1-1]-ts[is1-1][js-1]);
phase2(&pt[it-1][jt-1],&xt[it-1][jt-1],&ft[it-
    1],&at,&di[ib-1],&rt[it-1][jt-1],&fi[it-1][jt-
    1],&dpdx,&dt,&hi[it-1][jt-1],&slip,&al[it-1][jt-
    1],&ug[it-1][jt-1],&fp[it-1][jt-1],&ier);
if(ier)
    error(ier);
dpf=dpdx*dlt;
/* determine the momentum pressure drop */
dpm=(ug[it-1][jt-1]-ug1)/gsubc/2./3600./3600./144.;
/* determine the pressure exiting the cell */
pt[it-1][jt-1]=pt1+dpf+dpm;
```

```
/* determine the heat transfer coefficient on the
   inside of the tubes */
hai=hi[it-1][jt-1]*ai;
/* determine the heat transfer coefficient for the
   tube wall including the scale (crud) on both sides */
wall=2.*tk[ib-1]*(ao-ai)/log(ao/ai)/(d0[ib-1]-di[ib-
   1]);
hc=750.;
haw=1./(1./wall+1./(ao*hc));
/* determine the maximum mass flux across the tubes in
   the shell */
gm=fs[is-1][js-1]/dlt/wb[ib-1];
gmax=gm*smin[ib-1]/(smin[ib-1]-d0[ib-1]);
/* calculate the maximum local velocity in the shell
   */
us[is-1][js-1]=gmax*vs[is-1][js-1]/3600.;
/* compute thermodynamic and transport properties in
   the shell */
tsat=fTsat(ps[is-1][js-1]);
tg=TKgofT(tsat);
cpg=CpgofT(tsat);
dvg=MUgofT(tsat);
pr=cpg*dvg/tg;
/* calculate the frictional pressure drop by the
   method of Briggs and Young, Chem. Eng. Prog. Symp.
   Series No. 41, 59, 1965.
 * (see McGraw-Hill Handbook of Heat Transfer, p. 18-
   81) */
rs[is-1][js-1]=gmax*dr[ib-1]/dvg;
fo[is-1][js-1]=18.93*pow(st[ib-1]/smin[ib-
   1],0.515)/pow(rs[is-1][js-1],0.316)/pow(st[ib-
   1]/dr[ib-1],0.927);
dpf=-fo[is-1][js-1]*sq(gmax/3600.)*vs[is-1][js-
   1]/gsubc/144.;
/* calcualte the momentum pressure drop */
dpm=(sq(us1)/vs1-sq(us[is-1][js-1])/vs[is-1][js-
   1])/2./gsubc/144.;
/* calculate the pressure exiting the cell */
ps[is-1][js-1]=ps1+dpf+dpm;
/* determine the heat transfer coefficient on the
   outside of the tubes by the method of Robinson and
   Briggs,8th ASME/AIChE
 * National Heat Transfer Conference, No. 20,1965.
   (see McGraw-Hill Handbook of Heat Transfer, p.18-81)
   */
hdk=0.134*pow(rs[is-1][js-
   1],0.681)*pow(pr,0.333333)*pow(sp[ib-1]/fh[ib-
   1],0.2)*pow(sp[ib-1]/fk[ib-1],0.113);
ho[it-1][jt-1]=hdk*tg/dr[ib-1];
hao=ho[it-1][jt-1]*ao;
```

```
/* determine the overall conductance "ua" */
ua=1./(1./hai+1./haw+1./hao);
if(jt==1||jt==NC)
  {
  hi[it-1][jt-1]=0.;
  ho[it-1][jt-1]=0.;
  ua=0.;
  fo[is-1][js-1]=0.;
  ps[is-1][js-1]=ps1;
  }
u0[it-1][jt-1]=ua/a0;
/* determine the temperature difference (using a
  backward difference as this is conservative and never
  violates the Second
  * Law of Thermodynamics) and the heat transfer for
  this cell */
qmin=0.;
qmax=fmax(0.,ua*(tt[it-1][jt1-1]-ts[is1-1][js-1]));
/* use bisection method to determine heat transfer */
for(ibis=1;ibis<=20;ibis++)
  {
  qt[it-1][jt-1]=(qmin+qmax)/2.;
  if(ibis>1&&qmax-qmin<=.001*qt[it-1][jt-1])
    break;
  /* determine the exiting enthalpy from the first law
  */
  ht[it-1][jt-1]=ht1-qt[it-1][jt-1]/ft[it-1];
  hs[is-1][js-1]=hs1+qt[it-1][jt-1]/fs[is-1][js-1];
  /* check for gross overshoot in the heat transfer if
  there is overshoot divide the heat transfer by two
  and go back to the
  * First Law */
  if(!(ht[it-1][jt-1]>1499.||hs[is-1][js-1] <
    1.))
    {
    if(!(ht[it-1][jt-1]<1.||hs[is-1][js-1] >
      1499.))
      {
      /* determine the temperature,specific volume,and
  quality from the pressure and enthalpy */
      TofPH(pt[it-1][jt-1],ht[it-1][jt-1],&tt[it-
1][jt-1],&vt[it-1][jt-1],&xt[it-1][jt-1]);
      TofPH(ps[is-1][js-1],hs[is-1][js-1],&ts[is-
1][js-1],&vs[is-1][js-1],&xs[is-1][js-1]);
      /* bisection algorithm */
      if(!(qt[it-1][jt-1]>ua*(tt[it-1][jt-1]-ts[is-
1][js-1])||tt[it-1][jt-1]<=ts[is-1][js-1]))
        goto L_100;
      }
    qmax=qt[it-1][jt-1];
```

```
            continue;
        }
L_100:
    qmin=qt[it-1][jt-1];
    }
    /* calculate the average tube wall temperature,thermal
       expansion,and clamped stress */
    if(jt==1||jt==NC)
        {
        et[it-1][jt-1]=0.;
        cs[it-1][jt-1]=0.;
        }
    else
        {
        tw[it-1][jt-1]=(tt[it-1][jt-1]/hai+ts[is-1][js-
        1]/hao)/(1./hai+1./hao);
        bdt=bt[ib-1]*(tw[it-1][jt-1]-tw0);
        et[it-1][jt-1]+=bdt*dlt*12.;
        cs[it-1][jt-1]=bdt*es[ib-1];
        }
    }
```

Appendix C: Monte Carlo Codes

In order to create a Monte Carlo model within Excel® two functions are necessary: one to return a normally distributed random number and a second to count the number of occurrences between a range of values. The following section of VBA® code provides these:

```
function random(mean As Double, std As Double) As Double
    Dim i As Integer, r As Double
    r = 0
    For i = 1 To 12
      r = r + Rnd()
    Next i
    random = mean + std * (r / 6 - 1)
End Function
Function CountBetween(X As Range, Xmin As Double, Xmax
    As Double) As Long
    Dim i As Long, n As Long
    n = X.Count
    CountBetween = 0
    For i = 1 To n
      If (X(i) >= Xmin) Then
        If (X(i) < Xmax) Then
          CountBetween = CountBetween + 1
        End If
      End If
    Next i
End Function
```

The following code implements the Monte Carlo heat exchanger model from Chapter 17 and easily handles one million cases.

```
#define _CRT_SECURE_NO_DEPRECATE
#include <stdio.h>
#include <stdlib.h>
#include <memory.h>
#include <math.h>

void*allocate(unsigned count,unsigned siz)
  {
  void*ptr;
  if((ptr=calloc(count,siz))==NULL)
    {
    fprintf(stderr,"can't allocate memory\n");
    exit(1);
    }
  return(ptr);
  }

int nint(double d)
  {
  if(d>0.)
```

```
    return((int)(d+0.5));
  if(d<0.)
    return((int)(d-0.5));
  return(0);
  }

int urand()
  {
  int i,u;
  for(u=i=0;i<12;i++)
    u+=rand();
  return(u);
  }

double rnorm()
  {
  return(urand()/32767./6.-1.);
  }

double rdist(double a,double s)
  {
  return(a+6.*s*rnorm());
  }

double LMTD(double dT1,double dT2)
  {
  if(dT1<=0.||dT2<=0.)
    return(0.);
  if(fabs(dT1-dT2)<0.01)
    return(sqrt(dT1*dT2));
  return((dT1-dT2)/log(dT1/dT2));
  }

typedef struct{double avg,std;}RAN;

RAN mH ={4.00,0.40}; /* hot  side mass flow rate [kg/s]
   */
RAN CpH={3.50,0.05}; /* hot  side specific heat
   [kJ/kg/øC] */
RAN mC ={9.00,0.90}; /* cold side mass flow rate [kg/s]
   */
RAN CpC={2.50,0.04}; /* cold side specific heat
   [kJ/kg/øC] */
RAN THi={500.,0.67}; /* hot  side inlet temperature [øC]
   */
RAN THo={300.,0.67}; /* hot  side exit  temperature [øC]
   */
RAN TCi={250.,0.50}; /* cold side inlet temperature [øC]
   */
```

```c
RAN TCo={375.,0.50};   /* cold side exit  temperature [øC]
 */
double Area=100.;       /* surface area [mý] */

#define value(x) rdist(x.avg,x.std)

typedef struct{double
   mh,cph,mc,cpc,thi,tho,tci,tco,dt,qh,qc,uh,uc;}RES;

RES heatx(int lst)
  {
  static RES res;
  res.thi=value(THi);
  res.tho=value(THo);
  res.tci=value(TCi);
  res.tco=value(TCo);
  res.mh =value(mH);
  res.mc =value(mC);
  res.cph=value(CpH);
  res.cpc=value(CpC);
  res.qh=res.mh*res.cph*(res.thi-res.tho);
  res.qc=res.mc*res.cpc*(res.tco-res.tci);
  res.dt=LMTD(res.thi-res.tco,res.tho-res.tci);
  res.uh=1000.*res.qh/Area/res.dt;
  res.uc=1000.*res.qc/Area/res.dt;
  if(lst)
    printf("%4.0lf %4.0lf %5.2lf %5.1lf
    %5.1lf\n",res.qh,res.qc,res.dt,res.uh,res.uc);
  return(res);
  }

typedef struct{int p[50];double a,s,x[51];}STA;

STA stats(double*X,int n,char*name)
  {
  int i,j,m;
  double A,Xm,S,Xx;
  static STA sta;
  memset(&sta,0,sizeof(sta));
  Xm=Xx=X[0];
  A=S=0.;
  for(i=0;i<n;i++)
    {
    if(X[i]<Xm)
      Xm=X[i];
    if(X[i]>Xx)
      Xx=X[i];
    A+=X[i];
    S+=X[i]*X[i];
    }
```

```
  A/=n;
  S=sqrt((S-n*A*A)/(n-1));
  printf("%s: min=%1G, avg=%1G, max=%1G,
    std=%1G\n",name,Xm,A,Xx,S);
  sta.a=A;
  sta.s=S;
  m=sizeof(sta.x)/sizeof(sta.x[0]);
  for(i=0;i<m;i++)
    sta.x[i]=Xm+i*(Xx-Xm)/(m-1);
  for(i=0;i<n;i++)
    {
    j=nint((X[i]-Xm)*((double)(m-1))/(Xx-Xm));
    if(j<0)
       j=0;
    else if(j>m-2)
       j=m-2;
    sta.p[j]++;
    }
  return(sta);
  }

int main(int argc,char**argv,char**envp)
  {
  char fname[]="monte.csv";
  int i,m,n=1000000;
  double*Qh,*Qc,*dT,*Uh,*Uc;
  FILE*fp;
  RES res;
  STA sQh,sQc,sdT,sUh,sUc;
  printf("Monte Carlo Heat Exchanger Model\n");
  printf("%i cases\n",n);
  Qh=allocate(n,sizeof(double));
  Qc=allocate(n,sizeof(double));
  dT=allocate(n,sizeof(double));
  Uh=allocate(n,sizeof(double));
  Uc=allocate(n,sizeof(double));
  for(i=0;i<n;i++)
    {
    res=heatx((i%100000)==0);
    Qh[i]=res.qh;
    Qc[i]=res.qc;
    dT[i]=res.dt;
    Uh[i]=res.uh;
    Uc[i]=res.uc;
    }
  printf("done\n");
  sQh=stats(Qh,n,"Qh");
  sQc=stats(Qc,n,"Qc");
  sdT=stats(dT,n,"dT");
  sUh=stats(Uh,n,"Uh");
```

```
        sUc=stats(Uc,n,"Uc");
        if((fp=fopen(fname,"wt"))==NULL)
           {
           fprintf(stderr,"can't create output file
        %s\n",fname);
           exit(1);
           }
        fprintf(fp,"Qh,pQh,Qc,pQc,dT,pdT,Uh,pUh,Uc,pUc\n");
        m=sizeof(sQh.x)/sizeof(sQh.x[0]);
        for(i=0;i<m-1;i++)
           {
           fprintf(fp,"%lG,%lG,"
           ,(sQh.x[i]+sQh.x[i+1])/2.,((double)sQh.p[i])/((double
           )n));
           fprintf(fp,"%lG,%lG,"
           ,(sQc.x[i]+sQc.x[i+1])/2.,((double)sQc.p[i])/((double
           )n));
           fprintf(fp,"%lG,%lG,"
           ,(sdT.x[i]+sdT.x[i+1])/2.,((double)sdT.p[i])/((double
           )n));
           fprintf(fp,"%lG,%lG,"
           ,(sUh.x[i]+sUh.x[i+1])/2.,((double)sUh.p[i])/((double
           )n));

           fprintf(fp,"%lG,%lG\n",(sUc.x[i]+sUc.x[i+1])/2.,((dou
           ble)sUc.p[i])/((double)n));
           }
        fclose(fp);
        printf("see %s for results\n",fname);
        return(0);
        }
```

The output looks like this:

```
Monte Carlo Heat Exchanger Model
1000000 cases
3164 2701 81.09 390.1 333.1
2970 2953 82.39 360.5 358.4
2630 2680 81.65 322.1 328.3
3106 2940 81.99 378.9 358.6
2559 3091 81.63 313.5 378.7
3128 2953 81.56 383.5 362.1
3271 3303 82.19 398.0 401.9
2357 2451 81.42 289.5 301.0
2646 2611 82.02 322.6 318.4
2916 3186 81.99 355.7 388.6
done
Qh: min=1546.8,  avg=2800.4,  max=4099.52, std=283.679
Qc: min=1576.23, avg=2812.87, max=4008.61, std=285.542
dT: min=78.5398, avg=81.8496, max=84.9284, std=0.660767
Uh: min=191.825, avg=342.165, max=496.741, std=34.8006
```

```
Uc: min=191.586, avg=343.683, max=494.223, std=34.9743
see monte.csv for results
```
The graphics are in the second section of Chapter 17.

Appendix D: Steam Properties

The thermodynamic and transport properties of steam are defined in the various reports of the International Association for the Properties of Water and Steam. This is a link to their web site http://www.iapws.org/

Three primary versions are relevant in this case: IF-67[79], SF-95[80], and IF-97[81]. The designations IF and SF indicate Industrial and Scientific Formulations, respectively. The industrial formulations are in terms of pressure and temperature, while the scientific formulation is in terms of temperature and density. While the industrial formulations are less accurate and mathematically sloppy, they have been embraced throughout the industry. The notion that the scientific formulation is burdensome and unnecessarily complicated is, of course, made irrelevant by modern computers and well-crafted software. The IF-67 is used by some manufacturers (most notably G.E.) because it has been built into the basis of some archaic software and empirical correlations. There is precious little difference between the '67 and '97 industrial formulations when it comes to HRSG performance. The choice is most often driven by the steam turbine manufacturer and not by the HRSG manufacturer.

Excel® Add-In for Academic (Non-Commercial) Use Only

The most extensive text on Excel® Add-Ins is that of Steve Dalton.[82] This excellent document can be found in PDF form at several locations on the Web. Add-Ins (and, for that matter software) development is beyond the scope of this text. The source code provided can be compiled for either bitness. In fact, the batch file provided produces both. I have provided these steam properties and associated Excel® Add-In for academic (i.e., non-commercial) use only. While I have contributed considerable effort to translate these property functions into C and build them into the Add-In wrapper, the underlying code (mostly in FORTRAN) is the work of others (as noted) and can be found on the Web. There are several agents selling similar code for commercial. I do not want to compete with these. I also assume no liability for the use of this software. See the folder examples\AllSteam. You will also find a spreadsheet that generates a Mollier diagram in either English or SI units for any of the five formulations.

---

[79] Meyer, C. A., McClintock, R. B., Silvestri, G. J., and Spencer, R. C., Jr., *Thermodynamic and Transport Properties of Steam*, American Society of Mechanical Engineers, 1967.

[80] Wagner, W., and Pruß, A., "The IAPWS Formulation 1995 for the Thermodynamic Properties of Ordinary Water Substance for General and Scientific Use," Journal of Physical Chemistry, Ref. Data 31, pp. 387-535, 2002.

[81] Research and Technology Committee on Water and Steam in Thermal Power Systems, *ASME Steam Properties for Industrial Use*, The American Society of Mechanical Engineers.

[82] Dalton, S., Excel Add-in Development in C/C++: Applications in Finance, John Wiley & Sons, Ltd., Chichester, England, 2005.

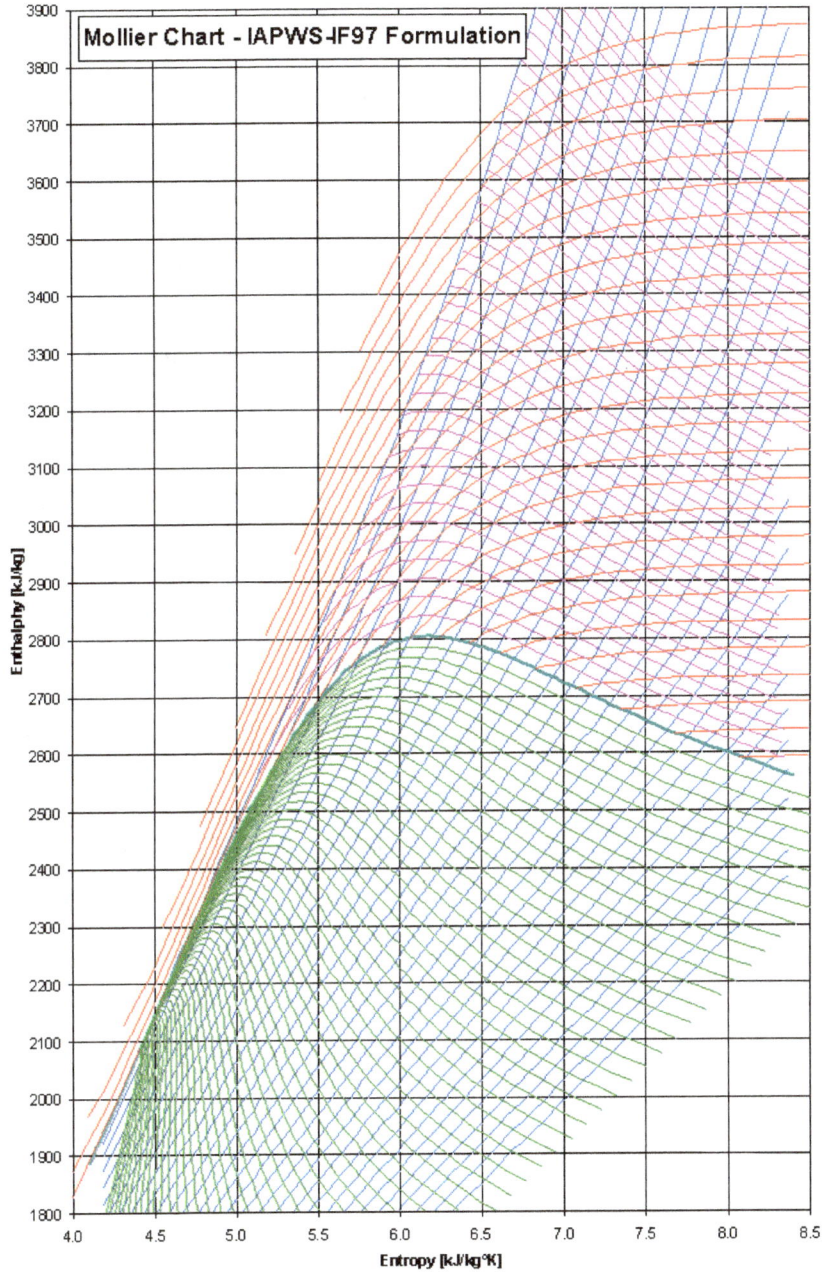

Appendix E: Exhaust Properties

We will only consider complete combustion, that is, combustion with more than ample excess air, which is characteristic of gas turbine exhaust and duct burner operation. While there are small levels of NOx and SOx present in some HRSGs, these contributions have negligible impact on the thermal performance, which is the only aspect of HRSGs with which we are concerned. We are not, for instance considering corrosion (material) or mechanical (structural) factors.

Gas properties are defined in the NASA Glenn report.[83] Any other references (e.g., ASME PTC22) are derived from these. You will find spreadsheets and source code in folder examples\exhaust. Note that specific heat and enthalpy are mass-weighted, not mole-weighted, and GT exhaust composition is most often specified in mole fractions. The simplest calculations for specific heat and enthalpy are listed on the next page after the figure.

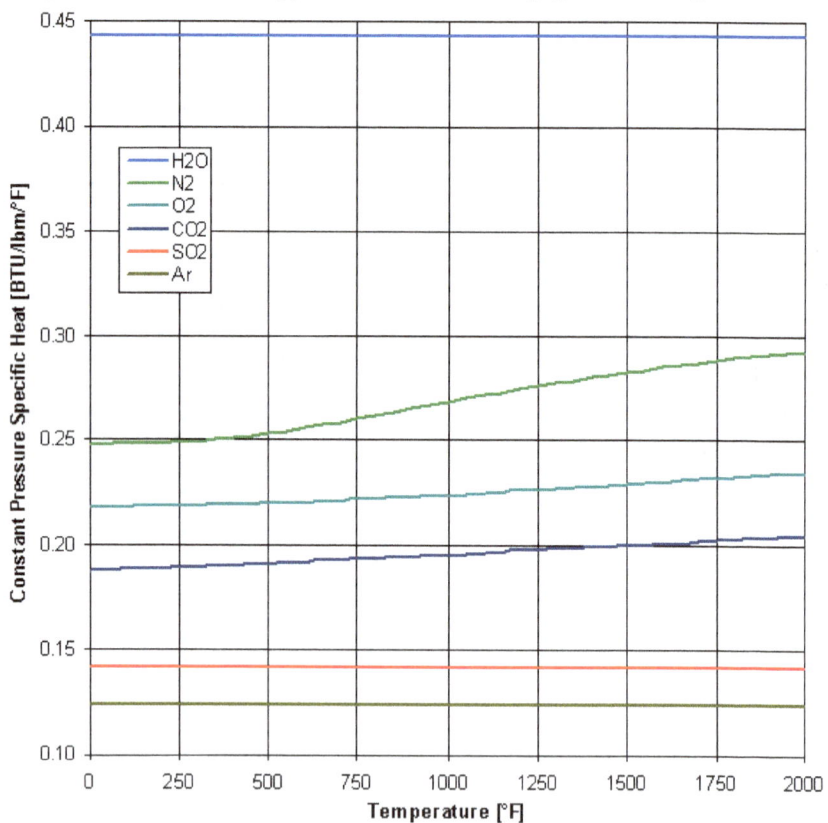

---

[83] McBride, B. J., Zehe, M. J., Gordon, S., "NASA Glenn Coefficients for Calculating Thermodynamic Properties of Individual Species," NASA Report No. 211556, 2002.

```c
typedef struct{double a,b,c,d,MW;}GAS;
GAS N2 ={-1.823844E-01,2.450080E-01,
9.515373E-06,1.278920E-09,28.013};
GAS O2 ={-1.173314E-02,2.177264E-01,
1.570874E-06,1.001867E-09,31.998};
GAS CO2={ 1.374508E-03,1.873375E-01,
3.953234E-06,1.303025E-10,44.010};
GAS H2O={ 9.393189E-05,4.431567E-01,
6.169783E-08,9.216964E-12,18.015};
GAS Ar ={ 7.922561E-14,1.242788E-01,
-2.526427E-18,9.458965E-22,39.948};
GAS SO2={ 1.495539E-05,1.417236E-01,
1.931810E-08,1.479716E-12,64.066};

double Hgas(MF mf,double T)
  {
  double HN2,HO2,HCO2,HH2O,HAr,HSO2,X,Y;
  HN2 =(( N2.d*T+ N2.c)*T+ N2.b)*T+ N2.a;
  HO2 =(( O2.d*T+ O2.c)*T+ O2.b)*T+ O2.a;
  HCO2=((CO2.d*T+CO2.c)*T+CO2.b)*T+CO2.a;
  HH2O=((H2O.d*T+H2O.c)*T+H2O.b)*T+H2O.a;
  HAr =(( Ar.d*T+ Ar.c)*T+ Ar.b)*T+ Ar.a;
  HSO2=((SO2.d*T+SO2.c)*T+SO2.b)*T+SO2.a;
  X= HN2* N2.MW*mf.N2
   + HO2* O2.MW*mf.O2
   +HCO2*CO2.MW*mf.CO2
   +HH2O*H2O.MW*mf.H2O
   + HAr* Ar.MW*mf.Ar
   +HSO2*SO2.MW*mf.SO2;
  Y= N2.MW*mf.N2
   + O2.MW*mf.O2
   +CO2.MW*mf.CO2
   +H2O.MW*mf.H2O
   + Ar.MW*mf.Ar
   +SO2.MW*mf.SO2;
  return(X/Y);
  }
double Tgas(MF mf,double H)
  {
  int iter;
  double T,T1,T2;
  T1=0.;
  T2=2000.;
  for(iter=0;iter<32;iter++)
    {
    T=(T1+T2)/2.;
    if(Hgas(mf,T)<H)
      T1=T;
    else
      T2=T;
```

```
        }
        return(T);
}
```

These calculations and additional spreadsheets can be found in the online archive in folder examples\exhaust. The most common gases are:

| | | | | NASA Glenn Properties | | | | | | | |
| | | | | specific heat [kJ/kg/°C] | | | | | enthalpy [kJ/kg] | | | |
	°K	°C	°F	N2	O2	CO2	H2O	Ar	N2	O2	CO2	H2O	Ar
ambient	275	2	35	1.04	0.92	0.82	1.86	0.52	-24	-21	-8961	-13466	-12
	300	27	80	1.04	0.92	0.85	1.86	0.52	2	2	-8940	-13420	1
	325	52	125	1.04	0.92	0.87	1.87	0.52	28	25	-8918	-13373	14
HRSG gas	350	77	170	1.04	0.93	0.89	1.88	0.52	54	48	-8896	-13326	27
	375	102	215	1.04	0.93	0.92	1.89	0.52	80	71	-8874	-13279	40
	400	127	260	1.04	0.94	0.94	1.90	0.52	106	95	-8850	-13232	53
	425	152	305	1.05	0.95	0.96	1.91	0.52	132	118	-8827	-13184	66
	450	177	350	1.05	0.96	0.98	1.93	0.52	158	142	-8803	-13136	79
	475	202	395	1.05	0.96	1.00	1.94	0.52	185	166	-8778	-13088	92
	500	227	440	1.06	0.97	1.01	1.96	0.52	211	190	-8753	-13039	105
	525	252	485	1.06	0.98	1.03	1.97	0.52	237	215	-8727	-12990	118
	550	277	530	1.06	0.99	1.05	1.99	0.52	264	239	-8701	-12940	131
	575	302	575	1.07	1.00	1.06	2.00	0.52	291	264	-8675	-12891	144
	600	327	620	1.07	1.00	1.08	2.02	0.52	317	289	-8648	-12840	157
	625	352	665	1.08	1.01	1.09	2.03	0.52	344	314	-8621	-12790	170
	650	377	710	1.09	1.02	1.10	2.05	0.52	372	339	-8594	-12739	183
	675	402	755	1.09	1.02	1.11	2.06	0.52	399	365	-8566	-12687	196
	700	427	800	1.10	1.03	1.13	2.08	0.52	426	391	-8538	-12636	209
GT exhaust	725	452	845	1.10	1.04	1.14	2.10	0.52	454	416	-8510	-12583	222
	750	477	890	1.11	1.04	1.15	2.12	0.52	481	443	-8481	-12531	235
	775	502	935	1.12	1.05	1.16	2.13	0.52	509	469	-8452	-12478	248
	800	527	980	1.12	1.05	1.17	2.15	0.52	537	495	-8423	-12424	261
	825	552	1025	1.13	1.06	1.18	2.17	0.52	565	521	-8394	-12370	274
	850	577	1070	1.13	1.06	1.19	2.18	0.52	593	548	-8364	-12316	287
	875	602	1115	1.14	1.07	1.20	2.20	0.52	622	575	-8334	-12261	300
	900	627	1160	1.15	1.07	1.20	2.22	0.52	650	601	-8304	-12206	313

The offset in the gas enthalpies is the heat of formation, which we don't want in our HRSG enthalpy but do want in the combustion calculations.

Appendix F: Combustion Calculations

We will only cover those aspects of combustion essential to HRSG modeling, which also means that we will only cover oil- and gas-fired duct burners. We will not cover gas turbine combustion, as the products of combustion are provided by the engine manufacturer and may account for inlet humidification (evaporative cooling or fogging), steam power augmentation, and water injection to control NOx formation.

Complete combustion of hydrocarbon fuels with moist air having excess oxygen is typical of gas turbines. The chemical reactions are often generalized in terms of the average fuel composition and the H/C molar ratio (m/n in this case). This reaction can be written:

$$0.7808 N_2 + 0.2095 O_2 + 0.0004 CO_2 + 0.0093 Ar + w H_2 O$$
$$+ a(Cn + Hm) = \alpha N_2 + \beta O_2 + \gamma CO_2 + \delta Ar + \varepsilon H_2 O \quad \text{(F.1)}$$

$$\alpha = 0.7808$$
$$\delta = 0.0093 \quad \text{(F.2)}$$

$$\beta = 0.2095 - a\left(n + \frac{m}{2}\right)$$
$$\gamma = 0.0004 + an \quad \text{(F.3)}$$
$$\varepsilon = w + \frac{m}{2}$$

The mole fractions of the constituents can be calculated directly from these parameters. The humidity ratio must be modified to account for the differing molecular weights of dry air and water vapor. The ambient and combustion functions are listed below:

```
typedef struct{double N2,O2,CO2,H2O,Ar;}MF;

MF Ambient(double baro,double Tdb,double RH)
    {
    double W,x,y;
    static MF mf;
    W=fWdbrh(baro,Tdb,RH);
    x=W/(1.+W);
    y=x*28.9645/18.01534;
    mf.H2O=y;
    mf.N2 =0.7808*(1.-y);
    mf.O2 =0.2095*(1.-y);
    mf.CO2=0.0004*(1.-y);
    mf.Ar =0.0093*(1.-y);
    return(mf);
    }
```

```
MF Combustion(MF mf1,double Fair,double Ffuel,double
   HCratio)
{
double C,H,S;
static MF mf2;
H=HCratio/(1.+HCratio);
C=1.-H;
mf2.N2 =Fair*mf1.N2;
mf2.O2 =Fair*mf1.O2 -Ffuel*(C+H/2.);
mf2.CO2=Fair*mf1.CO2+Ffuel*C;
mf2.Ar =Fair*mf1.Ar;
mf2.H2O=Fair*mf1.H2O+Ffuel*H/2.;
S=mf2.N2+mf2.O2+mf2.CO2+mf2.Ar+mf2.H2O;
mf2.N2 /=S;
mf2.O2 /=S;
mf2.CO2/=S;
mf2.Ar /=S;
mf2.H2O/=S;
return(mf2);
}
```

These calculations and additional spreadsheets can be found in the online archive in folder examples\exhaust and also examples\CCPP.

Appendix G. Moist Air Properties

The only properties of moist air considered are those developed by Hyland & Wexler[84,85,86], and refined by Nelson & Sauer.[87] The more recent formulation developed by Hermann, Kretzschmar, and Gatley[88,89] do not constitute a substantive improvement, merely an academic one; so there is little point implementing these. Moist air properties appear in various editions of the *ASHRAE Handbook of Fundamentals*. Beware that the equations in many editions of this otherwise excellent reference are wrong in that the equations contained therein don't produce the tabulated results.

In 1984 this author was part of a Cooling Technology Institute (CTI) task force investigating discrepancies in the published properties of moist air. The National Bureau of Standards (NBS)—now the National Institute of Standards and Technology (NIST)—lost Hyland & Wexler's original reports; however, a copy still existed in the Library of Congress (LoC). A colleague, Al Feltzin, went to the LoC and made a photocopy of the original reports. The tabulated values in the ASHRAE handbook are correct, but not all of the equations are, especially before 1993.

Formulations consistent with Hyland & Wexler, along with code plus an Excel® Add-In can be found in my book, *Evaporative Cooling*, and on my web site listed in the foreword. The code, calculations, and spreadsheets can be found in the online archive in folder examples\moistair. The humidity ratio, W, is the ratio of the mass of water to mass of dry air and is given by:

$$W = \left(\frac{MW_{H2O}}{MW_{AIR}}\right)\left(\frac{fP_{SAT}}{P_{BARO} - fP_{SAT}}\right) \quad (G.1)$$

Where MW_{H2O} and MW_{AIR} are the molecular weights of water and air, respectively. P_{SAT} and P_{BARO} are the saturation and barometric pressures,

---

[84] Hyland, R. W., Wexler, A., and Stewart, R., "Thermodynamic Properties of Dry Air, Moist Air and Water and SI Psychrometric Charts," ASHRAE RP-216 and RP-25, 1983.

[85] Hyland, R. W. and Wexler, A., "Formulations for the Thermodynamic Properties of the Saturated Phases of H2O from 173.15 K to 473.15 K," ASHRAE Trans., Vol. 89, pp. 500-519, 1983.

[86] Hyland, R. W. and Wexler, A., "Formulations for the Thermodynamic Properties of Dry Air from 173.15 K to 473.15 K, and of Saturated Moist Air from 173.15 K to 372.15 K, at Pressures to 5 MPa," ASHRAE Trans., Vol. 89, pp. 520-535, 1983.

[87] Nelson, H. F. and Sauer, H. J., "Formulation of High-Temperature Properties for Moist Air," HVAC&R Research Vol. 8, pp. 311-334, 2002.

[88] Herrmann, S., Kretzschmar, H.-J., and Gatley, D. P., "Thermodynamic Properties of Real Moist Air, Dry Air, Steam, Water, and Ice," HVAC&R Research, 2009.

[89] Herrmann, S., Kretzschmar, H.-J., and Gatley, D. P., "Thermodynamic Properties of Real Moist Air, Dry Air, Steam, Water, and Ice - Final Report," ASHRAE RP-1485, 2009.

respectively. The enhancement factor, f, is the ratio of the effective partial pressure of water vapor in air at the saturation point to the saturation pressure of steam alone (no air present). The enhancement factor varies with temperature and barometric pressure and is shown in the figure below for one atmosphere:

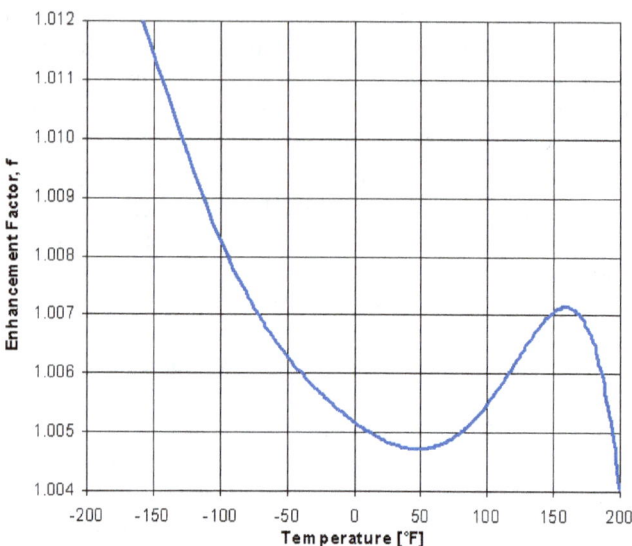

The saturation pressure for water vapor (against water liquid without air present) is shown in this next figure:

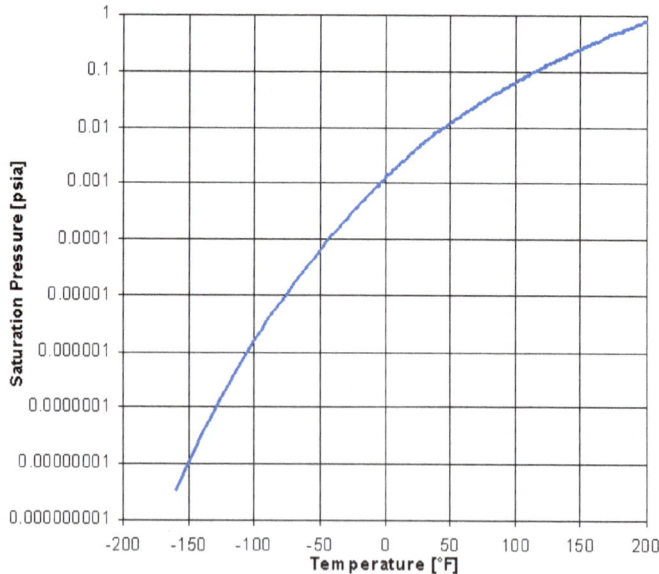

The resulting humidity ratio is then:

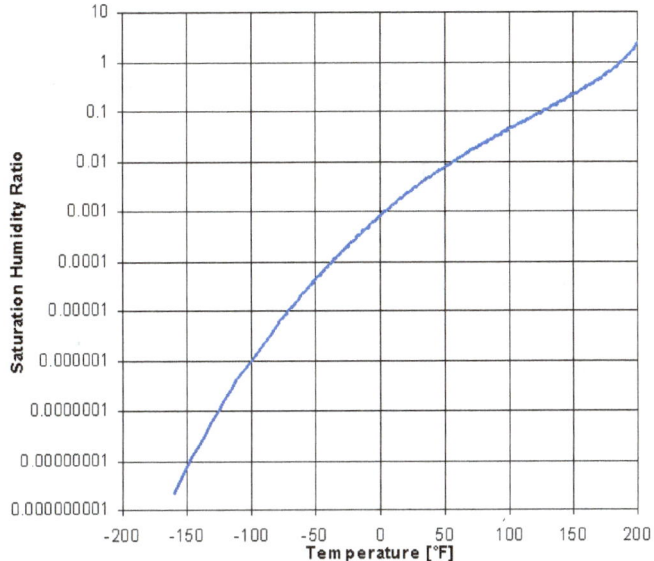

The enthalpy of moist air (per pound of dry air) is given by:

$$h = h_A + W h_G = 0.24\,T + W(1061 + 0.444\,T) \qquad (G.2)$$

where h_A and h_G are the enthalpies of dry air and water vapor, respectively. The enthalpy of water vapor is given by:

$$h_G = 1061 + 0.444\,T \qquad (G.3)$$

Note that this is the enthalpy of saturated water vapor, not the latent heat (h_{FG}) as is sometimes thought and even cited in the literature.[90]

Note that you will need rigorous psychrometric calculations at the gas turbine (GT) inlet, but DO NOT use these in the HRSG, as they are meaningless above the boiling point (212°F/100°C). In that case, use NASA Glenn, but be aware that the references are different. ASHRAE (Hyland & Wexler) reference liquid water at the triple point and NASA Glenn reference vapor water at standard conditions. If you do not consider this difference in the energy balance around a GT, you will get erroneous results.

---

[90] Need proof? Consider this… if you were to add steam at the critical point to dry air, you would significantly increase the enthalpy of the mixture; however, at the critical point h_{FG} is zero.

Appendix H. Heat Transfer Coefficients

Aside from tube wall resistance, which is trivial to calculate, the most important contributors to the overall heat transfer coefficient, U, are the inside and outside convective transports plus boiling in the case of evaporators. Fouling resistance is generally assumed, based on typical observed values, which are readily available. These conductances are combined as reciprocal resistances using Equation 20.1.

Convective Heat Transfer Coefficients

The convective contributions inside and outside the tubes are calculated similarly, only different correlations and properties are used for the gas and steam sides. These correlations all take the form of the classic equation, as developed by Dittus&Boelter[91]:

$$Nu = \frac{hD}{k} = 0.023 \, Re^{\frac{4}{5}} \, Pr^n \quad (H.1)$$

where n=0.4 for heating and n=0.3 for cooling. A more recent correlation has been developed by Sieder&Tate[92]:

$$Nu = 0.027 \, Re^{\frac{4}{5}} \, Pr^{\frac{1}{3}} \left(\frac{\mu_{BULK}}{\mu_{SURFACE}} \right)^{0.14} \quad (H.2)$$

Yet another correlation has been developed by Rabas&Cane[93]:

$$Nu = 0.0158 \, Re^{0.835} \, Pr^{0.462} \quad (H.3)$$

Resistance of the tube wall can be found in any heat transfer text:

$$R_W = \frac{D_O}{2k_W} \ln\left(\frac{D_O}{D_I}\right) \quad (H.4)$$

There are numerous correlations for the shell side heat transfer coefficients. There is considerable discussion in the literature on which correlation should be preferred and I suspect that each manufacturer has their own equation that has been fine-tunde. These vary with the fluids, baffles, and tube spacing. Fouling is often used as a *fudge* factor. The overall heat transfer coefficient is found by summing the resistances:

---

[91] Dittus, P. W. and L. M. Boelter, L.M., University of California Publications in Engineering, Vol. 1, No. 13, pp. 443-461 1930 (reprinted in *International Communications in Heat and Mass Transfer*, Vol. 12, pp. 3-22, 1985).
[92] Sieder, E. N. and G. E. Tate, "Heat Transfer and Pressure Drop of Liquids in Tubes," Industrial Engineering Chemistry, Vol. 28, p. 1429, 1936.
[93] Rabas, T. J., and D. Cane, "An Update of Intube Forced Convection Heat Transfer Coefficients of Water," *Desalinization*, Vol. 44, pp. 109-119, 1983.

$$\frac{1}{U} = \left(\frac{D_O}{D_I}\right)(R_{FI} + R_T) + R_M + R_S + R_{FO} \quad (H.5)$$

Here R_{FI} and R_{FO} are the inside and outside fouling resistance, respectively. R_T, R_W, and R_S are the tube side, tube material, and shell side resistances, respectively. The D_O/D_I term accounts for the fact that the inner and outer surface areas of a tube aren't the same per unit length.

There are also several methods for estimating the shell side heat transfer coefficient. One of the earliest methods was developed by Kern[94], based on industrial heat exchangers and is quite similar to the Sieder-Tate.

$$Nu = 0.36 \, Re^{0.55} \, Pr^{\frac{1}{3}} \left(\frac{\mu_{BULK}}{\mu_{SURFACE}}\right)^{0.14} \quad (H.6)$$

This correlation could also be cast in the same form as Dittus-Boelter for convenience:

$$Nu = 0.36 \, Re^{0.55} \, Pr^n \quad (H.7)$$

Tube Bundles

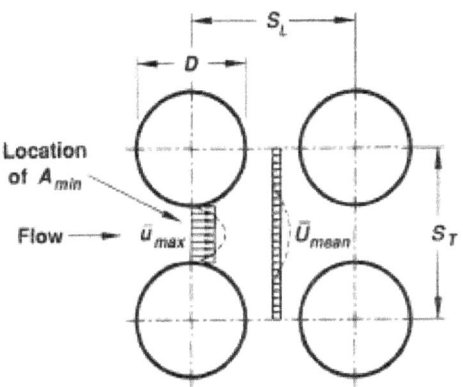

Inline Tube Bank

The length in both the Nusselt and Reynolds number is an equivalent diameter (D_E), that takes into account the tube spacing, pitch (P), and bundle alignment. This first equation (E.7) is for a square tube arrangement:

$$D_E = \frac{4P^2}{\pi D_O} - D_O \quad (H.8)$$

and this second (E.8) is for a triangular arrangement:

---

[94] Kern, D. Q., *Process Heat Transfer*, McGraw-Hill, 1950.

$$D_E = \frac{2\sqrt{3}P^2}{\pi D_O} - D_O \tag{H.9}$$

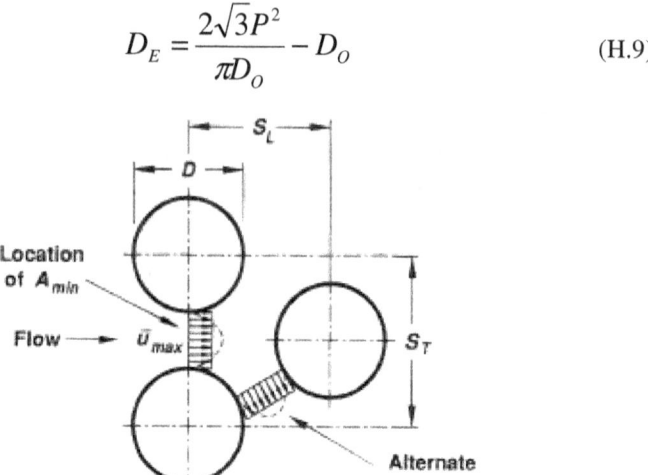

Staggered Tube Bundle

The shell-side velocity, V_s, is given by:

$$V_S = \frac{\dot{m}_S}{\rho N_R (P - D_O) \left(\dfrac{L_T}{N_B}\right)} \tag{H.10}$$

where N_R is the number of tubes per row (across the flow), L_T is the tube length, and N_B is the number of baffles. You may want to adjust the effective shell-side area to better represent the actual tube bundle and baffle arrangement in your heat exchanger. The Reynolds and Prandtl numbers are given by:

$$\text{Re} = \frac{DV}{\nu} \tag{H.11}$$

$$\text{Pr} = \frac{\mu C}{k} \tag{H.12}$$

You can devote considerable effort to calculating and refining these heat transfer coefficients, but always remember that the end result should be in the range discussed in Chapter 20. Years of performance testing experience have proven that there are limits to what may be practically achieved.

Various Tube Bundles
Finned Tubes

An excellent source of information for finned tubes is the Wolverine Engineering Data Book III.[95] This reference contains a variety of correlations, which you may find useful. Many correlations for convection over finned tubes are inspired by the Reynolds Analogy, which states that the Stanton Number is equal to the friction factor divided by two:

$$St = \frac{f}{2} \qquad (H.13)$$

The Stanton number is also equal to the Nusselt number divided by the Reynolds and Prandtl numbers.

$$St = \frac{Nu}{\text{Re Pr}} = \frac{h}{\rho u C_p} \qquad (H.14)$$

Fins do increase the heat transfer coefficient and also the pressure drop. Fin designs vary considerably. One common design is shown on the next page.

---

[95] Thome, J. R. *Engineering Data Book III*, Wolverine Tube, Inc. 2009.

Finned Tubes

Appendix I. Flow Measurement

The only accurate flow measurement for HRSG testing is either a Coriolis meter or sharp-faced orifice for fuel gas flow or an ASME-grade calibrated nozzle for feedwater. Performing a test based on an uncalibrated and/or unverifiably clean nozzle is a waste of time and money. Do not bother measuring steam flow in the vapor phase. The applicable standards for such calculations are either ASME MFC-3M (1989 or 2004), "Measurement of Fluid Flow in Pipes Using Orifice, Nozzle, and Venturi" and ASME PTC-19.5 (2004), "Flow Measurement." These standards contain various corrections and correlations for each of the pertinent devices. There are also ranges of applicability for physical dimensions, for instance, diameter ratios. Do not exceed these. The equations will not be duplicated here, as there are numerous considerations and figures that must be considered and these details are beyond the scope of this text.

Make sure that the range of operating conditions are within the applicability for each flow device. More than once I have encountered an otherwise well-designed system that could not be tested within the range of calibration of the feedwater flow nozzle and still demonstrate the guarantees. I have also encountered nozzles that were calibrated at Reynolds numbers far removed from the intended installation. Yet other cases exist where a very expensive nozzle ($15K) was never calibrated. Cleanliness is also important. It is often presumed that feedwater nozzles are clean, but this may not be true and these should be inspected. Physical damage (e.g., scraping, gouging, dents, etc.) can be sustained during shipping, storage, and installation so that scaling and corrosion are not the only things to consider. More than once I have seen a sharp-faced orifice installed backwards. The sharp face goes into the flow, as shown below:

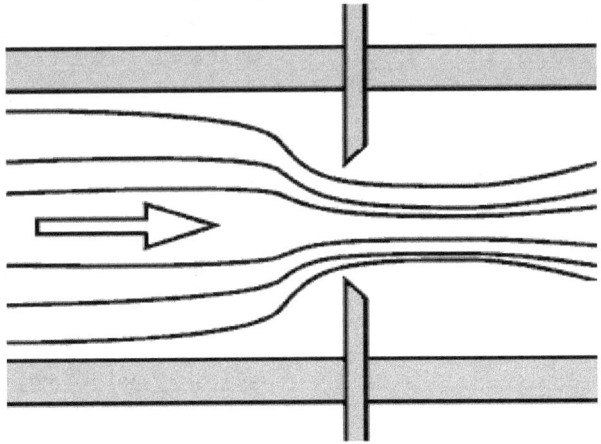

The calculation of flow can be quite involved. The following is typical for fuel gas through an orifice.

Input Data	Units	Test1	Test2	Test3	Avg.

Meter Type		ORIFICE	ORIFICE	ORIFICE	ORIFICE
Pipe Material		SS3	SS3	SS3	SS3
Throat Material		SS3	SS3	SS3	SS3
Pipe Diameter	inches	7.9810	7.9810	7.9810	7.9810
Throat Diameter	inches	4.7886	4.7886	4.7886	4.7886
Beta Ratio		0.6000	0.6000	0.6000	0.6000
Flowing Pressure	psia	483.40	483.32	483.17	483.30
Flowing Temperature	Deg F	25.69	25.41	26.21	25.77
Differential Pressure	in. H2O	120.63	120.13	120.24	120.34
Frequency	pulse	120.63	120.13	120.24	120.34
Fluid Type		GAS	GAS	GAS	GAS
Pipe Alpha Coefficient		9.000E-06	8.999E-06	9.001E-06	9.000E-06
Throat Alpha Coefficient		9.000E-06	8.999E-06	9.001E-06	9.000E-06
Pipe Reynolds Number		5,431,835	5,420,046	5,406,809	5,419,563
Corrected Pipe Diameter	inches	7.9775	7.9775	7.9776	7.9775
Corrected Throat Diameter	inches	4.7865	4.7865	4.7865	4.7865
Corrected Beta Ratio		0.6000	0.6000	0.6000	0.6000
Tap Term (for use in AGA 3 extrap. Cd)		-0.000209	-0.000209	-0.000209	-0.0002
A Term (for use in AGA 3 extrap. Cd)		0.007204	0.007216	0.007231	0.0072
C Term (for use in AGA 3 extrap. Cd)		0.553055	0.553476	0.553950	0.5535
Discharge Coefficient (ASME MFC 3M)		0.6045	0.6045	0.6045	0.6045
Discharge Coefficient (AGA3)		0.6041	0.6041	0.6041	0.6041
Discharge Coefficient (Cal. AGA R3)		0.6031	0.6031	0.6031	0.6031
Discharge Coefficient (Cal. PTC 19.5)		0.6041	0.6041	0.6041	0.6041
Discharge Coefficient		0.6045	0.6045	0.6045	0.604
Critical Pressure, Pcr	psia	667.95	668.29	666.98	667.74
Critical Temperature, Tcr	Deg F	346.64	346.49	345.89	346.34
Critical Volume, Vcr	ft^3/lb	0.0980	0.0980	0.0981	0.0980
Specific Gravity Rel. To Air, Gideal	ratio	0.5726	0.5717	0.5719	0.5721
Base Pressure, Pbase	psia	14.69	14.69	14.69	14.69
Base Temperature, Tbase	Deg F	70.00	70.00	70.00	70.00
AGA 8 Calculated Fuel Density	lb / ACF	1.6826	1.6802	1.6760	1.6796
AGA 8 Calculated Base Fuel Density	lb / SCF	0.0439	0.0438	0.0438	0.0439
Gas Viscosity	lbm/ft-sec	7.411E-06	7.407E-06	7.419E-06	7.412E-06
Natural Gas Flow Rate using NX-19 Density	KSCFH	1,761.83	1,760.00	1,757.99	1,759.94
Natural Gas Flow Rate using AGA8 Density	KSCFH	1,723.94	1,722.00	1,720.12	1,722.02
Natural Gas Flow Rate using AGA8 Density	KPPH	75.67	75.46	75.40	75.51

Nozzle flow calculations can also be quite involved, as illustrated in the following table:

Description	units	Design	Test1	Test2	Test3	Avg/
INPUTS						
Meter Type		NOZZLE	NOZZLE	NOZZLE	NOZZLE	NOZZLE
Pipe Material		CS	CS	CS	CS	CS
Throat Material		SS3	SS3	SS2	SS3	SS4
Tap Set		1.00	1.00	1.00	1.00	1.00
Pipe Diameter	inches	18.3771	18.3771	18.3771	18.3771	18.3771
Throat Diameter	inches	8.8875	8.8875	8.8875	8.8875	8.8875
Flowing Temperature	Deg F	270.0	270.0	270.0	276.0	272.0
Flowing Pressure	psia	350.0	350.0	350.0	348.0	349.3
Differential Pressure	in. H2O	377.7	398.8	378.1	372.0	383.0
Frequency	pulse	0.00	0.00	-1.00	0.00	-0.33
Fluid Type		WATER	WATER	WATER	WATER	WATER
Viscosity	lbm/ft-sec	1.631E-06	1.631E-06	1.631E-06	1.588E-06	1.617E-06
Fluid Density (1997)	lb/ft^3	58.3106	58.3106	58.3106	58.1301	58.2504
Reynolds Number, Re	-	6.16E+08	6.33E+08	6.15E+08	6.27E+08	6.25E+08
Gas Expansion Factor, Y1	-	1.0000	1.0000	1.0000	1.0000	1.0000
Beta @ Flowing Conditions	-	0.4839	0.4839	0.4836	0.4839	0.4838
Disc. Coeff. - PTC 19.5	-	1.0025	1.0025	1.0025	1.0025	1.0025
Disc. Coeff. (manual calc.)	-	0.9973	0.9973	0.9973	0.9973	0.9973
RESULTS						
Flow Rate (method 1)	lb/hr	4,354,496	4,474,396	4,351,473	4,315,146	4,380,338
Flow Rate (method 2)	lb/hr	4,355,040	4,474,956	4,352,017	4,315,680	4,380,884

Appendix J. Cooling Tower Terms

approach: the difference between the cold water (leaving) temperature and the ambient wet-bulb (for a cooling tower)

area per unit volume: It is not practical to measure the interfacial contact area between air and water droplets or sheets in cooling tower fill, but this is needed for mass transfer calculations. The fill (or packing) volume is easily measured. This parameter always appears multiplied by the mass or sensible heat transfer coefficients, so that the actual value is never needed.

backpressure: When a steam turbine exhausts into a condenser, the pressure in the condenser is "felt" at the turbine exit (i.e., "pushes" back).

dew point: the temperature at which condensation begins to form

drift eliminators: devices (usually in the shape of chevrons) designed to remove water droplets from moving air, thus reducing liquid carry-over and possibly pressure drop

dry-bulb: the conventional ambient air temperature

equilibrium temperature: For a cooling pond this is the temperature the pond would eventually reach, if all of the influences were held constant. The intent of this is to account for ambient temperature, ambient relative humidity, wind speed, and solar heating. Of course, this isn't possible, if for no other reason, the Sun doesn't remain at the same zenith.

rain zone: in a cooling tower beneath the fill (or packing) where the water droplets fall through the incoming air

range: the difference between the hot water (entering) temperature and the cold water (leaving) temperature (for a cooling tower)

wet-bulb: roughly equivalent to the adiabatic (no heat transfer) saturation temperature; measured by a temperature instrument covered with a wetted wick

Appendix K: Moist Air Property Functions

Accurate and fast moist air property functions are essential to all calculations involving evaporative cooling. The following VBA code provides all of the thermodynamic properties of moist air you will need. This code may also be found in several of the spreadsheets contained in the on-line archive that accompanies this text.

```
Option Explicit
'These properties are based on the 1993 ASHRAE Handbook
    of Fundamentals
'Chapter 6 Table 2 which is derived from the work of
    Hyland & Wexler
'The temperature and pressure dependence of the
    enhancement factor, f, are from
'Table 4 of this Section as well as Table 2 of Section 5
    of the 1977 Handbook.
'!!!!!!!!!!!!!!!!!!!!!!!!!!!!!!!!!!!!!!!!!!!!!!!!!
'Note: relative humidity is 0 and 1, not 0 to 100!
'!!!!!!!!!!!!!!!!!!!!!!!!!!!!!!!!!!!!!!!!!!!!!!!!!
Function fPs(Ts As Double) As Double
'saturation pressure of water
    Dim T As Double
    T = Ts + 459.67
    If (Ts < 32#) Then
        fPs = Exp((((-9.0344688E-14 * T + 3.5575832E-10) * T
        + 0.00000019202377) * T _
        - 0.0053765794) * T - 4.8932428 - 10214.165 / T +
        4.1635019 * Log(T))
    Else
        fPs = Exp(((-2.4780681E-09 * T + 0.00001289036) * T
        - 0.027022355) * T _
        - 11.29465 - 10440.397 / T + 6.5459673 * Log(T))
    End If
End Function
Function fT(T As Double) As Double
'temperaure dependence of the enhancement factor, f
    fT = ((((-1.22884575824E-13 * T + 2.98501566007E-11) *
    T - 1.68742180749E-09) * T _
    + 1.64960460192E-07) * T - 1.37450341076E-05) * T +
    1.00432811563436
End Function
Function fP(T As Double, P As Double) As Double
'pressure dependence of the enhancement factor
    fP = 0.9999278133 + (0.00002200673657 + (6.729181914E-
    08 - 7.590092091E-10 * T) * T) * T _
    + (0.0002997989681 + (-0.000001718254802 +
    8.054776534E-09 * T) * T - 3.40732948E-09 * T * P) *
    P
End Function
Function fTP(T As Double, P As Double) As Double
```

```
'composite (both temperature and pressure) dependence of
    the enhancement factor
  fTP = fT(T) * fP(T, P) / fP(T, 14.696)
End Function
Function fWdrh(Pbaro As Double, Tdb As Double, RH As
    Double) As Double
'humidity ratio from the pressure, dry-bulb, and
    relative humidity
  Dim f As Double, Ps As Double, Pw As Double
  f = fTP(Tdb, Pbaro)
  Ps = fPs(Tdb)
  Pw = f * RH * Ps
  fWdrh = 0.62198 * Pw / (Pbaro - Pw)
End Function
Function fPw(Pbaro As Double, W As Double) As Double
'partial pressure of water vapor in air, Pw=f*RH*Ps
  fPw = W * Pbaro / (0.62198 + W)
End Function
Function fDew(Pw As Double) As Double
'dew point from partial pressure
  Dim iter As Integer, T1 As Double, T2 As Double
  T1 = -80
  T2 = 200
  For iter = 1 To 32
    fDew = (T1 + T2) / 2
    If (fT(fDew) * fPs(fDew) < Pw) Then
      T1 = fDew
    Else
      T2 = fDew
    End If
  Next iter
End Function
Function fTdprh(Pbaro As Double, Tdb As Double, RH As
    Double) As Double
'dew point from dry-bulb and relative humidity
  fTdprh = fDew(fPw(Pbaro, fWdrh(Pbaro, Tdb, RH)))
End Function
Function fWdwb(Pbaro As Double, Tdb As Double, Twb As
    Double) As Double
'humidity ratio from pressure, dry-bulb, and wet-bulb
  Dim Ws As Double
  Ws = fWdrh(Pbaro, Twb, 1#)
  fWdwb = ((1093 - 0.556 * Twb) * Ws - 0.24 * Tdb + 0.24
    * Twb) / (1093 + 0.444 * Tdb - Twb)
End Function
Function fTdpwb(Pbaro As Double, Tdb As Double, Twb As
    Double) As Double
'dew point from dry-bulb and wet-bulb
  fTdpwb = fDew(fPw(Pbaro, fWdwb(Pbaro, Tdb, Twb)))
End Function
```

```
Function fTwdrh(Pbaro As Double, Tdb As Double, RH As
    Double) As Double
'wet-bulb from dry-bulb and relative humidity
  Dim iter As Integer, T1 As Double, T2 As Double, W As
    Double
  W = fWdrh(Pbaro, Tdb, RH)
  T1 = ((4.94704224634E-06 * Tdb - 0.00303815378496) *
    Tdb + 0.85025907521) * Tdb - 2.85417545122
  T2 = Tdb
  For iter = 1 To 32
    fTwdrh = (T1 + T2) / 2
    If (fWdwb(Pbaro, Tdb, fTwdrh) < W) Then
      T1 = fTwdrh
    Else
      T2 = fTwdrh
    End If
  Next iter
End Function
Function fRHdwb(Pbaro As Double, Tdb As Double, Twb As
    Double) As Double
'relative humidity from dry-bulb and wet-bulb
  Dim iter As Integer, R1 As Double, R2 As Double, W As
    Double
  W = fWdwb(Pbaro, Tdb, Twb)
  R1 = 0
  R2 = 1
  For iter = 1 To 32
    fRHdwb = (R1 + R2) / 2
    If (fWdrh(Pbaro, Tdb, fRHdwb) < W) Then
      R1 = fRHdwb
    Else
      R2 = fRHdwb
    End If
  Next iter
End Function
Function fTwdp(Pbaro As Double, Tdb As Double, Tdp As
    Double) As Double
'wet-bulb from dry-bulb and dew point
  Dim iter As Integer, T1 As Double, T2 As Double, W As
    Double
  W = fWdrh(Pbaro, Tdp, 1)
  T1 = ((4.94704224634E-06 * Tdb - 0.00303815378496) *
    Tdb + 0.85025907521) * Tdb - 2.85417545122
  T2 = Tdb
  For iter = 1 To 32
    fTwdp = (T1 + T2) / 2
    If (fWdwb(Pbaro, Tdb, fTwdp) < W) Then
      T1 = fTwdp
    Else
      T2 = fTwdp
```

```
      End If
    Next iter
  End Function
  Function fHdbw(Tdb As Double, W As Double) As Double
  'enthalpy of moist air per pound of DRY air
      fHdbw = 0.24 * Tdb + W * (1061 + 0.444 * Tdb)
  End Function
  Function fHdrh(Pbaro As Double, Tdb As Double, RH As
        Double) As Double
  'enthalpy of moist air per pound of DRY air
      fHdrh = fHdbw(Tdb, fWdrh(Pbaro, Tdb, RH))
  End Function
  Function fSdbw(Pbaro As Double, Tdb As Double, W As
        Double) As Double
  'entropy of moist air per pound of DRY air
      Dim R As Double, Sg As Double
      R = 0.068686
      Sg = ((-4.33393735904E-08 * Tdb + 1.80117532949E-05) *
        Tdb - 0.00494503299302) * Tdb + 2.44380828045
      fSdbw = 0.24 * Log((Tdb + 459.67) / 459.67) + W * Sg -
        R * Log(Pbaro / 14.696)
  End Function
  Function fSdrh(Pbaro As Double, Tdb As Double, RH As
        Double) As Double
  'enthalpy of moist air per pound of DRY air
      fSdrh = fSdbw(Pbaro, Tdb, fWdrh(Pbaro, Tdb, RH))
  End Function
```

Appendix L. Runge-Kutta 2D for Crossflow Calculations

The 4th order Runge-Kutta method for solving two-dimensional partial differential equations (PDEs) is as follows:

$$k_1 = f(t_n, x_n, y_n)$$
$$l_1 = g(t_n, x_n, y_n)$$
$$k_2 = f(t_n + \tfrac{1}{2}h, x_n + \tfrac{1}{2}h\,k_1, y_n + \tfrac{1}{2}h\,l_1)$$
$$l_2 = g(t_n + \tfrac{1}{2}h, x_n + \tfrac{1}{2}h\,k_1, y_n + \tfrac{1}{2}h\,l_1)$$
$$k_3 = f(t_n + \tfrac{1}{2}h, x_n + \tfrac{1}{2}h\,k_2, y_n + \tfrac{1}{2}h\,l_2)$$
$$l_3 = g(t_n + \tfrac{1}{2}h, x_n + \tfrac{1}{2}h\,k_2, y_n + \tfrac{1}{2}h\,l_2)$$
$$k_4 = f(t_n + h, x_n + h\,k_3, y_n + h\,l_3)$$
$$l_4 = f(t_n + h, x_n + h\,k_3, y_n + h\,l_3)$$
$$k = \tfrac{1}{6}(k_1 + 2k_2 + 2k_3 + k_4),$$
$$l = \tfrac{1}{6}(l_1 + 2l_2 + 2l_3 + l_4)$$
$$x_{n+1} = x_n + h\,k$$
$$y_{n+1} = y_n + h\,l$$
$$t_{n+1} = t_n + h$$

The following code implements the method:

```
typedef void (*PDE)(double X,double Y,double
    U,double*dU,double V,double*dV);

void RungeKutta2D(PDE pde,double X,double dX,double
    Y,double dY,double*U,double*V)
{
double dU,dU1,dU2,dU3,dU4,dV,dV1,dV2,dV3,dV4;
pde(X,Y,U[0],&dU1,V[0],&dV1);
pde(X+dX/2.,Y+dY/2,U[0]+dU1*dX/2,&dU2,
    V[0]+dV1*dY/2,&dV2);
pde(X+dX/2.,Y+dY/2.,U[0]+dU2*dX/2.,&dU3,
    V[0]+dV2*dY/2.,&dV3);
pde(X+dX,Y+dY,U[0]+dU3*dX,&dU4,V[0]+dV3*dY,&dV4);
dU=(dU1+2.*dU2+2.*dU3+dU4)/6.;
dV=(dV1+2.*dV2+2.*dV3+dV4)/6.;
U[0]+=dU*dX;
V[0]+=dV*dY;
}
```

The PDE is supplied by a function, for example:

```
void Cell(double X,double Y,double Ha,double*dHa,double
    Tw,double*dTw)
{
```

```
double Hw,Q;
Hw=fHtwb(Pbaro,Tw);
Q=KaY*(Hw-Ha);
dHa[0]=Q*LG;
dTw[0]=-Q;
}
```

The process of stepping through the domain is illustrated in the following code:

```
for(y=0;y<Ny;y++)
   {
   for(x=0;x<Nx;x++)
      {
      H=Ha[(Nx+1)*y+x];
      T=Tw[Nx*y+x];
      RungeKutta2D(Cell,X,1./Nx,Y,1./Ny,&H,&T);
      Ha[(Nx+1)*y+x+1]=H;
      Tw[Nx*(y+1)+x]=T;
      }
   }
```

Appendix M. Falling Droplet Trajectory and Mass Transfer

The 4th order Runge-Kutta method for solving one-dimensional ordinary differential equations (ODEs) is as follows:

```
void RungKutta4( /* one step of 4th order Runge-Kutta
    integration */
void dydx(double,double*,double*), /* function
    returning dY/dX */
double*x, /* independent variable */
double dx, /* step in X */
double*y, /* dependent variable array */
double*dy, /* step in Y array */
int n) /* number of dependent variables */
{
int i,j;
double a[4]={0,.5,.5,1};
double b[4]={1/6.,1/3.1/3.,1/6.};
double*v; /* working array of dimension n */
double*w; /* working array of dimension 4n */
w=calloc(4*n,sizeof(double));
v=calloc( n,sizeof(double));
dydx(x[0],y,w);
for(j=1;j<4;j++)
  {
  for(i=0;i<n;i++)
    {
    dy[i]=a[j]*w[n*(j-1)+i];
    v[i]=y[i]+dx*dy[i];
    }
  dydx(x[0]+dx*a[j],v,w+n*j);
  }
for(i=0;i<n;i++)
  {
  dy[i]=0;
  for(j=0;j<4;j++)
    dy[i]+=b[j]*w[n*j+i];
  y[i]+=dx*dy[i];
  }
x[0]+=dx;
free(w);
free(v);
}
```

The function providing the derivatives is listed below:

```
void DropTrajectory(double t,double*Q,double*dQ) /*
    trajectory of a droplet falling */
{ /* in a moving stream of air */
double As; /* droplet surface area [ft²] */
double Av; /* drop surface area per fill volume [1/ft]
    */
```

```
double Ax; /* droplet cross-sectional area [ft²] */
double Cd; /* drag coefficient */
double d; /* drop diameter [ft] */
double Dc; /* diffusion coefficient [ft²/sec] */
double Dv; /* drop volume [ft^3] */
double Nu; /* Nusselt number */
double Re; /* Reynolds number */
double Sh; /* Sherwood number */
double Ud; /* horizontal drop velocity component
   [ft/sec] */
double Ur; /* relative horizontal velocity component
   [ft/sec] */
double Vd; /* vertical drop velocity component
   [ft/sec] */
double Vr; /* relative vertical velocity component
   [ft/sec] */
double Ws; /* absolute humidity at saturation */
double y; /* net vertical displacement of drop [ft] */
d=max(1E-10,pow(6*Q[7]/Rw/pi,1/3.));/* compute drop
   diameter from mass */
As=pi*d*d; /* drop surface area */
Ax=As/4; /* drop cross-sectional area */
Dv=As*d/6; /* drop volume */
y=max(.000001,-Q[1]); /* net vertical movement */
tH=t*Hdrop/y; /* compute hold-up time */
Td=32+Q[6]/Dv/Rw/Cp;/* compute drop temperature from
   energy */
Ws=fWs(Td); /* absolute humidity at saturation */
Ud=Q[2]; /* absolute horizontal velocity component */
Vd=Q[3]; /* absolute vertical velocity component */
Ur=Ua-Ud; /* relative horizontal velocity component */
Vr=Va-Vd; /* relative vertical velocity component */
Re=sqrt(Ur*Ur+Vr*Vr)*d*Ra/mU; /* Reynolds number */
Cd=Drag(Re); /* drag coefficient */
dP=(Q[4]*Ua/Vo+Q[5]*Va/Vo)*Fd/y/(Ra*Vo*Vo/2/g);/*
   pressure drop */
Dc=mU/Ra/Sc; /* diffusion coefficient */
Nu=Nusselt(Re,Pr); /* Nusselt number */
Sh=Sherwood(Re,Sc); /* Sherwood number */
Av=As*Fd*t/y; /* drop surface area per unit of fill
   volume */
KaYL=y*Sh*Ra*Dc/d*Av/(Fl/3600); /* mass transfer
   parameter (KaY/L) */
dQ[0]=Q[2]; /* diff horizontal position = horizontal
   velocity */
dQ[1]=Q[3]; /* diff vertical position = vertical
   velocity */
dQ[2]=3*Cd*Ra*Ur*fabs(Ur)/4/d/Rw; /* diff horizontal
   velocity from horizontal impulse */
```

```
    dQ[3]=(3*Cd*Ra*Vr*fabs(Vr)-4*d*(Rw-Ra)*g)/4/Rw/d; /*
      diff vertical velocity from vertical impulse */
    dQ[4]=Cd*Ra*Ur*fabs(Ur)*Ax/2/g; /* diff horizontal
      impulse from horizontal force balance */
    dQ[5]=Cd*Ra*Vr*fabs(Vr)*Ax/2/g; /* diff vertical
      impulse from vertical force balance */
    dQ[6]=(Tdb-Td)*Nu*Ka*As/d+Hfg*Ra*(Wa-Ws)*Sh*Dc*As/d;/*
      diff energy from heat and mass transfer */
    dQ[7]=Ra*(Wa-Ws)*Sh*Dc*As/d; /* diff mass from
      evaporation */
    }
```

The Nusselt number, Sherwood number, and drag coefficient are provided by the following code:

```
    double Drag(double Re) /* drag coefficient for a sphere
       */
    {
    return(0.22+(1+0.15*pow(Re,0.6))*24/Re);
    }

    double Nusselt(double Re,double Pr) /* Nusselt number
       for a sphere */
    { /* from Kreith page 473 */
    return(2+(.4*pow(Re,.5)+.06*pow(Re,.67))*pow(Pr,.4));
    }

    double Sherwood(double Re,double Sc) /* Sherwood number
       for a sphere */
    { /* from Treybal page 68 */
    return(2+(.4*pow(Re,.5)+.06*pow(Re,.67))*pow(Sc,.4));
    }
```

The solution process is carried out by the following code:

```
    int Droplet(double Do,double fL,double hD,double
       vO,double angle,double To,double tDB,double rh)
    {
    double As,d,dQ[8],dt,Dv,Q[8],t;
    Fl=fL;
    Hdrop=hD;
    Vo=vO;
    Tdb=tDB;
    approach=range=0;
    t=0; /* initial time */
    dt=.01; /* time step */
    Td=To; /* initial drop temperature */
    Do*=FeetPerMm; /* convert drop diameter to ft */
    As=pi*Do*Do; /* droplet surface area */
    Dv=As*Do/6; /* droplet volume */
    Fd=Fl/Rw/3600/Dv; /* droplet mass flux */
    memset(Q,0,8*sizeof(double));
    Q[6]=Dv*Rw*Cp*(To-32); /* initial energy of droplet */
```

```
Q[7]=Dv*Rw;  /* initial mass of droplet */
Wa=fWdbr(Tdb,rh);  /* ambient absolute humidity */
Twb=fBdbr(Tdb,rh);  /* ambient wet-bulb */
Ra=fDdbw(Po,Tdb,Wa);  /* ambient air density */
Ua=Vo*cos(angle*pi/180);  /* horizontal air velocity
    component */
Va=Vo*sin(angle*pi/180);  /* vertical air velocity
    component */
while(-Q[1]<Hdrop)  /* step until droplet falls Hd */
    RungKutta4(DropTrajectory,&t,dt,Q,dQ,8);
approach=Td-Twb;
range=To-Td;
return(Td>Twb+.5);
}
```

The following is typical output from this program:

```
DROPS/V3.10: water droplets falling in flowing air
             by Dudley J. Benton, Knoxville, Tennessee

Do .....  initial drop diameter [mm] (1 to 10)
L ......  mass flux of water droplets [lbm/hr/sq.ft] (100 to
100000)
H ......  vertical distance of drop fall [ft] (1 to 10)
Va .....  air velocity [ft/sec] (1 to 100)
angle ..  angle of inclination of airflow [degrees] (-180 to
180)
Tid ....  initial temperature of drop [F] (32 to 150)
Tdb ....  dry-bulb temperature [F] (-20 to 120)
RH .....  relative humidity [%] (0 to 100)

enter Do,L,H,Va,angle,Tid,Tdb,RH  3 5000 30 10 0 90 60 50

L/G=1.826, Twb=50.2, Wa=0.0055, Ra=0.0761
time    D    X     Y     U     V    KaY/L    dP    holdup Tdrop
0.00  3.00  0.0   0.0   0.0   0.0   *****   ******  ******  90.0
0.10  3.00  0.0  -0.0   0.2  -1.4  0.027    2.152  73.571  89.6
0.20  3.00  0.0  -0.1   0.4  -2.8  0.055    1.131  41.858  89.2
0.30  3.00  0.0  -0.3   0.6  -4.2  0.083    0.758  29.274  88.8
0.40  3.00  0.1  -0.5   0.8  -5.6  0.114    0.564  22.553  88.5
0.50  3.00  0.1  -0.8   1.0  -6.9  0.146    0.446  18.382  88.1
0.60  3.00  0.2  -1.1   1.1  -8.2  0.180    0.367  15.549  87.7
0.70  3.00  0.2  -1.5   1.3  -9.5  0.217    0.310  13.502  87.3
0.80  3.00  0.3  -1.9   1.5 -10.7  0.255    0.267  11.958  86.9
0.90  3.00  0.3  -2.4   1.6 -11.8  0.295    0.234  10.754  86.5
1.00  3.00  0.4  -3.0   1.7 -12.9  0.337    0.208   9.790  86.1
1.10  3.00  0.5  -3.6   1.9 -14.0  0.381    0.187   9.003  85.7
1.20  3.00  0.6  -4.2   2.0 -15.0  0.425    0.169   8.349  85.3
1.30  3.00  0.7  -4.9   2.1 -16.0  0.471    0.155   7.797  84.9
1.40  3.00  0.8  -5.6   2.2 -16.9  0.518    0.142   7.327  84.4
1.50  3.00  0.9  -6.4   2.3 -17.7  0.566    0.131   6.921  84.0
1.60  2.99  1.0  -7.2   2.4 -18.6  0.614    0.122   6.569  83.6
1.70  2.99  1.1  -8.0   2.5 -19.3  0.663    0.114   6.260  83.2
1.80  2.99  1.2  -8.9   2.6 -20.0  0.713    0.107   5.987  82.8
1.90  2.99  1.3  -9.8   2.7 -20.7  0.762    0.101   5.745  82.4
2.00  2.99  1.4 -10.7   2.8 -21.3  0.812    0.095   5.529  81.9
```

2.10	2.99	1.6	-11.7	2.9 -21.9	0.863	0.090	5.335	81.5
2.20	2.99	1.7	-12.7	3.0 -22.4	0.913	0.086	5.160	81.1
2.30	2.99	1.8	-13.7	3.1 -23.0	0.964	0.082	5.002	80.7
2.40	2.99	2.0	-14.7	3.2 -23.4	1.015	0.078	4.859	80.3
2.50	2.99	2.1	-15.7	3.3 -23.9	1.065	0.075	4.728	79.9
2.60	2.99	2.3	-16.8	3.3 -24.3	1.116	0.072	4.608	79.5
2.70	2.99	2.4	-17.9	3.4 -24.6	1.166	0.069	4.498	79.1
2.80	2.99	2.6	-19.0	3.5 -25.0	1.217	0.067	4.397	78.7
2.90	2.99	2.7	-20.1	3.6 -25.3	1.267	0.064	4.304	78.4
3.00	2.99	2.9	-21.2	3.6 -25.6	1.317	0.062	4.218	78.0
3.10	2.99	3.0	-22.3	3.7 -25.9	1.368	0.060	4.138	77.6
3.20	2.99	3.2	-23.5	3.8 -26.1	1.418	0.058	4.064	77.2
3.30	2.99	3.4	-24.6	3.8 -26.3	1.467	0.056	3.995	76.9
3.40	2.99	3.5	-25.8	3.9 -26.6	1.517	0.055	3.931	76.5
3.50	2.99	3.7	-27.0	4.0 -26.8	1.567	0.053	3.871	76.2
3.60	2.99	3.9	-28.2	4.0 -26.9	1.616	0.052	3.815	75.8
3.70	2.99	4.1	-29.3	4.1 -27.1	1.665	0.050	3.763	75.5
3.76	2.99	4.2	-30.1	4.1 -27.2	1.694	0.050	3.733	75.3

range=14.7, approach=25.1

NOTE: drop temperature applies only to solitary drops; whereas, KaY/L, dP, and holdup apply to a population.

Appendix N. Nuclear Plant Thermal Performance

The following functions may be used to model the thermal performance of a large (c. 1000 MWe) power plant. These particular functions are for a Westinghouse pressurized water reactor. The condenser is in three sections, producing three sequentially higher backpressures. The pump curve (head vs. flow) and the hydraulic resistance is used to calculate the condenser cooling water (CCW) flow for one, two, and three pumps, which is not linear.

```
/*****************************************************
**************************
    APPROXIMATE STEAM PROPERTIES
*****************************************************
**********************/

double fTsat(double P)            /* Tsat of steam in øF
from pressure in psia */
{
 double A;
 A=log(P);
 return((26.5029*A+101.692)/(-0.0683446*A+1.));
}

double fPsat(double T)            /* Psat of steam in
psia from Tsat in øF */
{
 return(exp((0.0377315*T-3.83698)/(0.00257868*T+1.)));
}

/*****************************************************
**************************
    HEAT EXCHANGE INSTITUTE STANDARDS FOR STEAM SURACE
CONDENSERS 1989 ED.
*****************************************************
**********************/

/* standard tubing wall thickness for gages 12-24 */
double
Wall[13]={0.109,0.095,0.083,0.072,0.065,0.058,0.049,0.04
2,0.035,0.032,0.028,0.025,0.022};

double WaterTemperatureCorrectionFactor(double T)    /*
Heat Exch. Inst. 1989 */
 {                                                   /*
empirical correction factor   */
  return(0.376769/((((2.15973E-9*T-1.13681E-
6)*T+0.000221828)*T-0.0195586)*T+1.));
 }
```

```
/****************************************************
***********************
   GENERIC PIPE FLOW FUNCTIONS
*****************************************************
**********************/

double Colebrook(double Re,double k)           /*
Colebrook's formula   */
   {                                            /*
Re=Reynolds number    */
   double f;                                    /*
k=relative roughness  */
   int iter;                                    /*
f=friction factor     */
   if(Re<2300.)                                 /*
dP=f*(L/D)*rho*V²/2/g */
      return(64./Re);
   f=0.01;
   for(iter=0;iter<5;iter++)
      {
      f=1.14-2.*(log10(k)+log10(1.+9.3/Re/k/sqrt(f)));
      f=1./f/f;
      }
   if(Re<5500.)
      f=64./Re+(log10(Re)-3.3617)/0.3372*(f-64./Re);
   return(f);
   }

double WoodsF(double Re,double k)       /* Wood's
formula for friction factor */
   {                                    /* Re>1E4 see
Colebrook() for details */
   double a,b,c;
   a=0.094*pow(k,0.225)+0.53*k;
   b=88.*pow(k,0.44);
   c=1.62*pow(k,0.134);
   return(a+b/pow(Re,c));
   }

/****************************************************
***********************
   PLANT-SPECIFIC THERMAL CALCULATIONS
*****************************************************
*********************/

int     gage       =   22; /* condenser tube gage */
int     Ntubes     =44764; /* number of tubes per
condenser */
double Dtube       =   1.; /* condenser tube outside
diameter [in] */
```

```
double Ltube      =91.73; /* total length of condenser
tubes [ft] */
double cleanliness=  90.; /* default condenser
cleanliness [%] */
double material   = 0.79; /* HEI empirical tube material
factor */

double HeatRateCorrectionFactor(double pl,double bp)
/* change in heatrate */
  {                                              /*
pl=power level (1=100%)   */
  double Ts;                                     /*
bp=back pressure [in.Hg]  */
  Ts=fTsat(bp*0.4911540775);                     /*
Ts=steam saturation temp.  */
  return(((-0.0337506*pl+(0.762123*Ts-76.9946))*pl+
    ((-0.00484629*Ts-0.641417)*Ts+114.355))*pl
    +((2.75779E-5*Ts-0.000100014)*Ts+0.12553)*Ts-
40.1616);
  }

double GeneratorOutput(double rh,double bp)   /*
electrical power output [MWe] */
  {                                              /*
rh=reactor heat input [MWt]   */
  double gen,hrcf,locf,pl;                       /* bp=back
pressure [in.Hg]     */
  pl=rh/3425.;                                   /* pl=power
level (1=100%)      */
  hrcf=HeatRateCorrectionFactor(pl,bp);          /*
hrcf=heatrate correction fact */
  locf=100./(100.+hrcf);                         /*
locf=load correction factor % */
  gen=1175.*((-0.10336*pl+1.256)*pl-.15264);/*
gen=uncorrected generator MWe */
  return(max(0,gen*locf));
  }

double CondenserHeatRejection(double rh,double bp)  /*
cond. ht. rej. [BTU/hr] */
  {                                              /*
rh=reactor heat input [MWt]  */
  double go,qr;                                  /*
bp=back pressure [in.Hg]     */
  go=GeneratorOutput(rh,bp);                     /*
go=generator output [MWe]    */
  qr=rh-go;                                      /*
qr=gross heat reject. [MWt]  */
  return(0.98*qr*3412140.);                      /* about
98% goes to condenser */
```

```c
}
double CondenserRise(double rh,double bp,double gpm)   /* water temp. rise [øF] */
{                                                       /*
rh=reactor heat input [MWt] */
   double qr;                                           /*
bp=back pressure [in.Hg]     */
   qr=CondenserHeatRejection(rh,bp);                    /*
gpm=cooling water flow       */
   return(qr/gpm/8.335/60.);                            /*
qr=cond. heat rejt [BTU/hr] */
}

double PumpCurve(double gpm)            /* CCW pump head [ft] from flow [gpm] */
{
   double x;
   x=gpm/247000.;
   return(45.*((((1.64822*x-6.75411)*x+8.2608)*x-4.28862)*x+2.13944));
}

double CondenserFlow(                                   /*
condenser flow [gpm]     */
   int     pumps,                                       /*
number of pumps operating */
   int     tubes,                                       /*
number of active tubes    */
   double Di,                                           /* tube
inside diameter [in] */
   double length)                                       /* tube
length [ft]          */
{
   int iter;
   double area,drop,friction,gpm,gpm1,gpm2,head,Reynolds,velocity;

   if(pumps<1||tubes<10000||Di<0.5||length<10)    /* check for null conditions */
      return(0);

   Di/=12.;                                       /*
convert diameter to feet */
   area=tubes*M_PI*Di*Di/4.;                      /*
inside area of tubes */
   gpm=pumps*247000.;                             /*
initial estimate of flow */
```

```
      gpm1=1000.;
/* lower bound on flow */
      gpm2=2.*gpm;
/* upper bound on flow */

      for(iter=0;iter<32;iter++)                          /*
use bisection search */
         {                                                /*
to match drop & head */
         gpm=(gpm1+gpm2)/2.;
         velocity=gpm/area/7.48052/60.;                   /*
water velocity [ft/sec] */
         Reynolds=velocity*Di/0.0000122;
/* Reynolds number */
         friction=WoodsF(Reynolds,0.00006);
/* friction factor */

         drop=12.
/* static head loss */
               +12.16569*pow(gpm/729350.,2)               /*
waterbox + etc. head loss */
               +friction*(length/Di)*                     /*
tube friction head loss */
                   velocity*velocity/2./32.174;
         head=PumpCurve(gpm/pumps);                       /*
pump head at this flow */

         if(drop<head)
/* adjust flow so that */
            gpm1=gpm;
/* drop and head match */
         else
            gpm2=gpm;
         }

      return(gpm);
      }

double BackpresGeneratorOutput(
      /*** function input parameters
*********************************************/
      double   Wt,       /* reactor heat input [MWt]      *    These
calculations are from    */
      double   gpm,      /* condenser flow [gpm]          *    the
Heat Exchange Institute  */
      double   length,   /* tube length [ft]              *
Standards for Steam Surface   */
      int      tubes,    /* number of active tubes        *
Condensers, 1989 Edition.     */
```

```c
    double  Cc,    /* cleanliness factor [%]
**********************************/
    double  Cm,    /* empirical material factor    */
    double  Do,    /* outside tube diameter [in]   */
    double  Di,    /* inside tube diameter [in]    */
    double  Ti,    /* inlet water temp [øF]        */
    /*** function output parameters ***********/
    double  bp[3], /* zone backpressures [in.Hg]   */
    double  Te[3]) /* exit water temps [øF]        */
{
    int iter,zone;
    double area,bp1,bp2,Ct,Cv,effectiveness,excess,mwe,rise,Th,Tsat,NTU,velocity;

    if(Wt<856.||gpm<100000.)
    {
        Te[0]=Te[1]=Te[2]=Ti;
        bp[0]=bp[1]=bp[2]=0.;
        return(0.);
    }

    Ct=0.107*Do/Di/Di;                  /* empirical tube diameter and gage factor */

    Di/=12.;                            /* convert diameter to feet */
    area=Ntubes*M_PI*Di*Di/4.;          /* flow area [ft²] */

    velocity=gpm/area/7.48052/60.;      /* water velocity [ft/sec] */

    Cv=263./sqrt(velocity);             /* HEI empirical velocity factor */

    NTU=Ct*Cv*Cm*(Cc/100.)              /* number of heat transfer units */
        *(length/3.)/7.48052/60.;       /* (simplified heat exchange model) */

    effectiveness=exp(-NTU*WaterTemperatureCorrectionFactor(Ti));
    excess=effectiveness/(1-effectiveness);

    bp[0]=bp[1]=bp[2]=fPsat(Ti)/0.4911540775;  /* initial estimate of backpres */

    for(iter=0;iter<32;iter++)          /* iterate to converge on */
```

```c
   {
/* zone backpressures */
   for(zone=0,Th=Ti;zone<3;zone++)
      {
      rise=CondenserRise(Wt,bp[zone],gpm)/3.;
      Th+=rise;
      Te[zone]=Th;
      Tsat=Th+rise*excess;
      bp1=bp[zone];
      bp2=fPsat(Tsat)/0.4911540775;
      bp[zone]=sqrt(bp1*bp2);                   /* use sqrt()
to dampen iterations */
      }
   if(iter>2&&fabs(bp1-bp2)<0.001)                         /*
check for convergence */
      break;
   }

 for(mwe=zone=0;zone<3;zone++)                             /*
compute generator output */
   mwe+=GeneratorOutput(Wt,bp[zone])/3.;
 return(mwe);
 }

void PlantTables()
 {
 int pumps;
 double bp[3],gpm,mwe,mwt,pl,qrej,Te[3],Ti;

 printf("Typical Nuclear Power Plant Performance\n");

 printf("\n CCW Flow\n");
 printf("pumps   gpm\n");
 for(pumps=0;pumps<=3;pumps++)
    {
    gpm=CondenserFlow(pumps,Ntubes,Dtube-2.*Wall[gage-
12],Ltube);
    printf("  %i %6.0lf\n",pumps,gpm);
    }

 printf("\n    Heat Rate Correction\n");
 printf(" bp    100%%    75%%    50%%    25%%\n");
 for(bp[2]=0.5;bp[2]<=5.5;bp[2]+=0.5)
    {
    printf("%3.1lf",bp[2]);
    for(pl=1.;pl>0.;pl-=0.25)
      printf("
%5.1lf%%",HeatRateCorrectionFactor(pl,bp[2]));
    printf("\n");
    }
```

```
printf("\nPower & Heat Reject\n");
printf("  MWt     MWe   MBTU/hr\n");
for(mwt=1000.;mwt<=3500.;mwt+=250.)
  printf("%4.0lf  %6.11f
%6.11f\n",mwt,GeneratorOutput(mwt,2.),CondenserHeatRejec
tion(mwt,2.)/1E6);

printf("\n        Generator Output [MWe]\n");
printf(" bp     100%%    75%%    50%%     25%%\n");
for(bp[2]=0.5;bp[2]<=5.5;bp[2]+=0.5)
  {
  printf("%3.11f",bp[2]);
  for(pl=1.;pl>0.;pl-=0.25)
    {
    mwt=pl*3425.;
    printf("  %6.11f",GeneratorOutput(mwt,bp[2]));
    }
  printf("\n");
  }

printf("\nCondenser Heat Reject [MBTU/hr]\n");
printf(" bp     100%%    75%%    50%%     25%%\n");
for(bp[2]=0.5;bp[2]<=5.5;bp[2]+=0.5)
  {
  printf("%3.11f",bp[2]);
  for(pl=1.;pl>0.;pl-=0.25)
    {
    mwt=pl*3425.;
    printf("  
%6.11f",CondenserHeatRejection(mwt,bp[2])/1E6);
    }
  printf("\n");
  }

printf("\n  3 Pump Condenser Rise [øF]\n");
printf(" bp     100%%    75%%    50%%     25%%\n");
for(bp[2]=0.5;bp[2]<=5.5;bp[2]+=0.5)
  {
  printf("%3.11f",bp[2]);
  for(pl=1.;pl>0.;pl-=0.25)
    {
    mwt=pl*3425.;
    printf("  %6.21f",CondenserRise(mwt,bp[2],gpm));
    }
  printf("\n");
  }

printf("\n  3 Pump Backpressure [in.Hg]\n");
printf("  Ti    100%%    75%%    50%%     25%%\n");
```

```
   for(Ti=35.;Ti<=125.1;Ti+=5.)
     {
     printf("%3.0lf",Ti);
     for(pl=1.;pl>0.;pl-=0.25)
       {
       mwt=pl*3425.;

mwe=BackpresGeneratorOutput(mwt,gpm,Ltube,Ntubes,cleanli
ness,material,Dtube,Dtube-2.*Wall[gage-12],Ti,bp,Te);
       printf(" %6.3lf",bp[2]);
       }
     printf("\n");
     }

   printf("\n   3 Pump Generator  Output  [MWe]\n");
   printf(" Ti    100%%     75%%     50%%     25%%\n");
   for(Ti=35.;Ti<=125.1;Ti+=5.)
     {
     printf("%3.0lf",Ti);
     for(pl=1.;pl>0.;pl-=0.25)
       {
       mwt=pl*3425.;

mwe=BackpresGeneratorOutput(mwt,gpm,Ltube,Ntubes,cleanli
ness,material,Dtube,Dtube-2.*Wall[gage-12],Ti,bp,Te);
       printf(" %6.1lf",mwe);
       }
     printf("\n");
     }

   printf("\nCondenser Heat Reject [MBTU/hr]\n");
   printf(" Ti    100%%     75%%     50%%     25%%\n");
   for(Ti=35.;Ti<=125.1;Ti+=5.)
     {
     printf("%3.0lf",Ti);
     for(pl=1.;pl>0.;pl-=0.25)
       {
       mwt=pl*3425.;

mwe=BackpresGeneratorOutput(mwt,gpm,Ltube,Ntubes,cleanli
ness,material,Dtube,Dtube-2.*Wall[gage-12],Ti,bp,Te);
       qrej=60.*8.335*gpm*(Te[2]-Ti)/1E6;
       printf(" %6.1lf",qrej);
       }
     printf("\n");
     }

   printf("\n    Condenser Rise [øF]\n");
   printf(" Ti    100%%     75%%     50%%     25%%\n");
   for(Ti=35.;Ti<=125.1;Ti+=5.)
```

```
        {
        printf("%3.0lf",Ti);
        for(pl=1.;pl>0.;pl-=0.25)
           {
           mwt=pl*3425.;
           mwe=BackpresGeneratorOutput(mwt,gpm,Ltube,Ntubes,cleanli
        ness,material,Dtube,Dtube-2.*Wall[gage-12],Ti,bp,Te);
           printf(" %5.2lf",Te[2]-Ti);
           }
        printf("\n");
        }
     }
```

The function PlantTables() lists the calculated performance for a range of input variables and also illustrates how to call the functions. The output is as follows:

```
Typical Nuclear Power Plant Performance

   CCW Flow
pumps   gpm
   0      0
   1 317758
   2 569330
   3 733382

        Heat Rate Correction
  bp   100%    75%    50%     25%
  0.5   0.1%  -0.7%  -5.6%  -14.6%
  1.0  -0.7%  -1.7%  -4.8%  -10.0%
  1.5  -0.6%  -1.2%  -2.6%   -5.0%
  2.0  -0.0%   0.0%  -0.0%   -0.0%
  2.5   0.9%   1.5%   2.8%    4.9%
  3.0   2.0%   3.2%   5.8%    9.7%
  3.5   3.3%   5.1%   8.8%   14.4%
  4.0   4.6%   7.1%  11.9%   19.0%
  4.5   6.1%   9.1%  14.9%   23.4%
  5.0   7.6%  11.2%  18.0%   27.8%
  5.5   9.1%  13.3%  21.0%   32.1%

  Power & Heat Reject
   MWt    MWe  MBTU/hr
  1000   241.2  2537.4
  1250   343.1  3032.6
  1500   443.7  3532.2
  1750   543.0  4036.1
  2000   641.0  4544.3
  2250   737.7  5056.9
  2500   833.2  5573.8
  2750   927.3  6095.0
  3000  1020.1  6620.5
  3250  1111.7  7150.3
  3500  1201.9  7684.5
```

```
Generator Output [MWe]
bp    100%    75%     50%     25%
0.5   1173.3  865.5   559.7   213.0
1.0   1182.8  873.6   554.5   202.1
1.5   1181.9  869.2   542.4   191.6
2.0   1175.0  859.2   528.2   182.0
2.5   1164.5  846.4   513.6   173.5
3.0   1151.8  832.1   499.2   165.9
3.5   1137.8  817.3   485.3   159.1
4.0   1122.9  802.2   472.1   153.0
4.5   1107.7  787.3   459.6   147.5
5.0   1092.2  772.5   447.8   142.4
5.5   1076.7  758.1   436.6   137.8

Condenser Heat Reject [MBTU/hr]
bp    100%    75%     50%     25%
0.5   7529.5  5695.6  3854.9  2150.9
1.0   7497.8  5668.4  3872.1  2187.3
1.5   7500.7  5683.1  3912.7  2222.7
2.0   7523.8  5716.7  3960.2  2254.6
2.5   7558.8  5759.5  4009.1  2283.0
3.0   7601.3  5807.0  4057.2  2308.4
3.5   7648.2  5856.7  4103.5  2331.1
4.0   7697.9  5907.0  4147.6  2351.6
4.5   7749.0  5957.1  4189.5  2370.1
5.0   7800.8  6006.4  4229.1  2387.0
5.5   7852.6  6054.7  4266.6  2402.5

3 Pump Condenser Rise [øF]
bp    100%    75%     50%     25%
0.5   20.53   15.53   10.51   5.86
1.0   20.44   15.46   10.56   5.96
1.5   20.45   15.50   10.67   6.06
2.0   20.51   15.59   10.80   6.15
2.5   20.61   15.70   10.93   6.22
3.0   20.73   15.83   11.06   6.29
3.5   20.85   15.97   11.19   6.36
4.0   20.99   16.11   11.31   6.41
4.5   21.13   16.24   11.42   6.46
5.0   21.27   16.38   11.53   6.51
5.5   21.41   16.51   11.63   6.55

3 Pump Backpressure [in.Hg]
Ti    100%    75%     50%     25%
35    0.736   0.548   0.403   0.298
40    0.838   0.633   0.473   0.356
45    0.957   0.733   0.555   0.425
50    1.096   0.849   0.652   0.505
55    1.259   0.984   0.765   0.599
60    1.447   1.142   0.897   0.709
65    1.665   1.325   1.050   0.838
70    1.918   1.537   1.228   0.987
75    2.209   1.781   1.434   1.160
80    2.543   2.063   1.672   1.360
85    2.926   2.386   1.945   1.590
90    3.364   2.757   2.259   1.854
95    3.864   3.180   2.618   2.155
```

```
100  4.432  3.662  3.027  2.499
105  5.077  4.210  3.492  2.889
110  5.807  4.831  4.019  3.331
115  6.630  5.532  4.614  3.831
120  7.557  6.322  5.285  4.393
125  8.596  7.208  6.038  5.025
```

3 Pump Generator Output [MWe]

```
Ti    100%    75%    50%    25%
 35 1175.6  863.5  557.7  216.8
 40 1177.6  866.3  558.8  216.1
 45 1179.5  868.9  559.5  215.0
 50 1181.1  871.0  559.7  213.6
 55 1182.2  872.5  559.2  211.8
 60 1182.6  873.1  557.9  209.6
 65 1182.1  872.8  555.6  206.9
 70 1180.4  871.3  552.3  203.8
 75 1177.2  868.3  547.9  200.2
 80 1172.3  863.8  542.3  196.2
 85 1165.5  857.5  535.4  191.7
 90 1156.6  849.3  527.2  186.8
 95 1145.3  839.1  517.7  181.5
100 1131.6  826.9  506.9  175.9
105 1115.4  812.6  494.9  170.0
110 1096.5  796.4  481.9  164.0
115 1075.2  778.2  467.9  157.7
120 1051.4  758.3  453.0  151.3
125 1025.3  736.8  437.4  144.9
```

Condenser Heat Reject [MBTU/hr]

```
Ti    100%    75%    50%    25%
 35 7521.7 5702.2 3861.6 2138.1
 40 7515.0 5692.7 3857.8 2140.7
 45 7508.6 5684.2 3855.5 2144.3
 50 7503.3 5677.2 3854.9 2149.0
 55 7499.6 5672.2 3856.6 2155.0
 60 7498.3 5669.9 3861.0 2162.4
 65 7500.0 5671.0 3868.5 2171.4
 70 7505.8 5676.2 3879.5 2181.8
 75 7516.4 5686.1 3894.2 2193.8
 80 7532.7 5701.3 3913.1 2207.3
 85 7555.4 5722.4 3936.2 2222.2
 90 7585.3 5749.8 3963.6 2238.6
 95 7623.0 5783.8 3995.4 2256.2
100 7668.8 5824.6 4031.4 2275.0
105 7723.2 5872.2 4071.4 2294.6
110 7786.2 5926.6 4115.0 2315.0
115 7857.6 5987.3 4161.9 2335.9
120 7937.2 6053.9 4211.6 2357.1
125 8024.4 6125.7 4263.6 2378.5
```

Condenser Rise [øF]

```
Ti   100%   75%   50%  25%
35  20.51 15.55 10.53 5.83
40  20.49 15.52 10.52 5.84
45  20.47 15.50 10.51 5.85
50  20.46 15.48 10.51 5.86
```

```
 55 20.45 15.47 10.52 5.88
 60 20.44 15.46 10.53 5.90
 65 20.45 15.46 10.55 5.92
 70 20.46 15.48 10.58 5.95
 75 20.49 15.50 10.62 5.98
 80 20.54 15.54 10.67 6.02
 85 20.60 15.60 10.73 6.06
 90 20.68 15.68 10.81 6.10
 95 20.78 15.77 10.89 6.15
100 20.91 15.88 10.99 6.20
105 21.06 16.01 11.10 6.26
110 21.23 16.16 11.22 6.31
115 21.42 16.32 11.35 6.37
120 21.64 16.51 11.48 6.43
125 21.88 16.70 11.62 6.49
```

Appendix O. Cooling Pond Performance

The following functions implement Langhaar's model for cooling ponds. These may be used to model the thermal performance of large ponds.

POND-GEOMETRY
```
double PondArea(double Elevation)/* pond area in acres
from elevation */
  {/* in feet above mean sea level */
  if(Elevation<=583.5)
    return(0);
  return(252.305*pow(Elevation-583.5,0.945033));
  }
double PondVolume(double Elevation)/* pond volume in
acre-feet from elevation */
  {/* in feet above mean sea level */
  if(Elevation<=583.5)
    return(0);
  return(252.305*pow(Elevation-
583.5,1.945033)/1.945033);
  }
```
LANGHAAR'S POND CALCULATIONS
```
double yRH(double RH)
  {
  return((81.3295615276*RH-
359.043847242)*RH+566.189533239);
  }
double yTX(double Tdb)
  {
  return((-0.00364208459429*Tdb-
0.938531836914)*Tdb+563.661164074);
  }
double yTY(double Tdb)
  {
  return((((-
0.000011002057396*Tdb+0.00312849291125)*Tdb-
0.363391922646)*Tdb+24.0624509974)*Tdb-268.381574621);
  }
double fTE(double y)
  {
  return(((-7.64388403125E-8*y+1.767707201209E-
4)*y+0.02270881300489)*y+21.22088945338);
  }
double Teq1(double RH,double Tdb)
  {
  double x1,x2,x3,y1,y2,y3;
  x1=116.;
  y1=yRH(RH);
  x2=yTX(Tdb);
  y2=yTY(Tdb);
  x3=659.;
```

```
    y3=y1+(y2-y1)*(x3-x1)/(x2-x1);
    return(fTE(y3-1.9073576336026));
}
double yTE(double Te)
{
    return(((2.445370284702E-4*Te-
.0728401209826)*Te+12.12846891621)*Te-77.3036975845);
}
double gTE(double y)
{
    return(((-1.514543864717E-7*y+3.173255575958E-4)*y-
0.02245149639398)*y+21.17697337907);
}
double yWS(double WS)
{
    return((0.113846153846*WS-7.4)*WS+485.153846154);
}
double fQS(double y)
{
    return(exp((3.28103282914E-7*y-
0.005219646991425)*y+7.629928327939));
}
double yQS(double Qs)
{
    double z;
    z=log(Qs);
    return((2.732299407884*z-
232.2979147965)*z+1614.086511135);
}
double fWs(double X)
{
    return((-0.06*X+4.34)*X+700.8);
}
double Teq2(double Teq1,double WS,double Qa)
{
    double Qs,x1,x2,x3,y1,y2,y3,y4,y5;
    x1=85.;
    x2=336.;
    x3=519.;
    y1=yTE(Teq1);
    y2=yWS(WS);
    y3=y1+(y2-y1)*(x3-x1)/(x2-x1);
    Qs=fQS(y3+1.70916334657522);
    y4=yQS(Qs+Qa);
    y5=y4+(y2-y4)*(x1-x3)/(x2-x3);
    return(gTE(y5+0.874316939823416));
}
double yE(double Te)
{
```

```
        return((((3.06219996606E-6*Te-
.00119039629931)*Te+0.169554749432)*Te-
4.71740208339)*Te-.703035058119);
    }
    double Ey(double y)
    {
        return(0.174879844254*y+41.0017947314);
    }
    double gWS(double WS)
    {
        if(WS<9.5)
            return((-
1.13687436159*WS+43.2414708887)*WS+23.2145045965);
        if(WS<13.)
            return(((-0.572918896334*WS-
31.496518252)*WS+310.040267259)/(1.-0.118033856491*WS));
        if(WS<16.)
            return((-
0.366371018114*WS+15.7249219238)*WS+224.693316677);
        if(WS<25.)
            return((((-0.00208301885643*WS+0.189635219107)*WS-
6.29716977817)*WS+93.6137012375)*WS-143.871291359);
        if(WS<35.)
            return((((0.000265697376144*WS-
0.0337665004286)*WS+1.58346148339)*WS-
28.5970835609)*WS+559.210936962);
        if(WS<40.)
            return((((0.00080000001463*WS-
0.102666668718)*WS+4.88000010623)*WS-
97.93333574)*WS+1100.00002011);
        return((((1.6690358468E-6*WS-
0.00859920943636)*WS+1.24997058155)*WS-
53.9210443316)*WS+1171.17549936);
    }
    double gQ(double y)
    {
        return(exp(-0.00514002689607*y+1.18818451182));
    double fQ(double Te,double WS)
    {
        double x1,x2,x3,y1,y2,y3;
        x1=60.;
        x2=225.;
        x3=452.;
        y1=yE(Te);
        y3=gWS(WS);
        y2=y1+(y3-y1)*(x2-x1)/(x3-x1);
        return(gQ(y2+5.10204089863748E-2));
    }
    double yDTi(double dTi)
    {
```

```c
  double z;
  z=log(dTi);
  return(((3.50768670016*z-16.5785425742)*z-
132.330053787)*z+664.85176737);
}
double yP(double P)
{
  return(-1.14828363384*P+312.083287101);
}
double fDTo(double y)
{
  return(exp((((4.89112665649E-11*y-4.85669595205E-
8)*y+0.000017963796707)*y+0.00360401184359)*y+0.47386888
4976));
}
double dTe(double dTi,double P)
{
  double x1,x2,x3,y1,y2,y3;
  x1=68.;
  x2=300.;
  x3=529.;
  y1=yDTi(dTi);
  y2=yP(P);
  y3=y1+(y2-y1)*(x3-x1)/(x2-x1);
  return(fDTo(y3+0.943965516979461));
}
void PondTables()
{
  double ep,P,Qa=100.,RH,Tdb,Teq,WS;
  printf("\n   Pond Geometry\n");
  printf(" Elev  Area  Volum\n");
  printf(">MSL  acre  ac-ft\n");
  for(ep=583.5;ep<590.1;ep+=0.5)
    printf("%5.1lf %4.0lf
%5.0lf\n",ep,PondArea(ep),PondVolume(ep));
  printf("\nLanghaar's Pond Performance\n");
  printf("\nEquilibrium Temperature\n");
  printf("Tdb    0%%   25%%   50%%   75%%  100%%\n");
  for(Tdb=20.;Tdb<101.;Tdb+=10.)
    {
    printf("%3.0lf",Tdb);
    for(RH=0.;RH<1.01;RH+=0.25)
      printf(" %4.1lf",Teq1(RH,Tdb));
    printf("\n");
    }
  printf("\nAdjusted Equilibrium Temperature\n");
  printf("Teq    0     5    10    15    20 mph\n");
  for(Teq=30.;Teq<141.;Teq+=10.)
    {
    printf("%3.0lf",Teq);
```

```
      for(WS=0.;WS<20.1;WS+=5.)
        printf(" %5.1lf",Teq2(Teq,WS,Qa));
      printf("\n");
    }
  printf("\n         Scaling Factor, Q\n");
  printf("Teq    0    5    10    15    20 mph\n");
  for(Teq=30.;Teq<141.;Teq+=10.)
    {
    printf("%3.0lf",Teq);
    for(WS=0.;WS<20.1;WS+=5.)
      printf(" %5.3lf",fQ(Teq,WS));
    printf("\n");
    }
  printf("\n              Approach\n");
  printf("Teq P=50   100   150   200   250\n");
  for(Teq=30.;Teq<141.;Teq+=10.)
    {
    printf("%3.0lf",Teq);
    for(P=50.;P<251.;P+=50.)
      printf(" %5.2lf",dTe(Teq,P));
    printf("\n");
    }
  }
```

The function PondTables() lists the calculated performance for a range of input variables and also illustrates how to call the functions. The output is as follows:

```
 Pond Geometry
 Elev Area Volum
>MSL  acre ac-ft
583.5    0    0
584.0  131   34
584.5  252  130
585.0  370  285
585.5  486  499
586.0  600  771
586.5  713 1099
587.0  824 1483
587.5  935 1923
588.0 1045 2418
588.5 1155 2968
589.0 1264 3573
589.5 1372 4232
590.0 1480 4945

Langhaar's Pond Performance

Equilibrium Temperature
Tdb  0%  25%  50%  75% 100%
 20 20.6 20.9 21.3 21.8 22.2
 30 24.6 26.1 27.5 28.9 30.1
 40 31.1 33.5 35.9 38.0 39.9
 50 38.0 41.4 44.5 47.4 49.9
 60 44.8 49.1 53.1 56.7 59.9
 70 51.5 56.8 61.7 66.1 69.9
```

```
 80  58.1  64.5  70.3  75.6  80.1
 90  64.0  71.6  78.5  84.6  89.9
100  68.1  76.9  84.9  92.0  98.1
```

Adjusted Equilibrium Temperature

Teq	0	5	10	15	20 mph
30	66.1	58.3	52.8	49.2	46.7
40	72.3	65.2	60.2	56.9	54.7
50	79.0	72.5	68.0	65.0	63.0
60	85.8	80.0	76.0	73.3	71.6
70	92.7	87.5	84.0	81.6	80.1
80	99.7	95.1	92.0	89.9	88.6
90	106.7	102.7	100.0	98.2	97.0
100	114.0	110.4	108.1	106.5	105.5
110	121.5	118.5	116.4	115.1	114.2
120	129.5	126.9	125.2	124.1	123.4
130	138.1	135.9	134.5	133.6	133.0
140	147.3	145.6	144.5	143.8	143.3

Scaling Factor, Q

Teq	0	5	10	15	20 mph
30	3.304	2.201	1.647	1.533	1.483
40	2.996	1.996	1.494	1.390	1.345
50	2.628	1.750	1.310	1.219	1.180
60	2.254	1.501	1.123	1.046	1.012
70	1.908	1.271	0.951	0.885	0.857
80	1.606	1.070	0.801	0.745	0.721
90	1.351	0.900	0.673	0.627	0.606
100	1.138	0.758	0.567	0.528	0.511
110	0.960	0.640	0.479	0.445	0.431
120	0.811	0.540	0.404	0.376	0.364
130	0.682	0.454	0.340	0.316	0.306
140	0.568	0.379	0.283	0.264	0.255

Approach

Teq	P=50	100	150	200	250
30	13.13	6.20	2.98	1.64	1.44
40	16.48	7.72	3.67	1.90	1.39
50	19.47	8.99	4.27	2.14	1.41
60	22.13	10.07	4.78	2.35	1.45
70	24.50	11.00	5.21	2.54	1.50
80	26.61	11.79	5.58	2.70	1.55
90	28.47	12.46	5.90	2.84	1.59
100	30.12	13.04	6.16	2.96	1.63
110	31.56	13.53	6.39	3.06	1.67
120	32.81	13.95	6.58	3.15	1.70
130	33.89	14.31	6.74	3.22	1.73
140	34.82	14.61	6.88	3.28	1.75

Appendix P. Power Plant Thermal Simulation

The functions presented in Appendix N and O are combined to simulate the thermal response of this nuclear power plant. The code is provided below and in the on-line archive (filename nuke.c). The results were shown in Chapter 38.

```
double Required(/* required pond area [ft²/gpm] */
   double Tdb,    /* dry-bulb temperature [øF] */
   double RH ,    /* relative humidity (0 to 1) */
   double GHI,    /* solar input [BTU/hr/ft²] */
   double mph,    /* wind speed [mph] */
   double Thw,    /* inlet (hot) water temp [øF] */
   double Tcw)    /* exit (cold) water temp [øF] */
{
   double dT1,dT2,P,Q,Teq;
   if(Tcw<=Teq1(RH,Tdb))
      return(FLT_MAX);
   if(Thw<=Tcw)
      return(0.);
   Teq=Teq2(Teq1(RH,Tdb),mph,GHI);
   if(Teq<33.)
      Teq=33.;
   dT1=Thw-Teq;
   dT2=Tcw-Teq;
   if(dT2<=0.)
      return(FLT_MAX);
   P=fPDT(dT1,dT2);
   Q=fQ(Teq,mph);
   return(P*Q);
}

double bp[3]; /* backpressures (zones 1, 2, 3) [in.Hg] */
double Tc[3]; /* condenser temperatures (zones 1, 2, 3) [øF] */
double Tpond; /* pond temperature [øF] */
double mwe  ; /* generator output [MWe] */

void Thermal( /* thermal operating point */
   double Tdb,    /* dry-bulb temperature [øF] */
   double RH ,    /* relative humidity (0 to 1) */
   double GHI,    /* solar input [BTU/hr/ft²] */
   double mph,    /* wind speed [mph] */
   double elev,   /* pond elevation [ft above MSL] */
   int pumps ,    /* number of CCW pumps on */
   double mwt)    /* reactor heat input [MWt] */
{
   double Apond; /* pond area [acres] */
   double Di  ;  /* condenser inside tube diameter [in] */
   double gpm ;  /* condenser flow [gpm] */
```

```
    int iter;
    double Tp1,Tp2;
    Apond=PondArea(elev); /* pond surface area */
    Di=Dtube-2.*Wall[gage-12]; /* inside diameter */
    gpm=CondenserFlow(pumps,Ntubes,Di,Ltube); /* condenser
flow */
    Tp1=max(32.,Teq1(RH,Tdb)); /* upper and lower bound */
    Tp2=150.; /* on pond temperature */
    for(iter=0;iter<32;iter++) /* use a bisection search
to determine pond temp */
      {
      Tpond=(Tp1+Tp2)/2.;

mwe=Electrical(mwt,gpm,Ltube,Ntubes,cleanliness,material
,Dtube,Di,Tpond,bp,Tc);

if(Required(Tdb,RH,GHI,mph,Tc[2],Tpond)*gpm/43560.>Apond
)
        Tp1=Tpond;
      else
        Tp2=Tpond;
      }
    }

void Simulation(char*fname)
    {
    char bufr[128];
    int pumps=3;
    double elev=590.,GHI,mph,mwt=3425.,RH,Tdb;
    FILE*fp;
    fp=fopen(fname,"rt");
    while(fgets(bufr,sizeof(bufr),fp))
        {
        if(sscanf(bufr,"%lf%*[ ,\t]%lf%*[ ,\t]%lf%*[ ,\t]
%lf%*[ ,\t]",&Tdb,&RH,&mph,&GHI)==4)
            {
            if(Tdb<10.)
                Tdb=10.;
            Thermal(Tdb,RH,GHI,mph,elev,pumps,mwt);
            printf("%lG,%lG,%lG,%lG\n",Tpond,Tc[2],bp[2],mwe);
            }
        }
    fclose(fp);
    }

int main(int argc,char**argv,char**envp)
    {
    if(argc<2)
        {
        PlantTables();
```

```
    PondTables();
    }
else
    Simulation(argv[1]);
return(0);
}
```

Appendix Q. Nitrogen Supersaturation Models

The following code and data were used in the nitrogen supersaturation model for Jennings Randolph Project.

```c
#include <stdio.h>
#include <stddef.h>
#include <stdlib.h>
#include <float.h>
#include <math.h>

/* USACE WES Nitrogen SuperSaturation Model */

/* constants and empirical coefficients */

double Cs=     100; /* saturation concentration [%] */
double Cu=     100; /* upstream concentration [%] */
double g =     9.8; /* gravitational acceleration
[m/sec^2] */
double Pa=101325; /* atmospheric pressure [N/m^2] */
double Sc=     0.6; /* Schmidt number */
double Vr=     0.25; /* bubble rise velocity [m/sec] */

double alpha = 4.69; /* mass transfer parameter */
double beta  =  2.2; /* effective depth parameter */
double eta   = 0.66; /* power on Reynolds number */
double Ks    = 1.77; /* surface transfer parameter */
double lambda= 0.19; /* air layer thickness [m] */
double mu    =.00078; /* dynamic viscosity of water
[kg/m/sec] */
double rho   = 1000; /* density of water [kg/m^3] */
double sigma =0.0720; /* surface tension [N/m] */

#define gamma (rho*g)  /* specific weight of water
[N/m^3] */
#define nu    (mu/rho) /* kinematic viscosity of water
[m^2/sec] */

double Supersaturation(
  double Hr, /* river depth [m] */
  double Hs, /* stilling basin depth [m] */
  double Ht, /* total head [m] */
  double Ls, /* stilling basin length [m] */
  double Dp, /* maximum plunge depth [m] */
  double q)  /* flow per unit width [m^2/sec] */
{
  double bH; /* dimensionless bH=beta*Hb/L */
  double Ce; /* effective saturation concentration [%]
*/
  double Ci; /* infinite flow concentration [%] */
  double Db; /* effective bubble depth [m] */
```

```c
  double Dr;   /* effective river depth [m] */
  double Ds;   /* effective stilling basin depth [m] */
  double Dj;   /* depth of plunging jet [m] */
  double Fi;   /* volumetric air concentration */
  double Hb;   /* bubble half-life distance [m] */
  double Kb;   /* bubble mass transfer coefficient */
  double Rb;   /* bubble rise velocity Reynolds number */
  double Re;   /* Reynolds number */
  double rK;   /* increase in mass transfer coefficient */
  double Vj;   /* jet velocity [m/sec] */
  double We;   /* Weber number */

  Db=2*Hs/3;
  Dr=Hr/2;
  Re=q/nu;
  Hb=log(2)*q/Vr;
  bH=beta*Hb/Ls;
  if(bH>1)
     Ds=Dr+(Db-Dr)*exp(1-bH);
  else
     Ds=Db;
  if(Ds>Dp)
    {
    rK=Ds/Dp;
    Ds=Dp;
    }
  else
    rK=1;
  Rb=2*Ds*Vr/nu;
  Vj=.15*pow(q/sqrt(g*Ht)/Ht,.23)*sqrt(g*Ht);
  Fi=Vj*lambda/(Vj*lambda+q);
  Dj=q/Vj;
  We=rho*q*q/sigma/Dj;
  Kb=alpha*Fi*sqrt(1-Fi)/pow(1-
pow(Fi,5/3.),.25)*pow(We,.6)*pow(Re,eta)/sqrt(Sc)/Rb;
  Ce=Cs*(1+Ds*gamma/Pa);
  Ci=(rK*Kb*Ce+rK*Ks*Cs)/(rK*Kb+rK*Ks);
  return(Ci-(Ci-Cu)*exp(-rK*Kb-rK*Ks));
  }

double Weir(double Cu,double e)
  {
  return(Cu+e*(Cs-Cu));
  }

double Downstream(double Cu,double X,double Q)
  {
  double L=15079+2.99*Q;
  return(Cu+(Cs-Cu)*(1-exp(-X/L)));
  }
```

```c
#define FEET_PER_METER 3.2808399

/* site specific data and computed parameters */

double Dp=12.0; /* maximum plunge depth [m] */
double Hr= 2.8; /* river depth [m] */
double Hs=18.0; /* stilling basin depth [m] */
double Ht= 5.0; /* total head [m] */
double Ls=95.4; /* stilling basin length [m] */
double Ws=19.1; /* width of spillway [m] (62'8") */

double Model(double e,double X,double q)
  {

return(Downstream(Weir(Supersaturation(Hr,Hs,Ht,Ls,Dp,q)
,e),X,q*pow(FEET_PER_METER,3)*Ws));
  }

struct{
  double Ce;
  double Cs;
  double Ks;
  double Kb;
  double Tb;
  }mT;

void MassTransfer(double t,double*C,double*dC)
  {
  dC[0]=mT.Kb*(mT.Ce-C[0])+mT.Ks*(mT.Cs-C[0]);
  if(t<mT.Tb)
    dC[0]=mT.Kb*(mT.Ce-C[0]);
  else
    dC[0]=mT.Ks*(mT.Cs-C[0]);
  }

void RungeKutta4(              /* one step of 4th order
Runge-Kutta integration */
  void dydx(double,double*,double*),  /* function
returning dY/dX */
  double*x,                    /* independent variable
*/
  double dx,                   /* step in X */
  double*y,                    /* dependent variable
array */
  double*dy,                   /* step in Y array */
  int n)                       /* number of dependent
variables */
  {
  int i;
```

```
    double*v,*w1,*w2,*w3,*w4;

    v=calloc(5*n,sizeof(double));
    w1=v+n;
    w2=w1+n;
    w3=w2+n;
    w4=w3+n;

    dydx(x[0],y,w1);

    for(i=0;i<n;i++)
        v[i]=y[i]+w1[i]*dx/2;
    dydx(x[0]+dx/2,v,w2);

    for(i=0;i<n;i++)
        v[i]=y[i]+w2[i]*dx/2;
    dydx(x[0]+dx/2,v,w3);

    for(i=0;i<n;i++)
        v[i]=y[i]+w3[i]*dx;
    dydx(x[0]+dx,v,w4);

    for(i=0;i<n;i++)
      {
        dy[i]=(w1[i]+2*w2[i]+2*w3[i]+w4[i])/6;
        y[i]=y[i]+dy[i]*dx;
      }
    x[0]+=dx;

    free(v);
    }

double Cx=260,Qx=45;

void Results(
    char*site,
    double Hs  , /* stilling basin depth [m] */
    double Hr  , /* river depth [m] */
    double Ht  , /* total head [m] */
    double Ls  , /* stilling basin length [m] */
    double*data, /* data pairs (q,Kb) */
    int n,     /* number of data pairs */
    int set)
    {
    int i;
    double q;
    double bX; /* dimensionless bX=beta*X2/L */
    double Db; /* effective bubble depth [m] */
    double De; /* effective depth [m] */
    double Dr; /* effective river depth [m] */
```

```
   double X2;  /* bubble half-life distance [m] */

   printf("\nmodel parameters for %s\n",site);
   printf("   Hs=%9.3lG stilling basin depth [m]\n",Hs);
   printf("   Hr=%9.3lG river depth [m]\n",Hr);
   printf("   Ht=%9.3lG total head [m]\n",Ht);
   printf("   Ls=%9.3lG stilling basin length [m]\n",Ls);
   printf("\ncomputed parameters for %s\n",site);

   Db=2*Hs/3;
   printf("   Db=%9.3lG effective bubble depth [m]\n",Db);

   Dr=Hr/2;
   printf("   Dr=%9.3lG effective river depth [m]\n",Dr);

   printf("\n    q    Cse\n");
   printf("%s measured\n",site);
   for(i=0;i<n;i++)
      {
      q=data[2*i];
      printf("%4.1lf %3.0lf\n",q,data[2*i+1]);
      printf("%i %lG %lG\n",set,q,data[2*i+1]);
      }

   printf("%s computed\n",site);
   for(q=Qx;q>FLT_EPSILON;q/=1.1)
      {
      X2=log(2)*q/Vr;
      bX=beta*X2/Ls;
      if(bX>1)
         De=Dr+(Db-Dr)*exp(1-bX);
      else
         De=Db;
      printf("0 %lG %lG\n",q,100*(1+De*gamma/Pa));
      }
   }

double IceHarbor[]={
   3.2,115,
   3.4,121,
   6.3,127,
   6.5,126,
   9.7,130,
   9.7,136,
   12.7,136,
   12.8,140,
   15.9,138,
   16.2,136,
   18.9,135,
   19.3,134,
```

```
      22.0,132,
      25.5,130,
      28.6,128,
      31.7,127,
      31.3,124,
      34.1,124,
      37.2,122};

double TheDalles[]={
       3.3,114,
       3.2,116,
       5.8,115,
       5.5,117,
       8.3,115,
       7.3,117,
       8.4,118,
       9.6,118,
      10.9,117,
      11.2,119,
      12.6,118,
      13.2,118,
      13.6,117,
      16.7,116,
      19.0,114,
      19.3,115,
      21.4,113};

double LittleGoose[]={
       3.5,105,
       7.1,113,
       7.0,116,
      10.7,122,
      10.6,128,
      14.1,128,
      14.2,131,
      17.8,130,
      17.7,132,
      17.7,134,
      21.3,135,
      24.8,138,
      28.2,139,
      31.4,140,
      34.6,139,
      35.2,137,
      37.9,137,
      41.3,138};

double JenningsRandolph[]={
       3.71,117,
       3.71,116,
```

```
    3.71,117,
    3.71,118,
    4.45,117,
    5.19,120,
    5.93,119,
    6.67,122,
    6.67,121,
    6.67,124,
    6.67,118,
    7.41,118,
    7.41,125,
    7.41,126,
    7.86,121,
    8.15,121,
    8.15,122,
    8.60,121,
    8.89,126,
    9.34,117,
   -10.1,128,  /* extrapolation */
   -14.1,132,
   -16.0,134,
   -18.1-12,135,
   -20.1-12,132,
   -22.0-12,130,
   -24.1-12,128,
   -26.0-12,127,
   -28.0-12,126,
   -29.9-12,124,
   -32.0-12,123,
   -34.1-12,122,
   -36.0-12,121,
   -38.1-12,120,
   -40.2-12,119,
   -42.0-12,118,
   -44.0-12,118,
   -46.1-12,117,
   -48.1-12,117,
   -50.2-12,116,
   -52.1-12,116,
   -54.1-12,116,
   -56.1-12,115,
   -58.1-12,115,
   -60.0-12,115};

double erfc(double X)                          /*
approximation for the error function */
   {                                           /*
from Abramowitz & Stegun */
   static double C0= 0.327591100;     /* Handbook
of Mathematical Functions */
```

```
       static double C1= 0.254829592;
       static double C2=-0.284496736;
       static double C3= 1.421413741;
       static double C4=-1.453152027;
       static double C5= 1.061405429;
       double Q,T,Y;

       Y=fabs(X);
       T=1/(1+C0*Y);
       Q=((((C5*T+C4)*T+C3)*T+C2)*T+C1)*T;

       if(X<0)
         return(2-Q/exp(X*X));
       else
         return(Q/exp(X*X));
       }

    #define erf(x) (1-erfc(x))

    double SuperSat(double Qx,double Cu,double Cx,double Q)
       {
       static double A=-4.33407,a=-
    0.00326463,B=4.84528,b=0.706924,c=0.945739;
       return(Cu+(Cx-Cu)*(A*exp(-
    a*Q/Qx)*erf(b*Q/Qx)+B*erf(c*Q/Qx)));
       }

    #undef gamma
    #undef nu

    int main(int argc,char**argv,char**envp)
       {
       double gamma,nu;

       printf("USACE WES Nitrogen SuperSaturation Model\n");

       printf("\nconstants and empirical coefficients\n");
       printf("  alpha   =%9.3lG mass transfer
    parameter\n",alpha);
       printf("  beta    =%9.3lG effective depth
    parameter\n",beta);
       printf("  Cs      =%9.3lG saturation concentration
    [%%]\n",Cs);
       printf("  eta     =%9.3lG power on Reynolds
    number\n",eta);
       printf("  g       =%9.3lG gravitational acceleration
    [m/sec^2]\n",g);
       printf("  Ks      =%9.3lG surface transfer
    parameter\n",Ks);
```

```c
    printf("   lambda=%9.3lG air layer thickness [m]\n",lambda);
    printf("   mu     =%9.3lG dynamic viscosity of water [kg/m/sec]\n",mu);
    printf("   Pa     =%9.0lf atmospheric pressure [N/m^2]\n",Pa);
    printf("   rho    =%9.0lf density of water [kg/m^3]\n",rho);
    printf("   Sc     =%9.3lG Schmidt number\n",Sc);
    printf("   sigma  =%9.3lG surface tension [N/m]\n",sigma);
    printf("   Vr     =%9.3lG bubble rise velocity [m/sec]\n",Vr);

    printf("\ncomputed parameters\n");
    gamma=rho*g;
    printf("   gamma  =%9.0lf specific weight of water [N/m^3]\n",gamma);

    nu=mu/rho;
    printf("   nu     =%9.3lG kinematic viscosity of water [m^2/sec]\n",nu);

    Results("Ice Harbor"  ,12.2, 4.3,29.0,59.4,IceHarbor,sizeof(IceHarbor)/sizeof(double)/2,1);
    Results("The Dales"   , 6.8, 2.8,24.7,42.7,TheDalles,sizeof(TheDalles)/sizeof(double)/2,2);
    Results("Little Goose",21.3,11.3,29.6,88.4,LittleGoose,sizeof(LittleGoose)/sizeof(double)/2,3);

    return(0);
}
```

also by D. James Benton

3D Articulation: Using OpenGL, ISBN-9798596362480, Amazon, 2021 (book 3 in the 3D series).

3D Models in Motion Using OpenGL, ISBN-9798652987701, Amazon, 2020 (book 2 in the 3D series).

3D Rendering in Windows: How to display three-dimensional objects in Windows with and without OpenGL, ISBN-9781520339610, Amazon, 2016 (book 1 in the 3D series).

A Synergy of Short Stories: The whole may be greater than the sum of the parts, ISBN-9781520340319, Amazon, 2016.

Azeotropes: Behavior and Application, ISBN-9798609748997, Amazon, 2020.

bat-Elohim: Book 3 in the Little Star Trilogy, ISBN-9781686148682, Amazon, 2019.

Boilers: Performance and Testing, ISBN: 9798789062517, Amazon 2021.

Combined 3D Rendering Series: 3D Rendering in Windows®, 3D Models in Motion, and 3D Articulation, ISBN-9798484417032, Amazon, 2021.

Complex Variables: Practical Applications, ISBN-9781794250437, Amazon, 2019.

Compression & Encryption: Algorithms & Software, ISBN-9781081008826, Amazon, 2019.

Computational Fluid Dynamics: an Overview of Methods, ISBN-9781672393775, Amazon, 2019.

Computer Simulation of Power Systems: Programming Strategies and Practical Examples, ISBN-9781696218184, Amazon, 2019.

Contaminant Transport: A Numerical Approach, ISBN-9798461733216, Amazon, 2021.

CPUnleashed! Tapping Processor Speed, ISBN-9798421420361, Amazon, 2022.

Curve-Fitting: The Science and Art of Approximation, ISBN-9781520339542, Amazon, 2016.

Death by Tie: It was the best of ties. It was the worst of ties. It's what got him killed., ISBN-9798398745931, Amazon, 2023.

Differential Equations: Numerical Methods for Solving, ISBN-9781983004162, Amazon, 2018.

Equations of State: A Graphical Comparison, ISBN-9798843139520, Amazon, 2022.

Evaporative Cooling: The Science of Beating the Heat, ISBN-9781520913346, Amazon, 2017.

Forecasting: Extrapolation and Projection, ISBN-9798394019494, Amazon 2023.

Heat Engines: Thermodynamics, Cycles, & Performance Curves, ISBN-9798486886836, Amazon, 2021.

Heat Exchangers: Performance Prediction & Evaluation, ISBN-9781973589327, Amazon, 2017.

Heat Recovery Steam Generators: Thermal Design and Testing, ISBN-9781691029365, Amazon, 2019.
Heat Transfer Examples: Practical Problems Solved, ISBN-9798390610763, Amazon, 2023.
The Kick-Start Murders: Visualize revenge, ISBN-9798759083375, Amazon, 2021.
Jamie2: Innocence is easily lost and cannot be restored, ISBN-9781520339375, Amazon, 2016-18.
Kyle Cooper Mysteries: Kick Start, Monte Carlo, and Waterfront Murders, ISBN-9798829365943, Amazon, 2022.
The Last Seraph: Sequel to Little Star, ISBN-9781726802253, Amazon, 2018.
Little Star: God doesn't do things the way we expect Him to. He's better than that! ISBN-9781520338903, Amazon, 2015-17.
Living Math: Seeing mathematics in every day life (and appreciating it more too), ISBN-9781520336992, Amazon, 2016.
Lost Cause: If only history could be changed..., ISBN-9781521173770, Amazon, 2017.
Mass Transfer: Diffusion & Convection, ISBN-9798702403106, Amazon, 2021.
Mill Town Destiny: The Hand of Providence brought them together to rescue the mill, the town, and each other, ISBN-9781520864679, Amazon, 2017.
Monte Carlo Murders: Who Killed Who and Why, ISBN-9798829341848, Amazon, 2022.
Monte Carlo Simulation: The Art of Random Process Characterization, ISBN-9781980577874, Amazon, 2018.
Nonlinear Equations: Numerical Methods for Solving, ISBN-9781717767318, Amazon, 2018.
Numerical Calculus: Differentiation and Integration, ISBN-9781980680901, Amazon, 2018.
Numerical Methods: Nonlinear Equations, Numerical Calculus, & Differential Equations, ISBN-9798486246845, Amazon, 2021.
Orthogonal Functions: The Many Uses of, ISBN-9781719876162, Amazon, 2018.
Overwhelming Evidence: A Pilgrimage, ISBN-9798515642211, Amazon, 2021.
Particle Tracking: Computational Strategies and Diverse Examples, ISBN-9781692512651, Amazon, 2019.
Plumes: Delineation & Transport, ISBN-9781702292771, Amazon, 2019.
Power Plant Performance Curves: for Testing and Dispatch, ISBN-9798640192698, Amazon, 2020.
Practical Linear Algebra: Principles & Software, ISBN-9798860910584, Amazon, 2023.
Props, Fans, & Pumps: Design & Performance, ISBN-9798645391195, Amazon, 2020.
Remediation: Contaminant Transport, Particle Tracking, & Plumes, ISBN-9798485651190, Amazon, 2021.

ROFL: Rolling on the Floor Laughing, ISBN-9781973300007, Amazon, 2017.
Seminole Rain: You don't choose destiny. It chooses you, ISBN-9798668502196, Amazon, 2020.
Septillionth: 1 in 10^{24}, ISBN-9798410762472, Amazon, 2022.
Software Development: Targeted Applications, ISBN-9798850653989, Amazon, 2023.
Software Recipes: Proven Tools, ISBN-9798815229556, Amazon, 2022.
Steam 2020: to 150 GPa and 6000 K, ISBN-9798634643830, Amazon, 2020.
Thermochemical Reactions: Numerical Solutions, ISBN-9781073417872, Amazon, 2019.
Thermodynamic and Transport Properties of Fluids, ISBN-9781092120845, Amazon, 2019.
Thermodynamic Cycles: Effective Modeling Strategies for Software Development, ISBN-9781070934372, Amazon, 2019.
Thermodynamics - Theory & Practice: The science of energy and power, ISBN-9781520339795, Amazon, 2016.
Version-Independent Programming: Code Development Guidelines for the Windows® Operating System, ISBN-9781520339146, Amazon, 2016.
The Waterfront Murders: As you sow, so shall you reap, ISBN-9798611314500, Amazon, 2020.
Weather Data: Where To Get It and How To Process It, ISBN-9798868037894, Amazon, 2023.

www.ingramcontent.com/pod-product-compliance
Lightning Source LLC
Chambersburg PA
CBHW071445220526
45472CB00003B/676